The Physics of Solar Cells

Perovskites, Organics, and Photovoltaic Fundamentals

The Physics of Solar Cells

Perovskites, Organics, and Photovoltaic Fundamentals

Juan Bisquert
Universitat Jaume I, Castello, Spain

CRC Press is an imprint of the
Taylor & Francis Group, an **informa** business

CRC Press
Taylor & Francis Group
6000 Broken Sound Parkway NW, Suite 300
Boca Raton, FL 33487-2742

Printed on acid-free paper

International Standard Book Number-13: 978-1-138-09996-8 (Paperback)
978-1-138-09997-5 (Hardback)

Library of Congress Cataloging-in-Publication Data

Names: Bisquert, Juan, author.
Title: The physics of solar cells : perovskites, organics, and photovoltaic fundamentals / Juan Bisquert.
Description: Boca Raton, FL : CRC Press, Taylor & Francis Group, [2018]
Identifiers: LCCN 2017029026| ISBN 9781138099968 (pbk. ; alk. paper) | ISBN 1138099961 (pbk. ; alk. paper)
Subjects: LCSH: Solar cells. | Photovoltaic cells.
Classification: LCC QC715.4 .B57 2018 | DDC 621.31/2440153--dc23
LC record available at https://lccn.loc.gov/2017029026

Visit the Taylor & Francis Web site at
http://www.taylorandfrancis.com

and the CRC Press Web site at
http://www.crcpress.com

Contents

Preface

For practical reasons, the treatise has been divided into three volumes:

1. Nanostructured Energy Devices: Equilibrium Concepts and Kinetics (ECK)
2. Nanostructured Energy Devices: Foundations of Carrier Transport (FCT)
3. The Physics of Solar Cells: Perovskites, Organics, and Fundamentals of Photovoltaics (PSC)

The books aim to form a useful utensil for the investigation of photovoltaic devices based on their scientific understanding. Therefore, although each volume is independent, there are cross citations (indicated by the acronym) and they converge into the PSC that explains the structure, principles, and applications of the solar cells.

The first volume, *Equilibrium Concepts and Kinetics* (*ECK*), examines fundamental principles of semiconductor energetics, interfacial charge transfer, basic concepts and methods of measurement, and the properties of important classes of materials such as metal oxides and organic semiconductors. These materials and their properties are important in the operation of organic and perovskite solar cells either as the bulk absorber or as a selective contact structure. Electrolytic and solid ionic conductor properties also play relevant roles in organic and perovskite solar cells.

The second volume, *Foundations of Carrier Transport* (*FCT*), presents a catalog of the physics of carrier transport in semiconductors with a view to energy device models. We systematically explain the diffusion-drift model that is central to solar cell operation, the different responses of band bending and electrical field distribution that occur when a voltage is applied to a device with contacts and the central issue of injection and mechanisms of contacts. We describe the carrier transport in disordered materials that often appear as good candidates for easily processed solar cells. There are also excursions into other important topics such as the transistor configuration and the frequency domain techniques as Impedance Spectroscopy that produce central experimental tools for the characterization of the devices.

This volume, PSC, completes the goal of the collection: to provide an explanation of the operation of photovoltaic devices from a broad perspective, which embraces concepts from nanostructured and highly disordered materials to highly efficient devices such as the lead halide perovskite solar cells. The volume starts with three background chapters of essential physical and material properties about electromagnetic radiation fields, semiconductors, and the interaction of light with the latter causing the generation of carriers that are at the heart of photovoltaic action: creating energetic electrons that go round an external circuit. Our approach is to establish from the beginning a simple but very rich model of a solar cell, in order to develop and understand step by step the photovoltaic operation according to fundamental physical properties and constraints. It allows us to focus on the aspects pertaining to the functioning of a solar cell and the determination of limiting efficiencies of energy conversion, by intentionally removing the many avoidable losses such as transport gradients, which are treated in the final chapters of the book.

The conceptual picture of a solar cell has evolved in the last two decades, when a broad landscape of candidate materials and devices were discovered and systematically studied and reported. New concepts and a rather powerful picture that embraces very different types of devices have been established based upon many discussions and sometimes also conceptual clashes. The resulting scientific consensus, according to my own view, is the story I want to tell. The approach and examples are centered on the types of solar cells I have investigated over these years, namely the dye-sensitized solar cells, the organic solar cells, and the lead halide perovskite solar cells. A broad and transversal perspective and a generalizing spirit have been adopted so that the general insights are not restricted to these types but may be useful for addressing any type of solar cell. This I believe is useful both for experts to achieve a summarizing scheme of their activities, and for starters to get an overall picture of what needs to be known.

In this volume, you get a broad view of the conversion of photons to electricity using light-absorbing semiconductors. In many textbooks, the solar cell operation is related to a diode element as the starting point. Here we wish to clarify the physical mechanisms determining the operation of the diode in photovoltaic conversion, as this is an essential point for the understanding and design of solar cell structures. The operation involves a set of concepts and tools that are not well developed inside the boundaries of a specific discipline of physics or chemistry, such as the Fermi levels and detailed balance of light absorption and emission. This model starts historically from the central insights of William Shockley, Hans-Joachim Queisser and Robert T. Ross in papers published 50 years ago (see chapter 6). Building the knowledge that has been gathered in the three volumes, step by step through ECK, FCT, and then PSC, can provide you with a strong foundation both for understanding solar cell operation and for exploring related materials and device operation. However, if you are already familiar with the basis of the knowledge, you can directly read the PSC ignoring many citations to the previous volumes, and go straight to the specific heart of the matter of the physics of solar cells.

In the pursuit of new photovoltaic conversion materials, one can start with a rather poor efficiency and the initial mystery is, how can we extract charge at all, how does charge separation occur, and why do charges arrive at the electrical contact? Therefore, historically, the explanation of the functioning of a solar cell relies on the transport of carriers and the action of electrical fields. It has now been recognized that this approach may be highly misleading about the core of the operation of photovoltaic conversion. When the material becomes better understood, carrier collection is usually not the main issue of photovoltaic cells. In most cases, the crucial issue is obtaining the maximal photovoltage allowed by the fundamental physical constraints; therefore, one begins to worry about reducing the recombination and finding the optimized structures for selective contacts.

If the technology progresses sufficiently and is blessed with evolution toward really high efficiencies, the electronic operation of the device becomes reproducible and proficient, and then the imperative to maximize the extraction of power from every photon that comes to the device surface presents itself to the researcher, who is then forced to attend to the photonic characteristics. Thereafter appear schemes to harvest the full solar spectrum, a task that requires a combination of absorbers. Therefore, to keep up with the pace of research, one needs to go through a pretty varied ensemble of problems. We would like to establish a holistic perspective that enables evaluation of the solar cell properties at any stage, since the surprise of birth, enjoying the joviality of adolescence, pausing in the calm maturity, before our understanding passes away forever into the realm of the money-making entities.

As already announced, we begin this book with background physical ideas in Chapters 1 through 3, providing the needed notions of properties of radiation fields, light absorption and emission, and carrier recombination. Then Chapters 4 through 10 present a progressive explanation of the properties of solar cells. We start the subject matter by formulating the physical basis of the recombination diode, and from the standpoint of the built-in asymmetry, we state the interaction of light and the semiconductors, the creation of the split of Fermi levels, and the production of output electrical power. Then we continue with physical limitations to optimal conversion obtaining more realistic specific effects, configurations, and shortcomings that provide a summary of the multitude of effects that may come into play at the time of experimental investigations and technological development. Chapter 11 presents a set of advanced solar energy conversion schemes.

I am very grateful to several colleagues who provided comments and pointed out improvements on parts of this book: Osbel Almora, Henk Bolink, Albert Ferrando, Pilar López Varo, Juan P. Martínez Pastor, Luis M. Pazos Outón, Iván Mora Seró, Andrey Rogach, Rafael Sánchez, and Greg Smestad. I am especially grateful to Mehdi Ansari Rad and Thomas Kirchartz, who commented on the preliminary manuscript and Sandheep Ravishankar, who made a general reading and editing. Any remaining mistakes are solely my own.

About the Author

Juan Bisquert is a professor of applied physics at the Universitat Jaume I de Castelló and the funding director of the Institute of Advanced Materials at UJI. He earned an MSc in physics in 1985 and a PhD from the Universitat de València in 1992. The research work is in perovskite solar cells, nanostructured solar cells (including dye-sensitized solar cells, organic solar cells, and quantum-dot sensitized solar cells), and solar fuel converters based on visible light and semiconductors for water splitting and CO_2 reduction. His most well-known work is about the mechanisms governing the operation of nanostructured and solution-processed thin film solar cells. He has developed insights in the electronic processes in hybrid organic–inorganic solar cells, combining the novel theory of semiconductor nanostructures, photoelectrochemistry, and systematic experimental demonstration. His contributions produced a broad range of concepts and characterization methods to analyze the operation of photovoltaic and optoelectronic devices. He is a senior editor of the *Journal of Physical Chemistry Letters*. He has been distinguished several times in the list of ISI Highly Cited Researchers. Bisquert created nanoGe conferences and is the president of the Fundació Scito del País Valencià.

1 Blackbody Radiation and Light

The fundamental properties of radiation are a central tool for the design and utilization of energy devices. Radiation from the sun is the principal source of energy exploited in solar cell devices. Even in the dark, ambient thermal radiation causes electronic processes in a photoactive device, and this fact is used in arguments based on detailed balance that provide fundamental information of semiconductors interaction with radiation. The properties of light and color are necessary for the characterization of light sources. This chapter provides a brief revision of the main features of the electromagnetic spectrum, and some fundamental quantities that are employed to quantify the radiation. The main part of the chapter describes the blackbody radiation for later reference, starting with the radiative properties of a blackbody, and formulating the properties of thermal radiation, consisting of the spectral distribution of number of photons and their energy. The next step is to characterize the flux of photons and the energy flux. Finally, we examine the main aspects of the solar spectrum, which is the energy source that we wish to utilize for the production of useful energy, and how it relates to the idealized blackbody spectrum.

1.1 PHOTONS AND LIGHT

Electromagnetic radiation can be viewed as being composed of *electromagnetic waves,* or as consisting of massless energy quanta called *photons*. Radiation, in either view, can be classified according to its wavelength, λ, or frequency ν. These quantities hold the relationship

$$\lambda = \frac{c_\gamma}{\nu} \tag{1.1}$$

where c_γ is the speed of light, which depends on the *refractive index* of the medium, n_r, through which the radiation travels, as

$$c_\gamma = \frac{c}{n_r} \tag{1.2}$$

Here, $c = 2.998 \times 10^8$ ms^{-1} is the speed of light in vacuum. Common units of wavelength are the micrometer, μm, and the nanometer, nm.
The energy of the photon is given by the expression

$$E = h\nu = \hbar\omega \tag{1.3}$$

The angular frequency is $\omega = 2\pi\nu$. h is Planck's constant and $\hbar = h/2\pi$. The photon energy can be converted to wavelength using the formula

$$\lambda = \frac{1240}{h\nu\ (\text{eV})}\text{nm} \tag{1.4}$$

The parts of the electromagnetic spectrum are indicated in Figure 1.1. The *optical region*, which is the wavelength region of interest for solar energy conversion and lighting, is the long-wave portion comprising the ultraviolet region (UV), the visible-light region extending from approximately

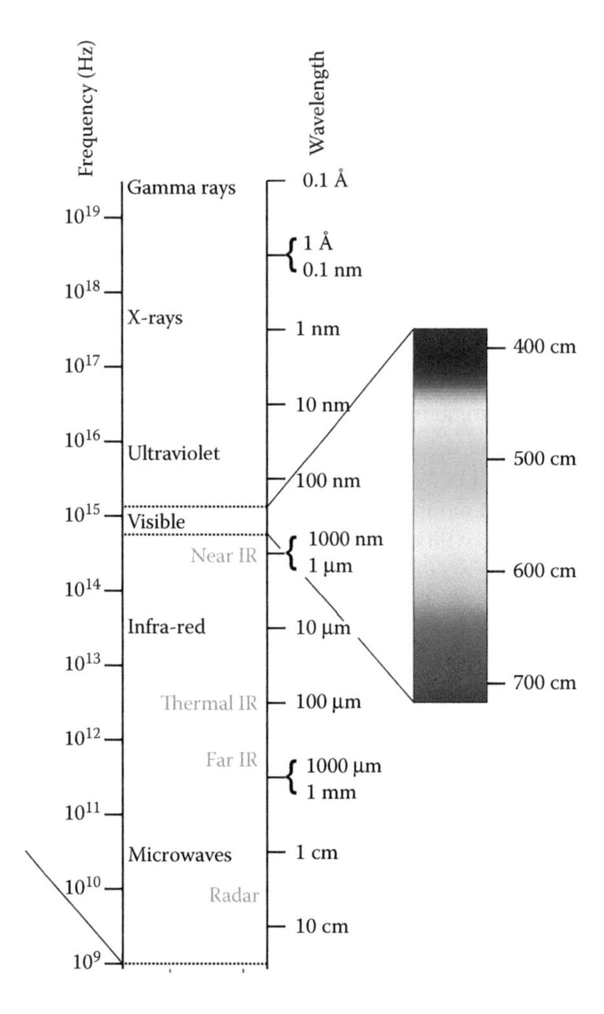

FIGURE 1.1 Electromagnetic spectrum in the range between microwaves and gamma rays. (Reprinted from the public domain from http://en.wikipedia.org/wiki/File:Electromagnetic-Spectrum.svg)

$\lambda = 0.4$–$0.7\ \mu$m, and the infrared region (IR) from beyond the red end of the visible spectrum to about $\lambda = 1000\ \mu$m. We denote *light* the part of the electromagnetic spectrum that provokes a visual response in humans. Figure 1.2 shows the solar spectrum in two different fashions: the number of photons arriving at the top of the earth's atmosphere, and at the earth's surface, per interval of wavelength. Approximately 50% of solar irradiation occurs outside the visible range, especially in the IR; therefore this part provides radiant energy and heat but not light. The spectral differences above and below the atmosphere are due to the filtering of certain wavelengths, as further discussed in Section 1.8.

1.2 SPREAD AND DIRECTION OF RADIATION

Light emitted at a point source propagates in the three dimensions of space, so that the intensity decreases with distance. The spread of light is defined by two factors: the area and the angle.

To quantify the direction of propagation, it is necessary to use a two-dimensional angle: the solid angle Ω, defined in terms of size and distance of a distant object. It corresponds to a fragment of a sphere that is invariant with the sphere's radius. In SI units the arc is measured in radians, $\theta = s/R$, where s is the arc and R is the radius of the circle. The solid angle is measured in *steradians*.

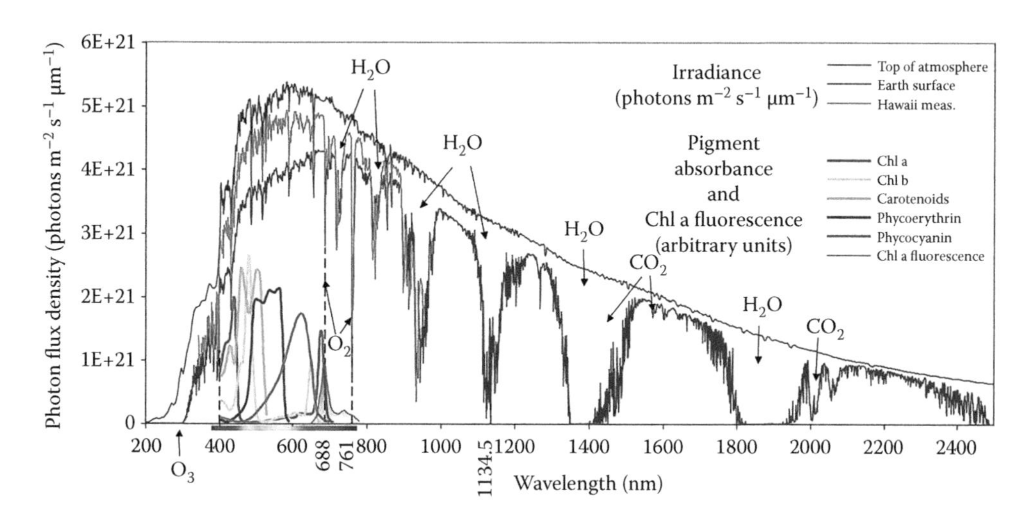

FIGURE 1.2 Solar spectral photon flux densities at the top of the earth's atmosphere and at the earth's surface, and estimated *in vivo* absorption spectra of photosynthetic pigments of plants and algae: Chl a and Chl b, carotenoid, phycoerythrin, phycocyanin absorption spectra. Chl a fluorescence spectrum, from spinach chloroplasts. (Reprinted with permission from Kiang, N. Y. et al. *Astrobiology* 2007, 7, 222–251.)

A steradian (sr) is defined as the solid angle subtended at the center of a sphere by an area on its surface numerically equal to the square of its radius. As shown in Figure 1.3a, for a surface element dS seen from a distance R the solid angle is given by

$$d\Omega = \frac{dS}{R^2} \tag{1.5}$$

The element of solid angle can be written in terms of angular coordinates as

$$d\Omega = \sin\theta \, d\phi \, d\theta \tag{1.6}$$

The total solid angle is 4π, and the solid angle of a hemisphere is 2π.

In geometrical optics étendue is a convenient concept to describe the propagation of light through an optical system. Consider a beam of radiation in a direction **s** in a medium of refractive index n_r shown in Figure 1.3b. dA is a cross-sectional element of the beam, and **n** is the normal to dA. The *étendue* ε is the product of the solid angle of the radiation and the projected area (Markvart, 2008).

$$d\varepsilon = n_r^2 \cos\theta \, dA \, d\Omega \tag{1.7}$$

An important property is that the étendue of a beam propagating in a clear and transparent medium is conserved.

Consider a cone of radiation of half-angle θ_m, impacting a surface element A, as shown in Figure 1.3c. The étendue of this light is

$$\varepsilon = n_r^2 A \int_0^{\theta_m} \cos\theta \sin\theta \, d\theta \int_0^{2\pi} d\phi = n_r^2 A\pi \sin^2\theta_m \tag{1.8}$$

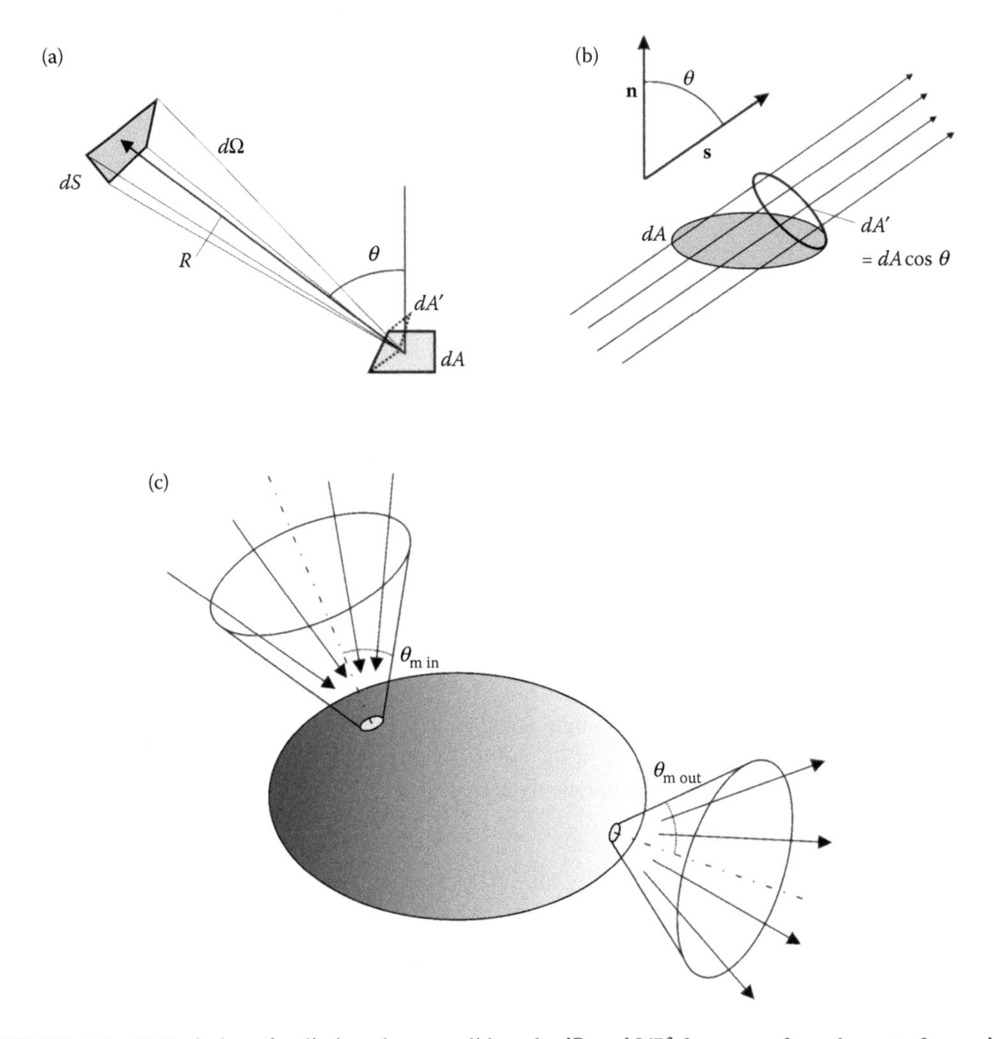

FIGURE 1.3 (a) Emission of radiation along a solid angle $d\Omega = dO/R^2$ from a surface element of area dA. The angle between the radiation and the normal to the surface is θ. (b) Representation of the flux in the direction **s** across an area dA with unit normal vector **n**. (c) Conical solid angles of incident and outgoing radiation on a solid object.

In the case of a hemisphere ($\theta_m = \pi/2$), it is $\varepsilon = n_r^2 A\pi$. This is the étendue for maximal concentration on a planar black absorber. The radiation incident on a solar cell carries the solid angle of the sun. The half-angle subtended by the sun is $\theta_S = 0.265°$, so that $\sin^2\theta_s = 2.139 \times 10^{-5}$ and $\varepsilon_S = n_r^2 A\, 6.8 \times 10^{-5}$ sr.

For an ideal diffusely reflecting surface or ideal diffuse radiator, the number of photons and the energy flux Φ across the small solid angle $d\Omega$ depends on the projected solid angle (the étendue)

$$\Phi(\theta) \propto (\cos\theta)\, d\Omega \tag{1.9}$$

Such a dependence is called *Lambert's* law. The factor $\cos\theta$ gives the reduction of apparent area of the emitting element dA seen from the receiving point, and the radiation is reduced by the same factor $\cos\theta$ with respect to the normal. Therefore, for the Lambertian reflector or radiator, the same number of photons is perceived when looking at dA from any angle. Conventional LEDs are approximately Lambertian. If we consider a curved surface that radiates in our direction, we obtain that for each element of the surface the emitted power is reduced by the same factor that the area is. Thus all elements of the Lambertian surface appear equally bright, and this is actually observed looking at the sun. Note that the Lambertian diffusor for light emitted from a surface is the reciprocal of a maximal concentration for the incoming light.

1.3 COLOR AND PHOTOMETRY

The visible spectrum ranges from 390 nm (violet) to 780 nm (red), see Figure 1.4. Our perception of color results from the composition of the light (the energy spectrum of the photons) that enters the eye. Cone cells in human eye are of three different types, sensible to three different ranges of frequency, which the eye interprets as blue (with a peak close to 419 nm), green (peaking at 531 nm), and red (with a peak close to 558 nm, which is more yellowish) (Figure 1.4). Naturally occurring colors are composed of a broad range of wavelengths. The wavelength that appears to be the most dominant in a color is the color's *hue*. The *saturation* is a measure of the purity of the color and indicates the amount of distribution in wavelengths in the color. A highly saturated color will contain a very narrow set of wavelengths.

The perception of color by the human eye is quantified by tristimulus values X, Y, Z defined by Commission Internationale de l'Éclairage (CIE). These parameters are derived from the spectral power distribution of the light emitted by a colored object, weighted by sensitivity curves, the standard colorimetric functions in Figure 1.4, that have been determined by actual measurement of the average human eye. Tristimulus values uniquely represent a perceivable hue. The two lowercase coordinates xy in the chromaticity diagram, Figure 1.5, are derived from the tristimulus values, and represent the relative contribution of the three primaries. The boundaries indicate maximum saturation, that is, the spectral colors. The diagram forms the boundary for all perceivable hues.

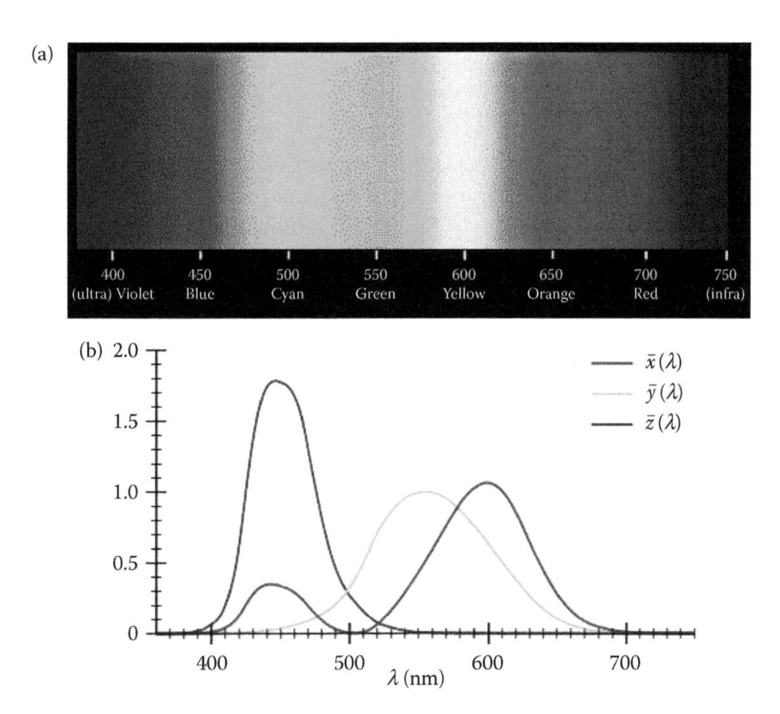

FIGURE 1.4 (a) Approximated wavelengths associated with the colors perceived in the visible spectrum. (b) CIE 1931 Standard Colorimetric Observer functions used to map blackbody spectra to XYZ coordinates. (Reprinted from the public domain https://en.wikipedia.org/wiki/File:CIE_1931_XYZ_Color_Matching_Functions.svg)

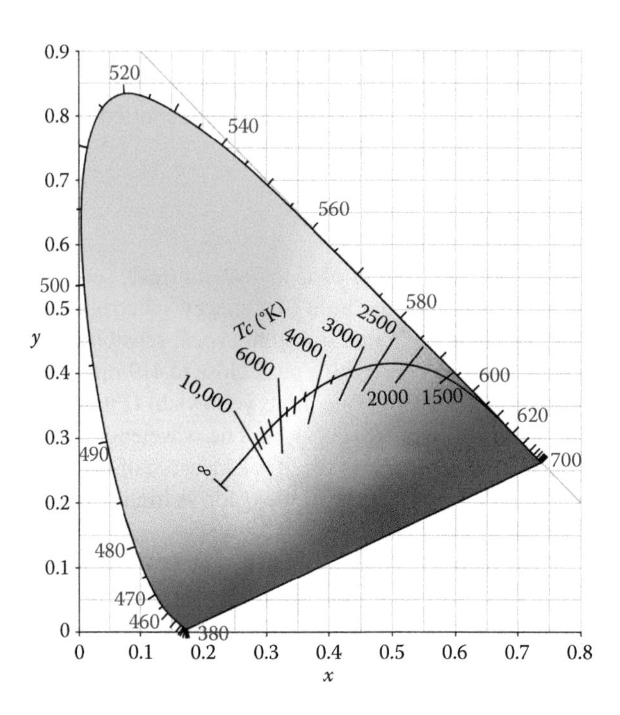

FIGURE 1.5 CIE *xy* 1931 chromaticity diagram including the Planckian locus. The Planckian locus is the path that a blackbody color will take through the diagram as the blackbody temperature changes. Lines crossing the locus indicate lines of constant CCT. Monochromatic wavelengths are shown at the boundary in units of nanometers. (Reprinted from the public domain from http://en.wikipedia.org/wiki/File:PlanckianLocus.png)

The emission of hot objects, described in Section 1.4, is called *incandescence*. Incandescent light sources emit a broad set of frequencies that covers a wide range of the visible spectrum. We normally denote white light, the light of the sun. One could also consider white light that of the tungsten filament bulb, which is similar to blackbody radiation at about 2900 K. In general, one can parametrize the white light emitted by a blackbody radiator by using the correlated color temperature (CCT). The trace of the points of the irradiating blackbodies in the chromaticity diagram, Figure 1.5, forms the Planckian locus. From low to very high temperatures, the emission changes from deep red, through orange, yellowish white, white, and finally bluish white.

There are three classes of lamps for lighting: incandescent, discharge, and solid state. Incandescent lamps heat a filament that glows. Discharge lamps and white LEDs (light-emitting diodes) are based on the emission of UV or blue photons, either by gas ionization or by electroluminescence. These high-energy photons are then used to excite phosphorescent atoms or molecules that emit lower energy photons in red or green. The downconversion process, further discussed in Section PSC.11.2, displaces the wavelength and allows the production of different colors with a single lamp.

White light can be created by a combination of red, green, and blue monochromatic sources. The emission of two complementary colors, at opposite sides of the Planckian line, also gives rise to white light. When we look at illuminated objects, we only see the reflected light. The spectrum of the light source affects the appearance of objects according to the color rendering. Incandescent light sources allow our visual system to easily distinguish the colors of objects. Discharge and fluorescent lamps may produce light that peaks strongly at certain emission wavelengths and lacks a significant part of the visible spectrum causing the perceived color of an object to be very different from that under natural light. An index denoted *Color Rendering Index* (CRI) compares the ability of a light source to reproduce the colors in comparison with a natural source. The reference

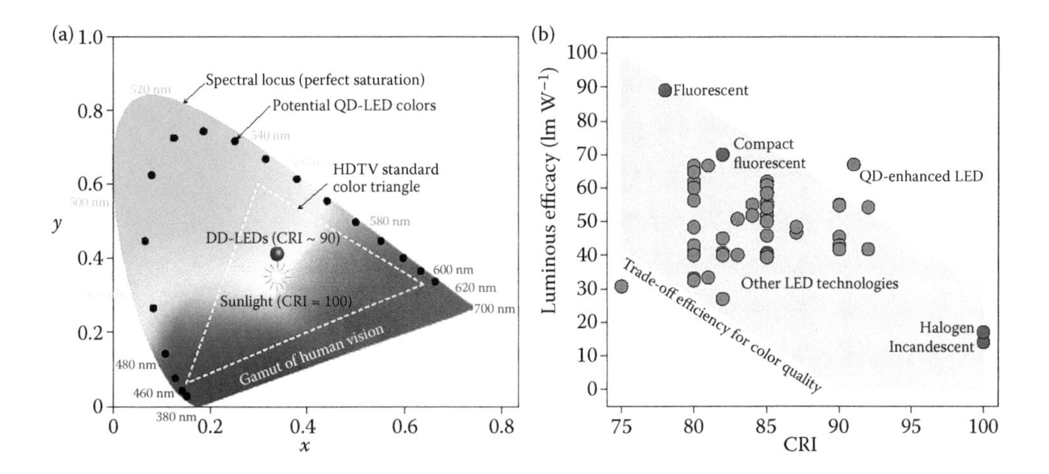

FIGURE 1.6 (a) CIE chromaticity diagram showing that the spectral purity of quantum dots (QDs) enables a color gamut (*dotted line*) larger than the high-definition television (HDTV) standard (*dashed line*). (b) Plot showing the luminous efficacy and CRI of various commercially available lighting solutions. (Adapted with permission from Shirasaki, Y. et al. *Nature Photonics* 2012, 7, 13–23.)

is a perfect blackbody radiator at the same nominal temperature, which is assigned the value 100. General lighting requires a CRI of 70 and some applications demand 80 or higher.

The intensity of the light is measured in SI in candela (cd). The *luminous intensity* per unit area of light, travelling in a given direction, is measured in cd m^{-2}. *Radiant intensity* is the amount of power radiated per unit solid angle, measured in W sr^{-1}. *Radiance* (in W m^{-2} sr^{-2}) is the energy flux per unit area per solid angle received by a surface.

The vision system of humans does not detect all wavelengths equally: UV and infrared light is not useful for illumination. The effect of production of light in a source for human vision therefore depends both on the efficacy of production of radiation and on how the produced spectrum adapts to the human eye. The response of vision of the average eye, in the visible spectrum, in conditions of bright illumination, is called the *photopic response*. The eye's response is maximum for green light of wavelength 555 nm and 1 W of irradiated power at this wavelength is defined as 683 lumen (lm). Solid state lighting and display applications require, respectively, a brightness of 10^3–10^4 and 10^2–10^3 cd m^{-2}. *Illuminance* measures the photometric flux per unit area, or visible flux density, in lux (lm m^{-2}). *Luminance* is the illuminance per unit solid angle, in lm m^{-2} sr^{-2}. Luminance is the density of visible radiation in a given direction. The spectrum of any light source can be measured and decomposed with the photopic response to provide a total production of lumens by the lamp. The luminous efficacy of a source is then the number of lumens produced per watt of electrical power supplied (lm W^{-1}).

High CCT sources (>5000 K) produce a bluish emission and low CRI in the range 80–85. For low CCT lights, in the range of the incandescent emitters (about 2700 K), higher CRI can be achieved but it is difficult to maintain high luminous efficiency due to losses in the infrared. Figure 1.6 compares the luminous efficiency and the CRI of different types of lighting solutions (Shirasaki et al., 2012).

1.4 BLACKBODY RADIATION

Sources of electromagnetic radiation can emit discrete wavelengths, as in phosphors, LEDs, or gases, by specific quantum transitions that radiate photons of energy equal to the difference between initial and final energy levels, Equation 1.3. On the other hand, many solids contain a near-continuum of

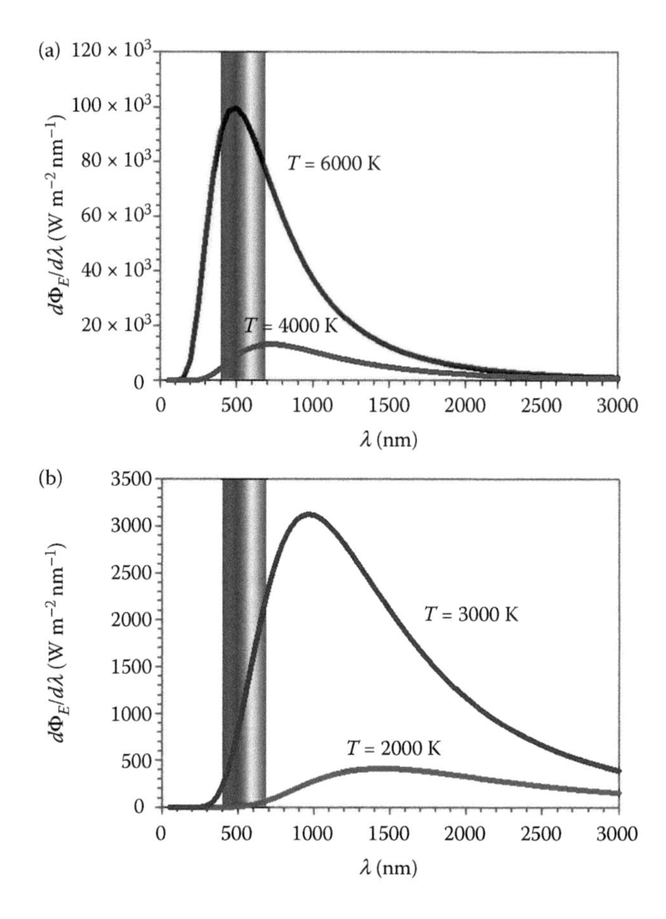

FIGURE 1.7 The spectrum of a blackbody at different temperatures with respect to wavelength. (a) and (b) show the spectra at different temperatures as indicated.

energy levels that allow for a very broad scope of excitations and consequently radiate a continuous electromagnetic spectrum extending over a wide range. The electromagnetic energy emitted from the surface of a heated body is called *thermal* radiation. Many solids radiate a spectrum that can be well approximated by *blackbody radiation*.

The spectrum of a blackbody is shown in Figure 1.7. Note that Figure 1.7 displays the power radiated while the solar spectrum in Figure 1.2 is the number of photons, per wavelength interval. The spectra in Figure 1.7 contain power at all wavelengths but display a characteristic distribution, consisting of a peak at the wavelength of maximum emission. Each spectrum shows a sharp decay toward the shorter wavelength region, while there is a longer and slower decay toward longer frequencies in the infrared. The wavelength of maximum emission is a function of temperature according to the expression

$$\lambda T = 2.9 \cdot 10^{-3}\,\text{mK} \tag{1.10}$$

This rule is known as Wien's displacement law. According to Equation 1.10, the wavelength of maximum emission moves to shorter wavelength when the temperature of the radiator rises. The total radiated power increases with the body temperature according to the Stefan–Boltzmann law discussed in Equation 1.39.

At room temperature, an object emits radiation in the infrared, and no visible light is generated. When an object is heated, the radiation spectrum approaches the visible wavelengths and the emission will change from dark red at low temperatures, through orange red, to white at the very high temperatures, as remarked before in the chromaticity diagram of Figure 1.5. To emit red light, a body must be heated above 850 K and at 3000 K it appears yellowish white, Figure 1.7b. The emission of a tungsten filament in a light bulb (at a power of 150 W) occurs at 2900 K. It is observed in Figure 1.7b that only a small fraction of the radiated energy (11%) is visible light, thus while the incandescent light bulb produces light containing all the frequencies of the visible spectrum, which is comfortable for human vision, the efficiency of conversion of electricity to light is small. Figure 1.7a shows the Planck spectrum at 6000 K, which is similar to the solar spectrum further discussed in Section 1.8.

1.5 THE PLANCK SPECTRUM

The *Planck spectrum* (*blackbody spectrum*) is a spectral distribution of energy per unit volume per unit frequency range $d\nu$. It is given by the form

$$\rho_{bb}(\nu) = \frac{8\pi h\nu^3}{c_\gamma^3} \frac{1}{e^{h\nu/k_BT} - 1} \tag{1.11}$$

A justification of Equation 1.11 will be presented below. It is derived from the concept of a blackbody, which is a body that absorbs all incoming radiation, so that it does not return, by reflection or scattering, any of the radiation falling on it. For a blackbody, the absorptivity, defined in Section PSC.2.1, is $a = 1$. Such a body appears *black* if its temperature is low enough such that it is not luminous by virtue of the radiation emitted on its own. In fact, the blackbody emits electromagnetic radiation, and such radiation, covering a broad spectral range, is the topic of interest here.

Although the blackbody is an idealized object, it is possible to build a radiation source with an emissivity close to that of the blackbody. Cavity radiation is a paradigmatic example of blackbody radiation and has actually been used for experiments in thermal radiation for more than a century. The cavity blackbody is an enclosure that has black internal walls, with a small hole drilled on it. Even though the internal wall may not be perfectly absorbing, photons entering the hole have a very low probability to bounce back out before being absorbed. The body is held at temperature T and the walls emit radiation so that the cavity is filled with isotropic radiation with the spectral distribution corresponding to the temperature T. In consequence, the hole in the cavity absorbs all incoming radiation and emits thermalized photons, so that *the hole* (not the cavity) can be considered a blackbody according to the above definition.

It can be rigorously shown that the radiation in equilibrium with the blackbody has a unique spectral distribution that is a function only of the absolute temperature of the blackbody, and not of its composition or the nature of the radiation enclosure (Landsberg, 1978). In addition, an ideal blackbody emits more energy than any other material at the same temperature, see Equation PSC.6.29. In general, however, blackbody radiation refers to the spectral distribution of the radiation, as given by Planck's formula, and we do not really need to think about cavities in most of the applications discussed here. The essential point is that all radiation coming from a blackbody must be regarded as *emitted* radiation.

The blackbody spectrum is a very important tool for a number of aspects of devices, physics and chemistry. First, it describes thermal radiation, which is the radiation of hot objects such as the sun, a hot filament or incandescent materials in general. Since our eyes are used to seeing objects under sunlight, the Planckian radiation is the landmark of white light. The radiation of artificial light sources is compared to the blackbody spectrum by attributing a temperature to the obtained light by comparison of its actual spectral distribution with the exact blackbody spectrum, which has a CRI of 100, as mentioned in Section 1.3. Historically, the quest for the Planckian spectrum was propelled at the end of the nineteenth century by the need to characterize the light emanating from a hot filament. For solar energy conversion, we need to know the spectral distribution of solar photons, their number, and energy in order to devise the transduction of the photon energy to electricity or chemical fuel. The blackbody radiation is a very good model for the spectrum of the sun, as already mentioned, although solar radiation contains additional features that are discussed in Section 1.8. Finally, thermalized radiation is also used to derive transition rates for light absorption and emission in semiconductors and molecules using the detailed balance arguments (Kennard, 1918). Fundamental rates of emission are obtained from the hypothesis of equilibrium with the ambient radiation, which is blackbody radiation at temperature $T = 300$ K in the normal environment of the earth surface.

1.6 THE ENERGY DENSITY OF THE DISTRIBUTION OF PHOTONS IN BLACKBODY RADIATION

We have commented in Chapter ECK.5 the thermalization of electrons in a semiconductor to a Fermi–Dirac distribution. We consider a similar situation concerning the *distribution of photons* that form electromagnetic radiation. Let us analyze a set of photons in an enclosement with perfect reflecting walls. These walls neither absorb nor emit radiation. Since the photons do not interact among themselves, the initial spectral distribution that was produced by the source of the radiation will be maintained. But, as remarked by Planck (1914), "as soon as an arbitrarily small quantity of matter is introduced into the vacuum, a stationary state of radiation is gradually established. If the substance introduced is not diathermanous for any color, for example, a piece of carbon, however small, there exists at the stationary state of radiation in the whole vacuum for all colors the intensity of black radiation corresponding to the temperature of the substance." Just a small piece of material that can absorb and emit all frequencies in the cavity (like a piece of carbon dust) will absorb and release photons which ultimately come to a thermal distribution, formed by isotropic radiation in equilibrium with the small body at temperature T. Such a thermal distribution of the photons is the Planck spectrum.

We analyze the spectral distribution of isotropic radiation, moving in all directions. This is composed of photons of all wavelengths (frequencies) that occupy the available states consisting of the volume density of electromagnetic modes, $D_{ph}(E)$, according to a thermal distribution defined by the occupation function $f_{ph}(h\nu)$.

The density of photons n_{ph} in a small range of energies around the energy $h\nu$ is given by

$$dn_{ph}(h\nu) = D_{ph}(h\nu) f_{ph}(h\nu) d(h\nu) \tag{1.12}$$

The Bose–Einstein distribution gives the occupancy of boson modes

$$f_{ph}(E) = \frac{1}{e^{(E-\eta_{ph})/k_B T} - 1} \tag{1.13}$$

Here η_{ph} is the electrochemical potential of the photons that coincides with their chemical potential $\mu_{ph} = \eta_{ph}$. The distribution of photons as function of their energy is shown in Figure 1.8. In contrast to the Fermi–Dirac function, there is no limit to the number of bosons that occupy an available mode so that the photons accumulate in the lowest energy available state (the ground state). For $E - \mu_{ph} \gg k_B T$, the distribution takes the form of the Boltzmann expression

$$f(E,T) \approx e^{-(E-\mu_{ph})/k_B T} \tag{1.14}$$

This term dominates the spectral luminescence as discussed later in Section PSC.6.4. The change of chemical potential of the photons (later to be identified with the voltage of the light-emitting device) produces a horizontal shift of the spectra, Figure 1.8a, while the temperature determines the slope, Figure 1.8b.

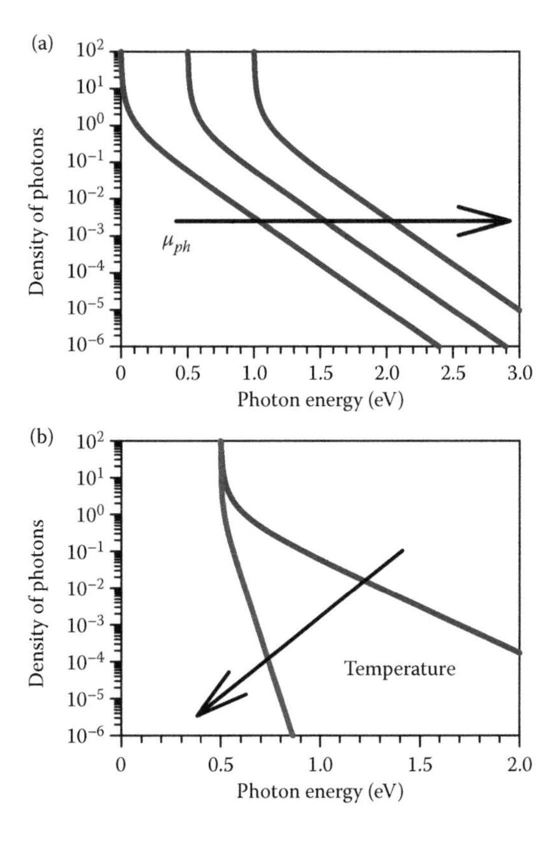

FIGURE 1.8 Number of photons at different electrochemical potentials (μ_{ph}) (a) and temperatures (b). The case $\mu_{ph} = 0$ is the thermal radiation.

Thermal radiation is characterized by $\eta_{ph} = 0$ and the distribution takes the form

$$f_{ph}(E) = \frac{1}{e^{E/k_BT} - 1} \tag{1.15}$$

The number of electromagnetic modes having energy lower than $h\nu$, per unit volume, is (Fowles, 1989)

$$N_{ph}(\nu) = \frac{(8\pi/3)(h\nu)^3}{\left(h^3 c_\gamma^3\right)} \tag{1.16}$$

Therefore, the density of states per volume and energy interval is

$$D_{ph}(h\nu) = \frac{dN_{ph}(\nu)}{d(h\nu)} = \frac{8\pi\nu^2}{hc_\gamma{}^3} \tag{1.17}$$

These states are isotropically distributed in a solid angle 4π, so that the density of states in a direction specified by the solid angle $d\Omega$ is $d\Omega/4\pi$. We include the index of refraction n_r, which causes an enhanced light intensity when the radiation is trapped in a medium (Yablonovitch and Cody, 1982), and we give the density of states per solid angle as follows:

$$D_{ph}(h\nu) = \frac{2n_r^3 d\Omega \nu^2}{hc^3} \tag{1.18}$$

Now *Planck's law of radiation* describes the number of photons, dn_{ph}, per volume per frequency interval. From Equation 1.12, we obtain

$$\frac{dn_{ph}(\nu)}{d\nu} = \frac{8\pi n_r^3 \nu^2}{c^3} \frac{1}{e^{h\nu/k_B T} - 1} \tag{1.19}$$

The photon density, per volume per frequency interval per solid angle, is

$$\frac{dn_{ph}(\nu)}{d\nu} = \frac{2n_r^3 \nu^2}{c^3} \frac{1}{e^{h\nu/k_B T} - 1} d\Omega \tag{1.20}$$

Figure 1.9a shows the distribution function for the number of photons at different temperatures for blackbody radiation. The energy per unit volume in a small range of wavelengths, de_{ph}, is obtained by the product of the photon density and their energy

$$de_{ph}(\nu) = (h\nu) dn_{ph}(\nu) \tag{1.21}$$

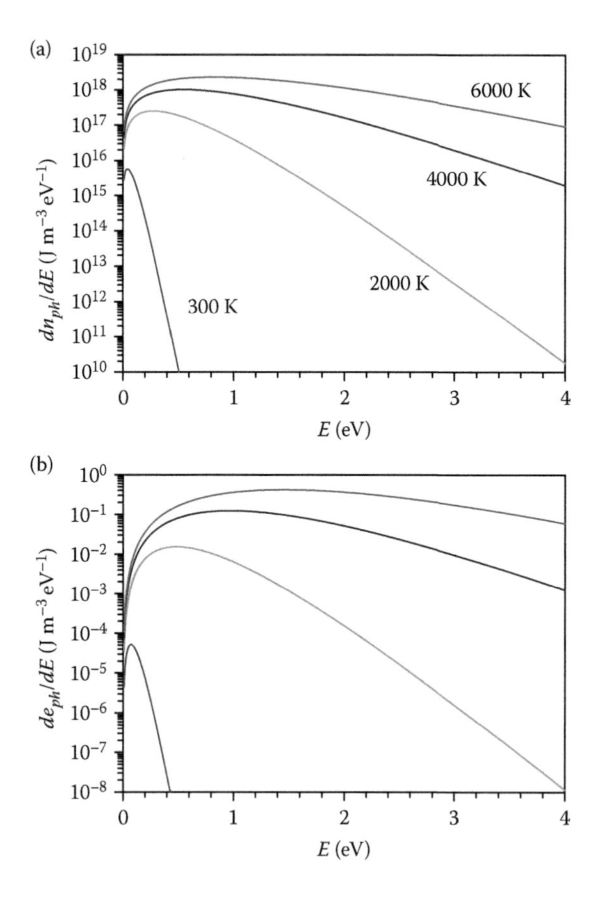

FIGURE 1.9 Representation of the photon and energy density distributions in the Planck radiation, as a function of the energy of the photons: (a) the number of photons per unit volume and (b) the energy density per unit volume.

As already anticipated in Equation 1.11, the Planck spectrum (blackbody spectrum) is normally given in terms of the radiant energy, that is, the spectral distribution of energy per unit volume per unit frequency range $d\nu$

$$\rho_{bb}(\nu) = \frac{de_{ph}(\nu)}{d\nu} = \frac{dn_{ph}(\nu)}{d\nu} h\nu = \frac{2n_r^3 h\nu^3 d\Omega}{c^3} \frac{1}{e^{h\nu/k_BT} - 1} \tag{1.22}$$

This function is shown in Figure 1.9b. We can also write the spectral distribution in terms of wavelength, noting that

$$\rho_{bb}(\nu)d\nu = \rho_{bb}(\lambda)d\lambda \tag{1.23}$$

Since

$$|d\nu| = \frac{c}{\lambda^2}|d\lambda| \tag{1.24}$$

we obtain

$$\rho_{bb}(\lambda) = \frac{de_{ph}(\lambda)}{d\lambda} = \rho_{bb}(\nu)\frac{d\nu}{d\lambda} = \frac{2n_r^3 hc\, d\Omega}{\lambda^5} \frac{1}{e^{hc/k_B\lambda T} - 1} \tag{1.25}$$

From Equation 1.25, it can be shown that the maximum of the spectrum is displaced to shorter wavelengths (higher energies) as the temperature of the radiation is raised, which is Wien's displacement law, discussed earlier in Equation 1.10.

1.7 THE PHOTON AND ENERGY FLUXES IN BLACKBODY RADIATION

Radiation in a cavity-type environment is isotropic but the radiation emitted from a source is directed, from the source to the observation point. For a radiating surface element dA with unit normal vector **n**, Figure 1.3b, the radiation can be emitted in any direction in the hemisphere characterized by the unit vector **s** and the solid angle $d\Omega$. The incoming or outgoing radiation is characterized by a number of photons $\Phi_{ph}(\nu_1, \nu_2, \mathbf{s})$ in the frequency interval (ν_1, ν_2) crossing the unit surface per second and they carry a flux of energy $\Phi_E(\nu_1, \nu_2, \mathbf{s})$. The variables used for the description of radiation are indicated in Table 1.1.

However, the *spectral flux* refers to the number of photons or photon energy per second in an interval in a small range of wavelengths $(\nu, \nu + d\nu)$ around a single frequency ν. The spectral quantity

$$\phi_E(\nu, \mathbf{s}) = \frac{d\Phi_E(0, \nu, \mathbf{s})}{d\nu} \tag{1.26}$$

TABLE 1.1
Physical Variables for Blackbody Radiation

Quantity	Symbol
Total photon flux	Φ_{ph}
Total energy flux	Φ_E
Spectral photon flux	ϕ_{ph}
Spectral energy flux	ϕ_E

describes the energy flux (power). The radiant flux density is called the spectral *irradiance* E_ν (Sizmann et al., 1991). The spectral photon flux is

$$\phi_{ph}(\nu,\mathbf{s}) = \frac{d\Phi_{ph}(0,\nu,\mathbf{s})}{d\nu} \tag{1.27}$$

The two previous magnitudes are connected by the relationship $\phi_E = h\nu\phi_{ph}$. The total flux of photons in a frequency interval is obtained as

$$\Phi_{ph}(\nu_1,\nu_2,\mathbf{s}) = \int_{\nu_1}^{\nu_2} \frac{d\Phi_{ph}(\nu,\mathbf{s})}{d\nu} d\nu = \int_{\nu_1}^{\nu_2} \phi_{ph}^{bb}(\nu,\mathbf{s}) d\nu \tag{1.28}$$

Similarly, we can obtain

$$\phi_E^{bb}(\lambda,\mathbf{s}) = \frac{d\Phi_E^{bb}(\lambda,\mathbf{s})}{d\lambda} \tag{1.29}$$

which is a very different quantity compared to Equation 1.26, due to the varying relationship between frequency and wavelength intervals, Equation 1.24. However, the integral of Equation 1.29 gives the same result as Equation 1.26 for the corresponding wavelength interval. The different flux distributions derived below are assembled in Table 1.2. Some of the distribution functions are represented in Figure 1.10, in the case of radiation at $T = 5800$ K with a total energy flux $\Phi_{E,tot}^{bb} = 1000\ \mathrm{Wm}^{-2}$ corresponding to that of the terrestrial solar spectrum, as further discussed in Section 1.8.

TABLE 1.2
Formulae of Thermal Blackbody Radiation ($n_r = 1$)

Quantity as Function of	Energy/Frequency	Wavelength
Photons per volume	$\frac{dn_{ph}(\nu)}{d\nu} = \frac{8\pi\nu^2}{c^3}\frac{1}{e^{h\nu/k_BT}-1}$	
Energy per volume	$\rho_{bb}(\nu) = \frac{de_{ph}(\nu)}{d\nu} = \frac{8\pi h\nu^3}{c^3}\frac{1}{e^{h\nu/k_BT}-1}$	$\rho_{bb}(\lambda) = \frac{de_{ph}(\lambda)}{d\lambda} = \frac{8\pi hc}{\lambda^5}\frac{1}{e^{hc/k_B\lambda T}-1}$
Spectral photon flux	$\phi_{ph}^{bb}(E,\Omega) = \frac{\varepsilon}{\pi} b_\pi \frac{E^2}{e^{E/k_BT}-1}$	$\phi_{ph}^{bb}(\lambda,\Omega) = \frac{2\varepsilon c}{\lambda^4}\frac{1}{e^{hc/k_B\lambda T}-1}$
	$\phi_{ph}^{bb,hemi}(\nu) = \frac{2\pi\nu^2}{c^2}\frac{1}{e^{h\nu/k_BT}-1}$	
Spectral energy flux	$\phi_E^{bb}(E,\Omega) = \frac{\varepsilon}{\pi} b_\pi \frac{E^3}{e^{E/k_BT}-1}$	$\phi_E^{bb}(\lambda,\Omega) = \frac{2\varepsilon hc^2}{\lambda^5}\frac{1}{e^{hc/k_B\lambda T}-1}$
	$\phi_E^{bb}(\nu,\Omega) = \frac{2\varepsilon h\nu^3}{c^2}\frac{1}{e^{h\nu/k_BT}-1}$	$\phi_E^{bb,hemi}(\lambda) = \frac{C_1}{\lambda^5}\frac{1}{e^{C_2/\lambda T}-1}$

$b_\pi = \frac{2\pi}{h^3c^2} = 9.883\times10^{26}\ \mathrm{eV^{-3}\,m^{-2}\,s^{-1}}$

$C_1 = 2\pi hc^2 = 3.742\times10^{-16}\ \mathrm{W\,m^2}$

$C_2 = \frac{hc}{k_B} = 1.439\times10^{-2}\ \mathrm{m\,K}$

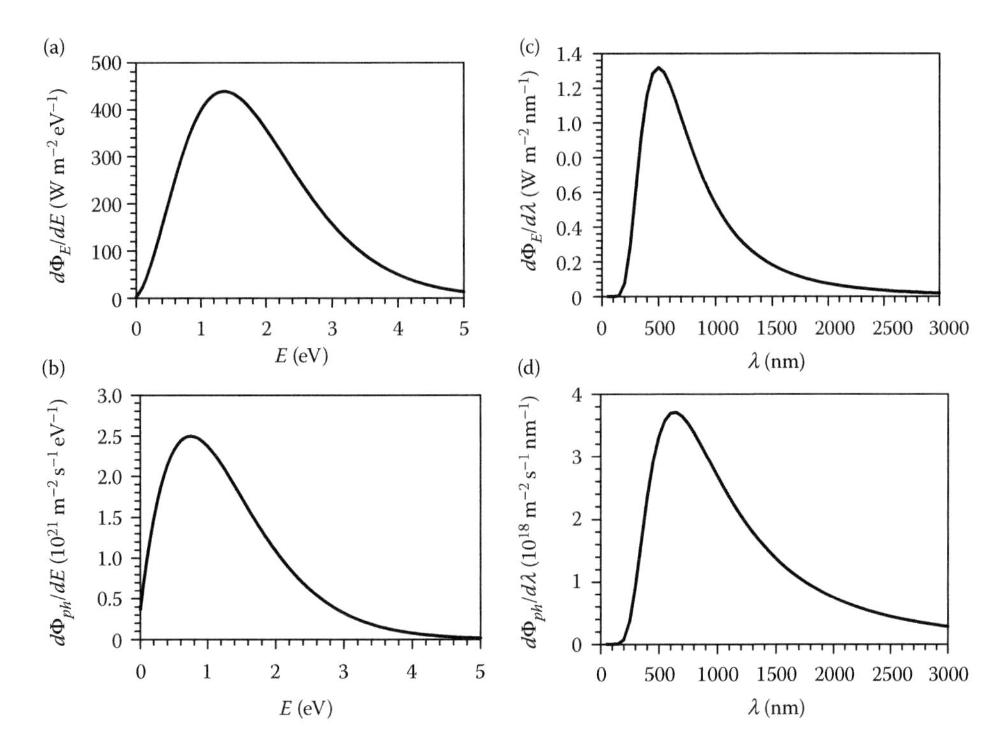

FIGURE 1.10 Energy and photon flux emitted by blackbody at $T = 5800$ K, as a function of energy (a, b) and wavelength (c, d). The energy flux is normalized to a total power density $\phi_E^{AM1.5G} = 1000\,\mathrm{W\,m^{-2}}$.

Let us consider the flux of photons in blackbody radiation that pass through a surface element, θ being the angle between **s** and **n**, as indicated in Figure 1.3b. It is given by

$$d\Phi_{ph}^{bb}(\nu,\mathbf{n}) = c_\gamma \frac{dn_{ph}(\nu)}{d\nu} \cos\theta\, d\Omega\, d\nu \tag{1.30}$$

Hereafter we consider a cone of radiation incoming or outgoing from a solid object as shown in Figure 1.3c. The projected solid angle in Equation 1.30 is integrated around the normal as indicated in Equation 1.8. Hence, the spectral photon number flux density in terms of étendue is

$$\phi_{ph}^{bb}(\nu,\mathbf{n}) = \frac{d\Phi_{ph}^{bb}(\nu)}{d\nu} = \frac{2\varepsilon\nu^2}{c^2} \frac{1}{e^{h\nu/k_BT} - 1} \tag{1.31}$$

If the spectral variable is the photon energy, the spectral distribution of photon flux in blackbody radiation at temperature T per unit energy interval is

$$\phi_{ph}^{bb}(E,\Omega) = \frac{\varepsilon}{\pi} b_\pi \frac{E^2}{e^{E/k_BT} - 1} \tag{1.32}$$

The constant for the radiation to a hemisphere ($\varepsilon = \pi$) is defined as

$$b_\pi = \frac{2\pi}{h^3c^2} = 9.883\times10^{26}\,\mathrm{eV^{-3}\,m^{-2}\,s^{-1}} \tag{1.33}$$

In the range of photon energy $h\nu \gg k_BT$, the term −1 in the denominator of Equation 10.32 can be neglected and the resulting distribution is Wien's approximation.

$$\phi_{ph}^{bb}(E,\Omega) = \frac{\varepsilon}{\pi} b_\pi E^2 e^{-E/k_BT} \tag{1.34}$$

The spectral distribution of energy flux per second, per energy interval dE, is

$$\phi_E^{bb}(E) = \frac{\varepsilon}{\pi} b_\pi \frac{E^3}{e^{E/k_BT} - 1} \tag{1.35}$$

This quantity is related to the *spectral radiance* L_ν (Sizmann et al., 1991)

$$L_\nu(\nu) = \frac{d\Phi_E(\nu)}{\cos\theta \, d\Omega \, d\nu} \tag{1.36}$$

as $\phi_E^{bb}(\nu) = (\varepsilon/\pi)L_\nu(\nu)$. The total energy flux is

$$\Phi_E^{bb} = \frac{\varepsilon}{\pi} b_\pi \int_0^\infty \frac{E^3}{e^{E/k_BT} - 1} dE \tag{1.37}$$

With the change $x = E/k_BT$ and applying

$$\int_0^\infty \frac{x^3}{e^x - 1} dx = \frac{\pi^4}{15} \tag{1.38}$$

we obtain

$$\Phi_{E,tot}^{bb} = \frac{\varepsilon}{\pi} b_\pi \frac{\pi^4}{15} (k_BT)^4 = \frac{\varepsilon}{\pi} \sigma T^4 \tag{1.39}$$

where the Stefan–Boltzmann constant is defined as

$$\sigma = \frac{2\pi^5 k_B^4}{15 h^3 c^2} = 5.6704 \times 10^{-8} \text{ W m}^{-2} \text{ K}^{-4} \tag{1.40}$$

The *Stefan–Boltzmann law* $\Phi_{E,tot}^{bb} = \sigma T^4$ gives the energy flux emitted by the blackbody radiator into a hemisphere, per unit area of the emitting surface.

1.8 THE SOLAR SPECTRUM

The solar spectrum is distributed over a range of wavelengths from 280 to 4000 nm. The total irradiation at the surface of the earth depends on the length of the path through the atmosphere, which is determined by the orientation of the sun with respect to the normal to the earth's surface at a given

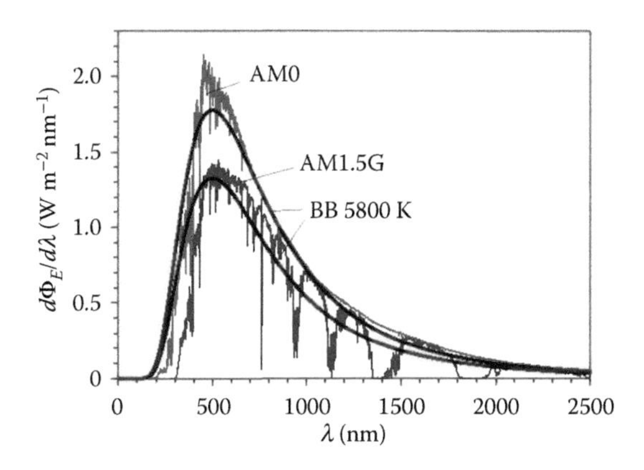

FIGURE 1.11 The spectral irradiance (energy current density, per wavelength interval) from the sun just outside the atmosphere (AM0 reference spectrum) and (AM1.5G) terrestrial solar spectrum. The lines are reference spectra of a blackbody at $T = 5800$ K, normalized to a total power density of $\phi_E^{AM0} = 1366\,\text{W m}^{-2}$ and $\phi_E^{AM1.5G} = 1000\,\text{W m}^{-2}$.

location. The length of the path is quantified by a coefficient denoted *air mass* (AM) that has the value 1 for normal incidence. The solar irradiation arriving at the earth fluctuates with the seasons and varies over extended periods of time. However, for energy conversion applications, it is convenient to refer the operation of devices and processes to a standardized spectrum. Reference spectra are the AM0 corresponding to the spectrum outside the atmosphere and Air Mass 1.5 Global (AM1.5G) that describes the radiation arriving at the earth's surface after passing through 1.5 times a standard air mass, with the sun at 48.2°. Both are shown in Figure 1.11.

The AM0 reference spectrum, representing the typical spectral solar irradiance measured outside the atmosphere, is given by the international standard ASTM E-490 spectrum of the American Society for Testing and Materials. Two reference spectra are defined for the terrestrial irradiance under absolute air mass of 1.5. The AM1.5G (ASTM G173) has an integrated power of

$$\Phi_E^{AM1.5} = 1000\,\text{W m}^{-2} = 100\,\text{mW cm}^{-2} \tag{1.41}$$

and an integrated photon flux of

$$\Phi_{ph}^{AM1.5} = 4.31\times 10^{21}\,\text{s}^{-1}\,\text{m}^{-2} \tag{1.42}$$

The AM1.5 Direct (+circumsolar) spectrum is defined for applications of solar concentrator and has an integrated power density of 900 W m^{-2}.

Consider the sun as a blackbody radiator of radius R_S and temperature T_S. The rate of energy emission from the whole surface is $\sigma T_S^4 4\pi R_S^2$. The flow of energy across a sphere of radius R, centered at the sun, is $f4\pi R$ where the *solar constant* f is a standard measure of the average energy received from sunlight and is defined as the energy received per unit time per unit area (perpendicular to the radiation) at the earth's mean distance from the sun. Therefore, the solar constant has the value

$$f = \left(\frac{R_S}{R}\right)^2 \sigma T_S^4 \tag{1.43}$$

Using $R = 1.5 \times 10^{11}$ m and the outer radius of the sun $R_S = 6.95 \times 10^8$ m, we have $(R_S/R)^2 = 2.147 \times 10^{-5}$. From the effective temperature $T_S = 5760$ K, one obtains the value $f = 1349$ W m^{-2} (Landsberg, 1978). The actual value of the solar constant is obtained by an average of satellite measurements, and the integrated spectral irradiance of ASTM E-490 is made to conform to the accepted value of the solar constant which is

$$f = \Phi_E^{\text{AM0}} = 1366.1 \,\text{W}\,\text{m}^{-2} \tag{1.44}$$

The terrestrial solar spectral irradiance on the surface actually applied in solar energy conversion differs from the extraterrestrial irradiation due to the effect of the filtering by the atmosphere. The scattering by the atmosphere disperses the blue light more than the red part of the spectrum, which is the cause of the red color of sunset. Therefore, the atmospheric extinction affects predominantly the shorter frequencies of the incoming radiation. In addition, selective absorption by low concentration gases causes a strong decrease or even full extinction of the radiation in certain specific ranges of wavelengths that are indicated in Figure 1.2. In the infrared, the terrestrial solar spectrum is especially highly structured by the principal agents of opacity that are carbon dioxide and water vapor.

Most bodies on the earth's surface emit thermal infrared radiation, and the interception by the atmosphere is a cause of greenhouse effect. The emission and absorption properties of the atmosphere are observed in detail in Figure 1.12. By looking at space from the surface, the atmosphere shows a transparency window (8–13 mm) that can be used to dissipate heat from the earth into outer space for different applications such as passive building cooling, renewable energy harvesting, and passive refrigeration in arid regions (Chen et al., 2016). For thermophotovoltaic applications, it is important to develop materials in which the normal radiative emission in the infrared is suppressed (Dyachenko et al., 2016).

It is sometimes practical to use an analytical function that approximates solar AM1.5G spectrum for evaluation of the efficiency of solar energy conversion. If the receiver measures only the spectral band of the visible and the close infrared, a blackbody spectrum of 5800 K approximately describes the spectrum in this wavelength range (Zanetti, 1984), see Figure 1.11. Though no single function can approach all details of the standard spectrum, we choose a blackbody spectrum at $T_S = 5800$ K (with the thermal factor $k_B T_S = 0.5$ eV) as shown in Figure 1.13, which indicates the quality of the approximation. To use the blackbody radiation expression, the photon flux at the earth's surface is reduced with respect to the emission at the sun's surface by the projected solid angle $\varepsilon_S^{\text{AM1.5}}$

$$\phi_{ph}^{\text{bb, 1 sun}}(E, T_S) = \frac{\varepsilon_S^{\text{AM1.5}}}{\pi} b_\pi \frac{E^2}{e^{E/k_B T_S} - 1} \tag{1.45}$$

The total energy flux is given in Equation 1.38. Comparing the total irradiated power of the blackbody at $T_S = 5800$ K to Air Mass 1.5 Global (AM1.5G) conditions ($\Phi_{E,\text{tot}} = 1000$ W m^{-2}), we obtain the value of the dilution factor (Smestad and Ries, 1992)

$$D = \frac{\varepsilon_S^{1.5\text{AM}}}{\pi} = 1.56 \times 10^{-5} \tag{1.46}$$

and the corresponding *equivalent* projected solid angle is $\varepsilon_S^{1.5\text{AM}} = D\pi = 4.8 \times 10^{-5}$. This is slightly different from the actual solid angle of the sun since $\varepsilon_S^{1.5\text{AM}}$ takes into account the extinction of the atmosphere.

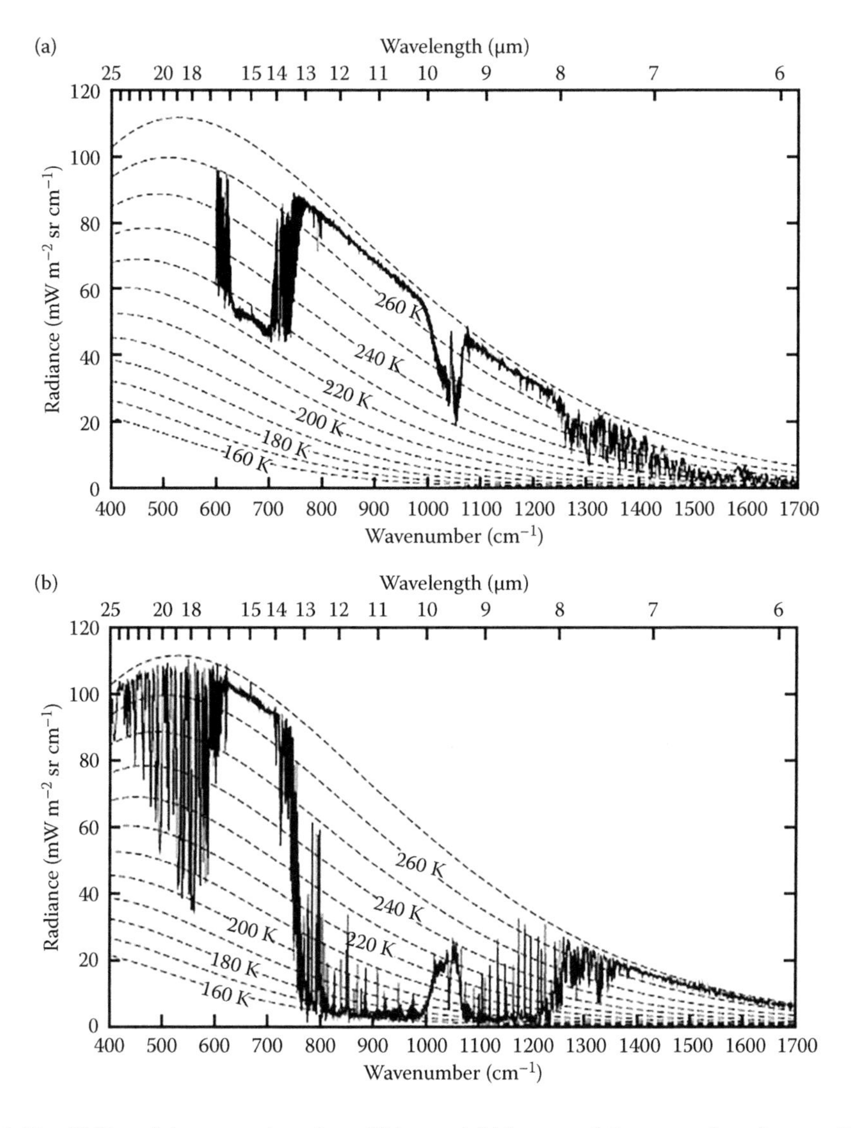

FIGURE 1.12 (a) Top of the atmosphere from 20 km and (b) bottom of the atmosphere from surface in the Arctic. Dashed curves represent the Planck function at different temperatures. The CO_2 *bite* is between the 14 and 17 μm, and the *bite* in the 9.5–10 μm area is due to O_2 and O_3 absorption spectra. In these bands, the atmosphere is opaque to outgoing radiation. Emission observed when looking up is from the lowest levels of the atmosphere, closest to the ground. However, the emission of the ground, at ≈270 K, is observed when looking down from a plane 20 km high. In (a) the emission from CO_2 and O bands comes from higher levels in the atmosphere and corresponds to the Planck spectrum of lower temperatures, ≈270 K. (Reproduced with permission from Petty, G. W. *Atmospheric Radiation*, 2nd edition; Sundog Publisher: Madison, 2006.)

The Standard Test Conditions (STC) for testing and verification of photovoltaic cell performance consist of air mass, sunlight intensity, and cell temperature at AM1.5G, 100 mW cm^{-2}, and 25°C, respectively. To test solar conversion devices, it is necessary to use a radiation source that approaches the AM1.5G spectrum as much as possible. Artificial light sources have important limitations to mimic the solar spectrum. It should be checked that the calibrated photon flux and energy flux matches the standard spectrum in the absorption range of the device (Murphy et al., 2006; Doscher et al., 2016). The spectra of widely used xenon lamp solar simulators are shown in Figure 1.14. Figure 1.15 shows the spectra of typical laboratory sources of white light, compared to the AM1.5G irradiance.

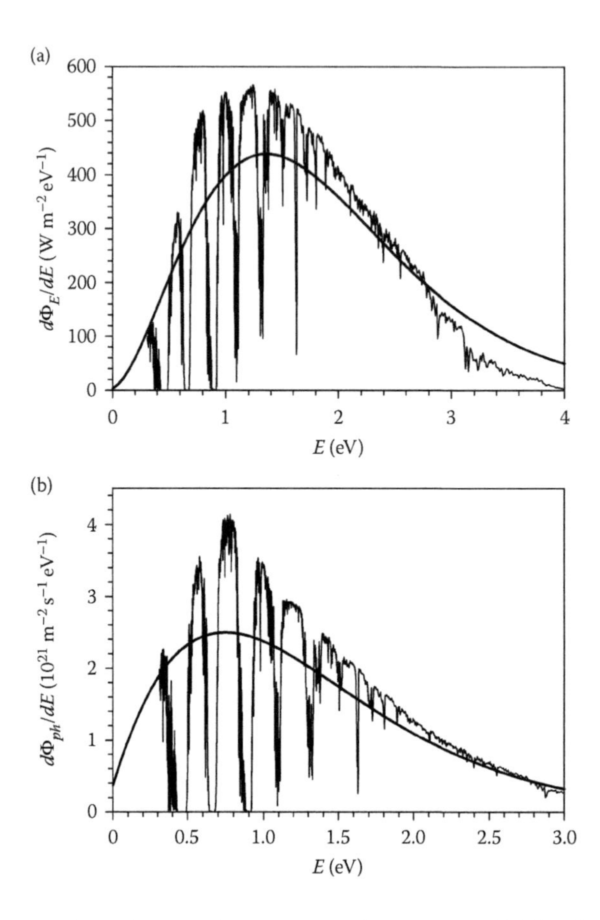

FIGURE 1.13 Energy and photon flux for the terrestrial solar spectrum (AM1.5G) as a function of photon energy. The lines are reference spectra of a blackbody at $T = 5800$ K, normalized to a total power density of $\phi_E^{AM1.5G} = 1000\,\mathrm{W\,m^{-2}}$.

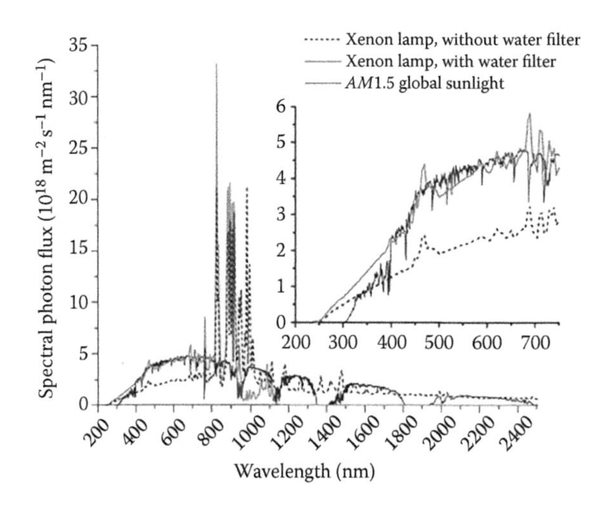

FIGURE 1.14 Spectral photon flux for the xenon lamp, with and without the water filter, compared to the AM1.5G solar spectrum. The lamp had logged 1250 h of operation when the spectrum was recorded. The inset shows detail of short wavelengths. In all cases, the total irradiance is normalized to 1000 W m^{-2}. (From Murphy, A. B. et al. *International Journal of Hydrogen Energy* 2006, 31, 1999–2017.)

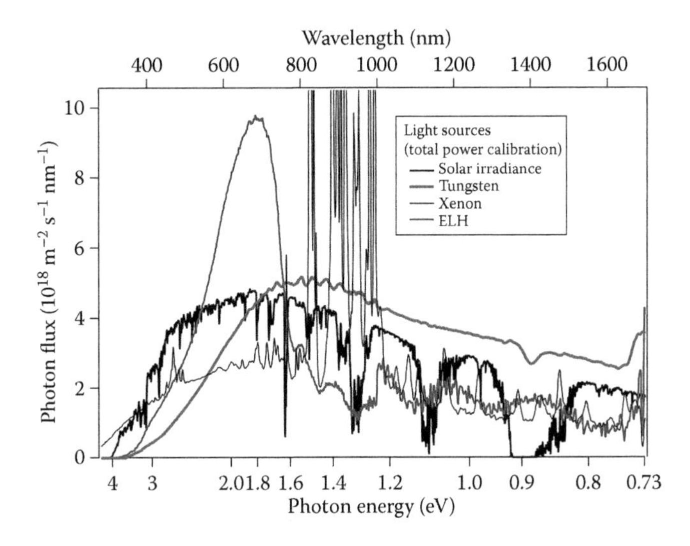

FIGURE 1.15 Spectral distribution of flux for AM1.5G irradiance compared to typical laboratory white light sources (adjusted to provide solar equivalent illumination power). ELH source is a tungsten halogen lamp with dichroic reflector. (Reproduced with permission from Doscher, H. et al. *Energy & Environmental Science* 2016, 9, 74–80.)

GENERAL REFERENCES

Blackbody radiation and the Planck spectrum: Kangro (1976), Waldman (1983), Sizmann et al. (1991), MacIsaac et al. (1999), Robitaille (2008), Castrejón-García et al. (2010), and Howell et al. (2010).

Distribution of photons in blackbody radiation: Ross (1967), Landsberg (1978), Yablonovitch (1982), Ries and McEvoy (1991), Overduin (2003), Markvart (2007), Würfel (2009), and Mooney and Kambhampati (2013).

REFERENCES

Castrejón-García, R.; Castrejón-Pita, J. R.; Castrejón-Pita, A. A. Design, development, and evaluation of a simple blackbody radiative source. *Review of Scientific Instruments* 2010, 81, 055106.

Chen, Z.; Zhu, L.; Raman, A.; Fan, S. Radiative cooling to deep sub-freezing temperatures through a 24-h day–night cycle. *Nature Communications* 2016, 7, 13729.

Doscher, H.; Young, J. L.; Geisz, J. F.; Turner, J. A.; Deutsch, T. G. Solar-to-hydrogen efficiency: Shining light on photoelectrochemical device performance. *Energy & Environmental Science* 2016, 9, 74–80.

Dyachenko, P. N.; Molesky, S.; Petrov, A. Y.; Störmer, M.; Krekeler, T.; Lang, S.; Ritter, M.; Jacob, Z.; Eich, M. Controlling thermal emission with refractory epsilon-near-zero metamaterials via topological transitions. *Nature Communications* 2016, 7, 11809.

Fowles, G. R. *Introduction to Modern Optics*; Dover: New York, 1989.

Howell, J. R.; Siegel, R.; Menguc, M. P. *Thermal Radiation Heat Transfer*; CRC Press: New York, 2010.

Kangro, H. *Early History of Planck's Radiation Law*; Taylor & Francis: London, 1976.

Kennard, E. H. On the thermodynamics of fluorescence. *Physical Review* 1918, 11, 29–38.

Kiang, N. Y.; Siefert, J.; Govindjee, B.; Blankenship, R. E. Spectral signatures of photosynthesis. I. Review of Earth organisms. *Astrobiology* 2007, 7, 222–251.

Landsberg, P. T. *Thermodynamics and Statistical Mechanics*; Dover: New York, 1978.

MacIsaac, D.; Kanner, G.; Anderson, G. Basic physics of the incandescent lamp (lightbulb). *The Physics Teacher* 1999, 37, 520 523.

Markvart, T. Thermodynamics of losses in photovoltaic conversion. *Applied Physics Letters* 2007, 91, 064102.

Markvart, T. The thermodynamics of optical étendue. *Journal of Optics A: Pure and Applied Optics* 2008, 10, 015008.

Mooney, J.; Kambhampati, P. Get the basics right: Jacobian conversion of wavelength and energy scales for quantitative analysis of emission spectra. *The Journal of Physical Chemistry Letters* 2013, 4, 3316–3318.

Murphy, A. B.; Barnes, P. R. F.; Randeniya, L. K.; Plumb, I. C.; Grey, I. E.; Horne, M. D.; Glasscock, J. A. Efficiency of solar water splitting using semiconductor electrodes. *International Journal of Hydrogen Energy* 2006, 31, 1999–2017.

Overduin, J. M. Eyesight and the solar Wien peak. *American Journal of Physics* 2003, 71, 219–219.

Petty, G. W. *Atmospheric Radiation*, 2nd edition; Sundog Publisher: Madison, 2006.

Planck, M. *The Theory of Heat Radiation*; P. Blakiston's Son & Co: Philadelphia, PA, 1914.

Ries, H.; McEvoy, A. J. Chemical potential and temperature of light. *Journal of Photochemistry and Photobiology A: Chemistry* 1991, 59, 11–18.

Robitaille, P.-M. Blackbody radiation and the carbon particle. *Progress in Physics* 2008, 3, 36–55.

Ross, R. T. Some thermodynamics of photochemical systems. *The Journal of Chemical Physics* 1967, 46, 4590–4593.

Shirasaki, Y.; Supran, G. J.; Bawendi, M. G.; Bulovic, V. Emergence of colloidal quantum-dot light-emitting technologies. *Nature Photonics* 2012, 7, 13–23.

Sizmann, R.; Köpke, P.; Busen, R. Solar radiation conversion. In *Solar Power Plants. Fundamentals, Technology, Systems, Economics*; Winter, C.-J., Sizmann, R. L., Vant-Hull, L. L. (Eds.); Springer: Berlin, 1991.

Smestad, G.; Ries, H. Luminescence and current-voltage characteristics of solar cells and optoelectronic devices. *Solar Energy Materials and Solar Cells* 1992, 25, 51–71.

Waldman, G. *Introduction to Light*; Dover Publications: New York, 1983.

Würfel, P. *Physics of Solar Cells. From Principles to New Concepts*, 2nd edition; Wiley: Weinheim, 2009.

Yablonovitch, E. Statistical ray optics. *Journal of the Optical Society of America* 1982, 72, 899–907.

Yablonovitch, E.; Cody, G. D. Intensity enhancement in textured optical sheets for solar cells. *IEEE Transactions on Electron Devices* 1982, 29, 300–305.

Zanetti, V. Sun and lamps. *American Journal of Physics* 1984, 52, 1127–1130.

2 Light Absorption, Carrier Recombination, and Luminescence

The interaction of photons with molecules and materials, resulting in the absorption and production of radiation, is a fundamental component of many energy devices. A range of spectral techniques based on light absorption and emission shows the physical properties of ions, molecules, and materials. Light is used to generate carriers on a large scale in a solar cell and conversely, radiative recombination produces light emission in chromophores and semiconductor LEDs. In this chapter, we discuss the general properties of light absorption and luminescence in solids, nanostructured materials, and molecules. We address the general concepts of quantum efficiency (QE) and quantum yield (QY) that govern the conversion between electricity and light. Finally, we review basic recombination mechanisms in semiconductors and the methods of determination of carrier lifetime.

2.1 ABSORPTION OF INCIDENT RADIATION

Incident radiation on a material results in different types of interactions. The incident light can be reflected at any surface. A part of the radiation is transmitted through the back surface. The *coefficient of reflection* R is defined as the ratio of reflected power to the incident power of the electromagnetic radiation and similarly, the *transmittance* T is the ratio of transmitted power to the incident power. For a beam of energy E propagating in the x direction, with an intensity (power per unit area) $\Phi_E(E, x)$, we have

$$T = \frac{\Phi_E(E,x)}{\Phi_E(E,0)} \tag{2.1}$$

In the absence of absorption, the conservation of energy states that

$$R + T = 1 \tag{2.2}$$

The laws of reflection and refraction can be deduced from Huygens' principle. The refractive index of a medium is related to the relative dielectric constant as

$$n_r = \sqrt{\varepsilon_r} \tag{2.3}$$

For normal incidence, the coefficient of reflection is

$$R = \left[\frac{n_r - 1}{n_r + 1}\right]^2 \tag{2.4}$$

The light passing through the material can also be absorbed or scattered. In addition, some of the absorbed light can be reemitted. In this section, we consider the description of the absorption processes. The light-emitting processes will be described in the next section. Different effects of

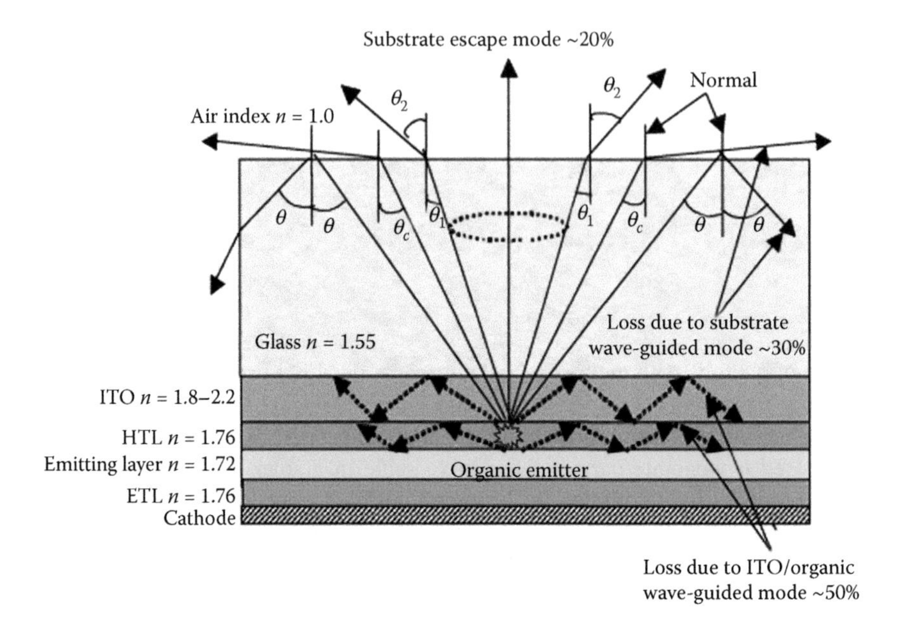

FIGURE 2.1 Schematic of multilayer OLED structure and optical ray diagram of light propagation via various modes, that is, substrate escape, substrate wave-guided mode, and ITO/organic wave-guided modes. (Reproduced with permission from Saxena, K. et al. *Optical Materials* 2009, 32, 221–233.)

light propagation via various reflection modes in multilayer organic light-emitted diodes (OLEDs) are shown in Figure 2.1.

The absorption of light by a medium is quantified by its linear absorption coefficient α (m^{-1}), which is defined as the fraction of the power absorbed per unit volume. In terms of the decrease of intensity in an incremental slice of thickness dx, we have

$$\alpha(E) = -\frac{1}{\Phi_E(E,x)}\frac{d\Phi_E}{dx} \tag{2.5}$$

The integral of Equation 2.5 gives Beer–Lambert's law that describes the attenuation of light in a medium

$$\Phi_E(x) = \Phi_E(0)e^{-\alpha x} \tag{2.6}$$

In solar cells, efficiently absorbing most of the solar photons is a crucial aspect of materials and device design. The quantity $L_\alpha = \alpha^{-1}$ is denominated as the absorption length. It indicates the characteristic size of the absorber layer needed to collect all the solar spectral photons. Increasing the thickness of the active layer improves the device absorptivity but involves other drawbacks in device operation. To enhance the probability to convert photons to electricity, the light in the solar cell should be internally reflected until it is absorbed.

Consider the passage of the incident radiation through a medium across a length z. The dimensionless product

$$A = \alpha z \tag{2.7}$$

is called the *absorbance* or the *optical density*. The intensity decreases as

$$\Phi_E = \Phi_E(0)e^{-A} \tag{2.8}$$

and we have

$$\log T = -A \tag{2.9}$$

When the absorption is a function of the absorbing centers in the medium, the absorbance is expressed as

$$\Phi_E(z) = \Phi_E(0)10^{-\varepsilon cz} \tag{2.10}$$

where c is the molar concentration of absorbing species and the quantity ε, in units M^{-1} cm^{-1}, is called the decadic *molecular extinction coefficient.*

We now consider the light absorption property of a slab of thickness d. The fraction of incoming radiation absorbed at a given wavelength or photon energy E and not reflected or emitted is described by the function $a(E)$, the spectral absorptivity or *absorptance* ($0 \leq a \leq 1$). It depends on the absorption coefficient $\alpha(E)$, the thickness of the layer, and the reflection properties of the surfaces. If reflection can be neglected, the absorptance relates to the absorption coefficient as

$$a(E) = \frac{\Phi_E(0) - \Phi_E(d)}{\Phi_E(0)} = 1 - e^{-\alpha d} \tag{2.11}$$

For an absorber layer with a backside mirror, the optical path length is doubled, and we have

$$a(E) = 1 - e^{-2\alpha d} \tag{2.12}$$

Considering the reflectivity at the front and back surface $R_f(E)$ and $R_b(E)$, a more general expression is obtained (Trupke et al., 1998):

$$a(E) = \frac{(1 - R_f(E)) \cdot (1 - e^{-\alpha(E)d})(1 + R_b(E)e^{-\alpha(E)d})}{1 - R_f(E)R_b(E)e^{-2\alpha(E)d}} \tag{2.13}$$

If both reflectivities are identical and the absorption is very weak, $\alpha d \ll 1$, then

$$a(E) = \alpha d \tag{2.14}$$

However, if $d \gg L_\alpha$, all photons are absorbed except those reflected at the front surface

$$a(E) = 1 - R_f(E) \tag{2.15}$$

We note that if we neglect reflection effects, the absorptance has two important limits. When the film is optically thick, it is always $a = 1$ and when it is thin, the absorptance is proportional to absorption coefficient as in Equation 2.14.

Scattering refers to the interaction of a light beam with small particles in the medium. The scattering coefficient α_{sc} depends on the ratio of the size of the scattering particles to the wavelength of the light. The reduction in the intensity of a beam of light, which has traversed a medium containing scattering particles, is given by an expression identical to Beer's law:

$$\Phi_E(x) = \Phi_E(0)e^{-\alpha_{sc}x} \tag{2.16}$$

Scattering is maximized when the particle size is somewhat less than the wavelength of the light. It also depends on the ratio of the refractive indices of the particle and the surrounding medium. Scattering layers can play a large role to enhance photon collection in nanostructured solar cells (Usami, 1997).

In a solar cell with planar surfaces, the absorptance is essentially given by Equation 2.12, and if a reflective coating is used in the backside surface the optical pathway is only doubled. Weakly absorbed light in the long wavelength region, therefore, has a low chance to contribute to carrier generation by absorption. The light generated or scattered inside the solar cell that impacts the surface will leave the sample if it is included in the *escape cone* of the surface, Ω_c. The solid angle subtended by the escape cone is

$$\Omega_c = \frac{1}{2n_r^2} 4\pi \tag{2.17}$$

Thus, for a planar surface, a randomly generated ray has a probability $p_e = \Omega_c/4\pi = 1/2n_r^2$ to escape from the active zone in an encounter with the surface (Yablonovitch, 1982). The high-refraction index of the semiconductor reduces the escape cone for the emission of the generated radiation. For silicon with a refractive index $n_r \approx 3.6$ in the wavelength range around $\lambda = 1150$ nm (Trupke et al., 2003a,b), a high-reduction factor $2n_r^2 \approx 26$ is obtained.

Randomly textured surfaces or regularly textured surfaces enhance the light absorption providing a light trapping effect due to the randomizing of the internal reflection angle avoiding the escape of internally reflected rays. This effect results in a much longer propagation distance of the light ray and, hence, a substantial increase of absorptance. A randomly textured surface also increases the probability of light emission, since the photons that hit the surface out of the narrow escape cone can be scattered into the cone by the random orientation. A Lambertian scatterer at the front surface disperses the light according to the law in Equation PSC.1.9, which gives the highest degree of randomization. The upper limit of light path enhancement for random scattering textures has been shown to be the quantity $4n_r^2$ (Yablonovitch, 1982), and this is called the Yablonovitch limit (or the Lambertian limit). In a textured cell, in the weak absorption limit, the absorptance is well described by the expression given by Tiedje et al. (1984):

$$a(E) = \frac{\alpha}{\alpha + \dfrac{1}{4n_r^2 d}} \tag{2.18}$$

Figure 2.2a shows the higher light emission of a textured silicon diode compared to planar diodes. The light emission properties of the textured cell compared with the blackbody radiation will be further discussed in Section 2.3 and PSC.7.4.

In solar cells made of very thin layers that use high-absorption materials, the active layers may be of the same order of size as the wavelength of the light, which leads to interference effects. The cell design must take these effects into account in order to optimize the light harvesting properties as well as the spatial distribution of photogeneration of charge.

2.2 LUMINESCENCE AND ENERGY TRANSFER

In Chapter PSC.1, we have described blackbody radiation, which is the emission of radiation resulting from heat, also denoted as incandescence. *Luminescence* generally denotes the emission of light by a material after it has absorbed energy. Luminescence requires the promotion of an atom, a molecule, or a solid to an excited state that is subsequently demoted, emitting a photon.

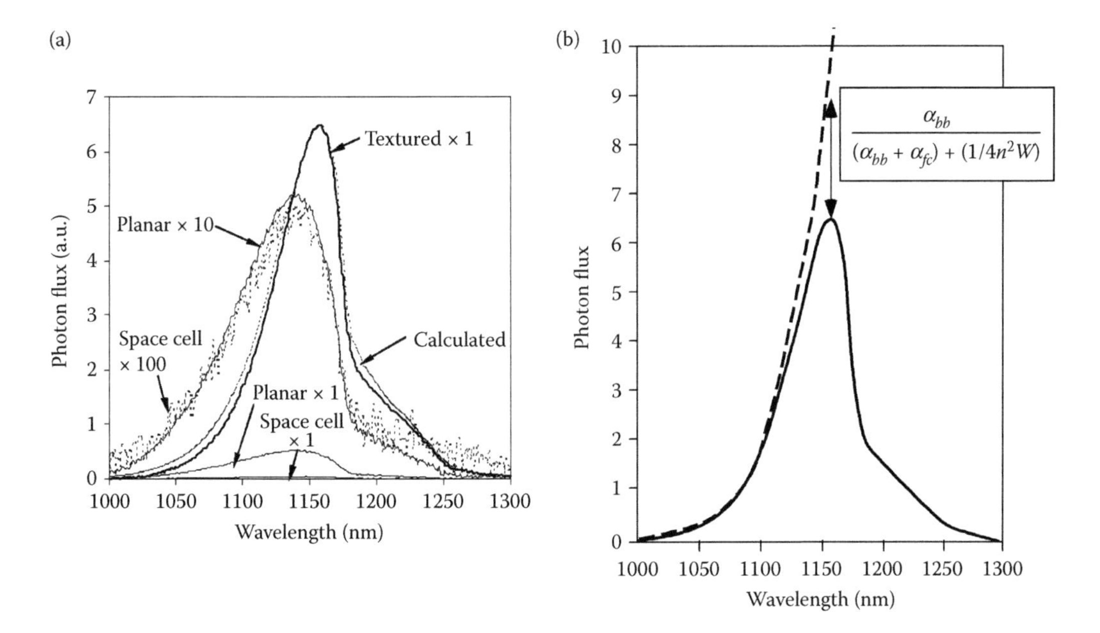

FIGURE 2.2 (a) Photon flux versus wavelength emitted by three silicon diodes. The lowest curve (shown multiplied 100 times) is for a high-performance silicon space cell. Planar cells perform 10 times better, due to reduction of parasitic nonradiative recombination. Textured cells perform 10 times higher again due to better optical properties, particularly much higher absorptance and hence emittance at the wavelengths shown. (b) Luminescence of a front-textured high-performance silicon solar cell compared to the blackbody emission (*dashed line*) for energies above the silicon bandgap (wavelengths shorter than 1102 nm). The expression in the inset (Equation 2.18) gives the ratio of these values, α_{bb} being the band-to-band absorption, α_{fc} the free carrier absorption, n the refraction index, and W the device thickness. (Reproduced with permission from Green, M. A. et al. *Physica E: Low-dimensional Systems and Nanostructures* 2003, 16, 351–358.)

Depending on the nature of the excitation source, there are different types of luminescence. If the excited state is obtained by photoexcitation, then the subsequent light emission is called photoluminescence (PL), Figure 2.3. The major forms of PL are *fluorescence* and *phosphorescence*. If the carriers are created by voltage injection to a solid, then the light emission process is called *electroluminescence* (EL), Figure 2.4.

In Section PSC.3.5, we show that the main PL emission in organic molecules is by spontaneous photon emission from the first singlet excited state and is called *fluorescence*. In general, fluorescence is a luminescence that occurs only during excitation or shows a very short decay time ($\tau < 10$ ms). On the other hand, *phosphorescence* persists for some time after the excitation has been disconnected, due to the long decay time ($\tau > 0.1$ s).

Photon absorption and emission in an ion or molecule are physically reciprocal processes. The respective probability coefficients are connected by a detailed balance relationship by Einstein (Strickler and Berg, 1962). If the absorption coefficient is large, so is the emission coefficient, and the radiative lifetime is short as shown in Equation PSC.6.23. However, in organic chromophores and in inorganic quantum dots, absorption and emission do not occur at the same wavelength due to the vibrational or thermal relaxation losses that occur after the absorption process, as shown in Figure 2.3a. This general principle is known as Stokes law. The displacement of the peak of the emission to longer wavelengths from the absorption peak is generally known as *Stokes shift*. It is normally measured as the difference between the maximum wavelengths in the excitation and emission spectra. The Stokes shift due to vibrational or orientational relaxation is discussed in Section PSC.3.5.

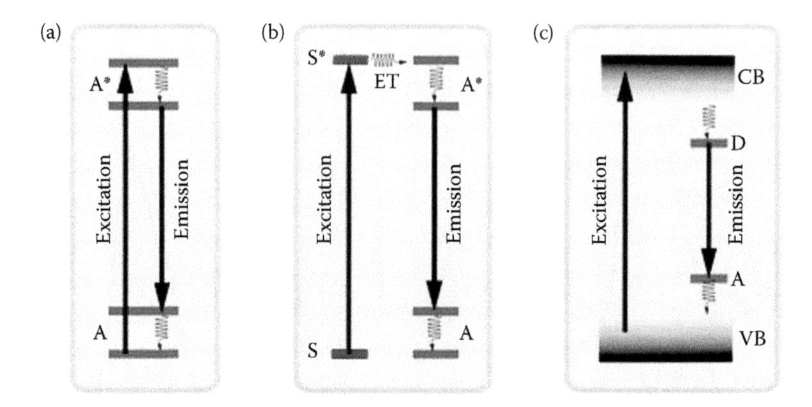

FIGURE 2.3 Luminescence processes. (a) Emission from a luminescence activator upon excitation. (b) Sensitized emission from an activator through energy transfer from a sensitizer to the activator upon excitation of the sensitizer. (c) Emission from a semiconductor after band-to-band excitation. A and A* represent the ground and excited states of the activator, respectively. S and S* represent the ground and excited states of the sensitizer, respectively. VB and CB represent the valence and conduction bands of the semiconductor while D and A represent the donor and acceptor energy levels, respectively. (Reproduced with permission from Huang, X. et al. *Chemical Society Reviews* 2013, 42, 173–201.)

Inorganic solids that give rise to luminescence are called *phosphors* or luminescent materials. The luminescence can be obtained from two different mechanisms. One is the emission of localized centers or activators embedded in a larger host crystalline structure that constitutes the bulk of the phosphors. The second is recombination in semiconductors, which may be band to band or via recombination centers. In semiconductors, most electrons and holes undergo rapid thermalization on a ps time scale, as discussed in Figure ECK.5.8, so that band-to-band recombination occurs between carriers close to the respective band edges, Figures 2.3 and 2.4. Therefore, the emitted photon spectrum peaks sharply at energy $h\nu \approx E_g$. A variety of recombination mechanisms are discussed in Section PSC.2.4. EL can also be influenced by the free-exciton, bound-exciton, free-to-bound, and donor–acceptor recombination mechanisms.

The main metal ions used in inorganic host materials to form phosphors are lanthanide ions, which are a family of 15 chemically similar elements from lanthanum (La) to lutetium (Lu). The lanthanides and the yttrium (Y) are generally denoted as "rare earths." The lanthanides show rich optical properties due to a large number of possible transitions between diverse $4f^n$ states ($0 < n < 14$). The transitions between the different $4f^n$ states are parity forbidden and become allowed through

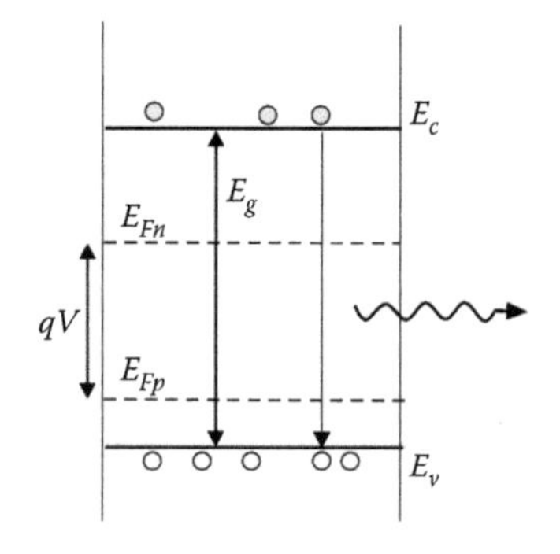

FIGURE 2.4 Emission of a photon by radiative recombination in an LED.

mixing with opposite parity states of crystal field components of vibrations. The absorption lines are very narrow because the coupling with the vibrations is weak. There is no Stokes shift from absorption to emission, which allows for avoiding thermal losses by vibrations, enhancing the luminescent QY.

The decay from an excited molecular state or the recombination of an electron–hole pair does not necessarily imply the emission of a photon, because the electron in the excited state may decay via other *nonradiative* recombination pathways. In this case, the energy of the excitation is dissipated in the lattice of the material as heat. Nonradiative recombination processes are mediated by phonons and are largely suppressed at low temperatures. Another way to prevent radiationless recombination is to use a dielectric environment in which the phonon frequency is small. This is because multiphonon processes become less likely when the number of phonons that is involved in the decay of the excited state is larger.

The mechanism of sensitization separates the absorption and emission processes in PL into two different ions or molecules, as shown in Figure 2.3b. In the terminology of phosphors, the *sensitizer* absorbs light and transfers the excitation to an *activator*, which emits a photon. The acceptor can also be termed the *annihilator* or *emitter*. Conservation of energy in the transfer process and energy relaxation in the vibrational modes of the activator implies that Stokes shift will normally occur.

There are several ways to transfer the energy from a sensitizer to a target molecule. In photovoltaic applications, one can extract electrons and holes from the sensitizer to metal oxide and electrolytic contacts (Gerischer et al., 1968), as in a dye-sensitized solar cell. For the transfer from one molecule to another, one normally distinguishes three mechanisms: radiative transfer, nonradiative transfer, and multiphonon-assisted energy transfer. In the *radiative transfer* mechanism, a real photon is emitted by the sensitizer and reabsorbed in the acceptor. *Resonant energy transfer involves* the excitation of a donor molecule that decays and passes its energy to the acceptor molecule by a suitable interaction, before the sensitizer is able to emit a quantum of fluorescence. *Förster resonant energy transfer* (FRET) occurs by Coulombic dipole–dipole interaction (Scholes, 2003). The rate constant for energy transfer is given by the expression

$$k_{ET} = k_D \left(\frac{R_0}{R} \right)^6 = \frac{1}{\tau_D^0} \left(\frac{R_0}{R} \right)^6 \tag{2.19}$$

Here $k_D = 1/\tau_D^0$ is the decay rate constant of the excited donor in the absence of energy transfer, R the distance between donor and acceptor, and R_0 is the *critical quenching radius* or *Förster radius*, that is, the distance at which the rate constants $k_{ET} = k_D$, so that energy transfer and spontaneous decay of the excited donor are equally probable (Braslavsky et al., 2008). This mechanism operates for large separations up to 2 nm. FRET gives rise to diffusion of excitons and is applied in antenna complexes in the natural photosystem, in which the excitation travels until it arrives to the reaction center.

Energy transfer based on higher multipole and exchange interaction is termed *Dexter energy transfer*. It depends on the spatial overlap of wave functions and operates only at very short distances (<0.5 nm). FRET has become a general denomination for energy transfer that does not involve fluorescence in the donor. Different types of excitonic states are discussed in Chapter PSC.3.

2.3 THE QUANTUM EFFICIENCY

Starting from the QY, we define several quantum efficiencies for LEDs and solar cells. The QY of a photo-induced process is the number of defined events (electron–hole generation, fluorescence, photochemical reaction) occurring per photon absorbed. The QY is defined for monochromatic radiation of frequency ν, hence the denominator contains the incident spectral photon flux $\phi_{ph}^{in}(\nu)$. Assuming $a(\nu) = 1$, we have

$$\eta_{QY}(\nu) = \frac{\text{Number of events}}{\phi_{ph}^{in}(\nu)} \tag{2.20}$$

The *QE* of a luminescent material is the number of photons emitted per input. The input can be an incident photon flux, as in PL, or the (current density)/q, in EL, as in an LED. The QE is defined for monochromatic radiation of frequency ν.

The *external quantum efficiency* (EQE) in PL is given by the ratio of incoming and outgoing photon flux

$$EQE_{PL}(\text{n}) = \frac{\phi_{ph}^{out}(\nu)}{\phi_{ph}^{in}(\nu)} \tag{2.21}$$

The EQE is a combination of several factors:

$$EQE_{PL} = \eta_{LHE} IQE \eta_{outco} \tag{2.22}$$

The light harvesting efficiency (LHE) is the number of photons that are absorbed by the converter, determined by absorption coefficient, thickness, and other optical features such as reflection, texturing, scattering layers, etc., which were discussed in Section PSC.2.1, so that, in terms of the absorptivity, $\eta_{LHE} = a(\nu)$. The *internal quantum efficiency* (IQE) is the probability that an excited molecule or an electron–hole pair recombines radiatively. Extraction or outcoupling efficiency η_{outco} is the fraction of internally generated photons that escape out of the material and contribute to the radiated emission. It depends strongly on the geometry and optical characteristics of the material, as indicated in Figure 2.1.

A semiconductor LED operates on the principle of injection of electrons and holes from separate contacts that recombine radiatively in the active layer. The EQE is defined as the ratio of output-emitted photons to electron–hole pairs injected by the current density j

$$EQE_{LED} = \frac{\Phi_{ph}^{out}}{j/q} \tag{2.23}$$

The EQE is related to two efficiency factors:

$$EQE_{LED} = IQE_{LED} \eta_{outco}^{LED} \tag{2.24}$$

The external voltage produces total recombination current j_{rec} that contains both radiative j_{rad} and nonradiative current j_{nrad}, hence the LED IQE is given by the relationship

$$IQE_{LED}(V) = \frac{j_{rad}(V)}{j_{rec}(V)} = \frac{j_{rad}(V)}{j_{rad}(V) + j_{nrad}(V)} \tag{2.25}$$

In organic LEDs, it is often necessary to establish the process of formation of excitons that subsequently produce the radiative emission. Hence, the EQE contains additional terms as follows:

$$EQE_{LED} = \eta_{inject}^{LED} \chi_{op} \eta_{rad}^{LED} \eta_{outco}^{LED} \tag{2.26}$$

The injection efficiency, η_{inject}^{LED}, is the fraction of injected carriers that become excited electron–hole pairs, or excitons. χ_{op} is the fraction of excitons whose states have spin-allowed optical transitions. For thermalized triplet and singlet states in organic LEDs, $\chi_{op} = 0.25$. η_{rad}^{LED} is the ratio of radiative to total recombination. The first three factors in Equation 2.26 form the IQE of the LED.

Internally trapped photons can be absorbed and reemitted in a cyclic fashion. *Photon recycling*, further described in Section PSC.7.4, refers to the process of reabsorption and reemission of photons that are not able to escape the LED in the first emission cycle produced by the voltage. Photon recycling allows photons to be reemitted several times before eventually escaping the LED provided that the nonradiative recombination rate is low. Under total internal reflection and photon recycling processes, the existence of nonradiative recombination pathways enhances the probability that multiple regenerated photons are finally recombined nonradiatively, decreasing the external quantum efficiency of the LED. In the presence of photon recycling and reabsorption effects, for a large IQE value the product in Equation 2.24 is not valid and a more general expression is given in Equation PSC.7.32.

The rate of external emission of either direct light radiation or regenerated photons in an LED also depends on the optical characteristics of the device as shown in Figure 2.1. As commented earlier in Equation 2.17, the escape probability of a photon is considerably reduced when the refractive index of the medium is high. Determination of η_{outco} can be made at low temperatures where nonradiative recombination is reduced and the IQE is close to unity. In practice, separating all the different factors involved in the EQE of an LED is a difficult task (Matioli and Weisbuch, 2011; Kivisaari et al., 2012).

The PL EQE of a textured 500 μm thick n-type silicon sample is shown in Figure 2.5. The EQE increases at low temperature because the absorption coefficient of silicon for band-to-band transitions decreases strongly in the spectral range 1000–1250 nm (1.0–1.2 eV), where significant luminescence is emitted from bulk silicon, see Figure 2.2. At lower temperatures the PL is enhanced, as the reabsorption of internally generated photons by band-to-band transitions is considerably reduced (Trupke et al., 2003a,b).

In a solar cell, the photovoltaic EQE is the ratio of electron flux, in the form of electrical current (at short circuit) to the incoming photon flux:

$$EQE_{PV}(\nu) = \frac{j(\nu)}{q\phi_{ph}(\nu)} \tag{2.27}$$

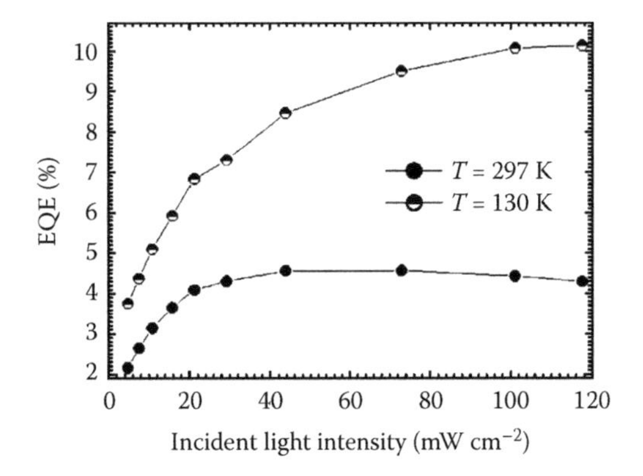

FIGURE 2.5 PL *EQE* of a textured 500 μm thick 30 Ω cm n-type silicon sample as a function of the incident light intensity. (Reproduced with permission from Trupke, T. et al. *Journal of Applied Physics* 2003a, 94, 4930–4937. Trupke, T. et al. *Applied Physics Letters* 2003b, 82, 2996–2998.)

The EQE_{PV} is obtained by measuring the short-circuit current under monochromatic light as a function of the frequency or wavelength. The photovoltaic EQE depends on the conditions of conductivity and recombination, which may depend on background illumination conditions, as discussed in Section PSC.10.10.

The EQE_{PV} can be separated into its optical and electrical parts.

$$\begin{aligned} EQE_{PV}(\nu) &= \eta_{LHE} IQE_{PV} \\ &= a(\nu) IQE_{PV}(\nu) \end{aligned} \tag{2.28}$$

IQE_{PV} is the flux of collected electrons per photon absorbed, given by

$$IQE_{PV} = \eta_{sep}\eta_{col} \tag{2.29}$$

The efficiency of charge separation of a generated electron-hole pair to free carriers is denoted η_{sep} (or injection efficiency, η_{inj}, in heterogeneous solar cells). This process must be considered in systems in which photogeneration creates a spatially localized electron hole pair, or an exciton, that has a probability to recombine, see the full discussion in Chapter PSC.9. The charge collection efficiency η_{col} is the probability that the separated electron and hole carriers are collected at the contacts and not lost by recombination or other processes, as discussed in Chapter PSC.10. In summary, the EQE can be expressed as a product

$$EQE_{PV}(\nu) = \eta_{LHE}\eta_{sep}\eta_{col} \tag{2.30}$$

The EQE_{PV} is also termed incident-photon-to-current-collected-electron-efficiency (η_{IPCE}) and IQE_{PV} is denoted as absorbed-photon-to-collected-electron-efficiency (η_{APCE}).

For a layer of reflectivity R of the front surface, absorption coefficient α, and no back layer reflection, the EQE relates to IQE as

$$EQE_{PV} = (1-R)(1-e^{-\alpha d}) IQE_{PV} \tag{2.31}$$

The normalized EQE of record photovoltaic devices of several technologies is shown in Figure PSC.10.2a.

It is important to remark that EQE_{LED} is a scalar quantity, which describes the whole outgoing photon flux, while EQE_{PV} is a function of photon energy or wavelength.

2.4 THE RECOMBINATION OF CARRIERS IN SEMICONDUCTORS

The excitation of a semiconductor by supra-bandgap light or the voltage injection process creates excess carrier densities with respect to equilibrium values n_0, p_0. In the recombination process, an electron in the conduction band of the semiconductor makes a downward transition to a valence band state resulting in annihilation of the electron–hole pair (Figure 2.6). In crystalline inorganic semiconductors such as silicon or GaAs, there are three principal recombination mechanisms: band to band, via defect levels, and Auger.

In band-to-band recombination, the electron makes a single transition from a state in the conduction band to an empty state in the valence band. Most of the energy of the excited carrier is released via photon emission, and this process is hence called radiative recombination. The density dependence of the recombination rate from band to band is well described by the expression

$$U_{np} = Bnp \tag{2.32}$$

where B is a recombination coefficient that depends on the temperature and other factors. The value for radiative recombination in Si at 300 K is $B \approx 1 \times 10^{-14}$ cm^3 s^{-1} (Michaelis and Pilkuhn, 1969;

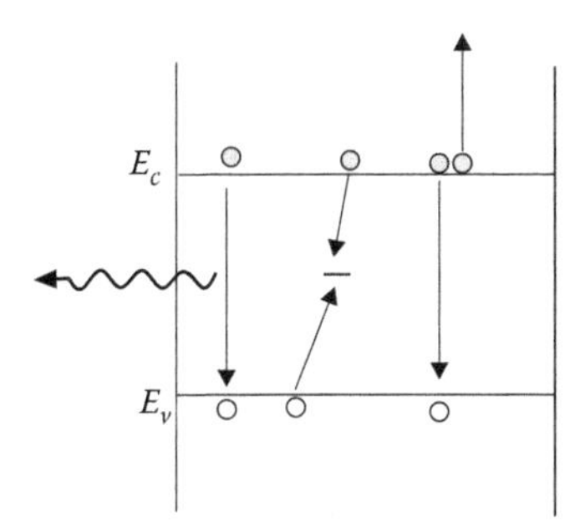

FIGURE 2.6 From left to right: Band-to-band, SRH, and Auger recombination.

Gerlach et al., 1972; Trupke et al., 2003a,b), and for GaAs it is $B = 7.2 \times 10^{-10}$ cm^3 s^{-1} (Varshni, 1967a,b). More generally, the radiative recombination coefficient can be extracted from the absorption coefficient by reciprocity arguments as described in Section PSC.6.2. From Equation ECK.5.56, we may write Equation 2.32 showing the dependence of recombination rate on applied voltage

$$U_{np} = Bn_i^2 e^{qV/k_BT} \tag{2.33}$$

The recombination of excess carriers is

$$U_{np} = B\left(np - n_i^2\right) \tag{2.34}$$

In a p-doped semiconductor, the majority carrier density p is basically constant, $p \approx p_0$ (except at very high-injection levels), and the recombination rate (2.34) depends only on the local electron density as follows:

$$U_n = k_{rec}(n - n_0) \tag{2.35}$$

where

$$k_{rec} = Bp_0 \tag{2.36}$$

The recombination rate is often described in terms of the *electron lifetime*. This quantity requires a careful definition as commented in Section PSC.2.5, but for a linear recombination rate indicated in Equation 2.35, it is given simply by

$$\tau_n = \frac{1}{k_{rec}} \tag{2.37}$$

see also Equation 2.46. If the recombination event occurs immediately after the generation of an electron–hole pair, before any charge separation occurs, it is called geminate recombination.

Recombination may occur as well via trap levels in the bandgap. Therefore, the drop of an electron from the conduction band to a bandgap level can produce the emission of a photon that is redshifted with respect to the bandgap photons. Recombination via a midgap state more often consists of a multiphonon process in which the bandgap energy is released to a number of phonons rather than by the emission of photons, hence called radiationless or *nonradiative* recombination.

Nonradiative recombination can be associated with defects located in the bulk or the surface of the semiconductor. The surface of the semiconductor where the contacts are located is often a source of defects and recombination sites. Recombination at these surface sites is called surface recombination.

Recombination via midgap states is often an important effect in semiconductor devices. For a nonradiative recombination process, the larger the energy between initial and final electron states, the greater the number of phonons required and the process becomes more unlikely. Thus, a midgap state, that is readily able to capture both electrons and holes, greatly increases the probability of radiationless recombination with respect to band-to-band recombination. A frequently used recombination model is the Shockley–Read–Hall (SRH) recombination that has been already described in Section ECK.6.3. It occurs at a localized state in the bandgap that receives both an electron from the conduction band and a hole from the valence band (Figure 2.6). The recombination rate of excess carriers is given by the formula ECK.6.35, often expressed as

$$U_{SRH} = \frac{\left(np - n_i^2\right)}{\tau_p(n+n_1)+\tau_n(p+p_1)} \tag{2.38}$$

where τ_n is the electron lifetime for a large density of holes, τ_p is the hole lifetime for a large density of electrons, and n_1, $p_1 = n_i^2/n_1$ are the electron and hole densities when the Fermi level coincides with the energy of the trap through which the recombination takes place:

$$\begin{aligned} n_1 &= N_c \exp((E_t - E_c)/k_B T) \\ p_1 &= N_v \exp((E_v - E_t)/k_B T) \end{aligned} \tag{2.39}$$

A value τ_n of 4 ms is typical for the low injection SRH electron lifetime measured in lightly doped silicon (Stephens et al., 1994).

Another important effect in heavily doped inorganic semiconductors is the Auger recombination, in which the energy released in band-to-band recombination of an electron and a hole is transmitted to another electron or hole (Figure 2.6). The resultant gain of kinetic energy of the third particle (electron) is normally lost by relaxation of the carrier to the band edge energy. It is, therefore, a three carrier nonradiative process usually expressed as

$$U_{Auger} = (C_n n + C_p p)\left(np - n_i^2\right) \tag{2.40}$$

where C_n and C_p are the Auger coefficients.

Figure 2.7 shows the minority carrier lifetime and luminescent IQE of different semiconductor solar cells. Auger recombination becomes the dominant effect at high-carrier density (Vossier et al., 2010).

Recombination in semiconductors is often investigated by observing luminescence dependence on incident light intensity. The identification of the dominant recombination processes as a function of the excitation rate can be established using an expression of the type

$$U_n = An + Bn^2 + Cn^3 \tag{2.41}$$

The first term is for monomolecular recombination, when the recombination is dominated by minority carriers under low-carrier injection, or for SRH when the capture of one carrier to traps governs the overall rate in Equation 2.38. The second term is for bimolecular recombination, which involves the densities of both electrons and holes. This is the case in band-to-band radiative

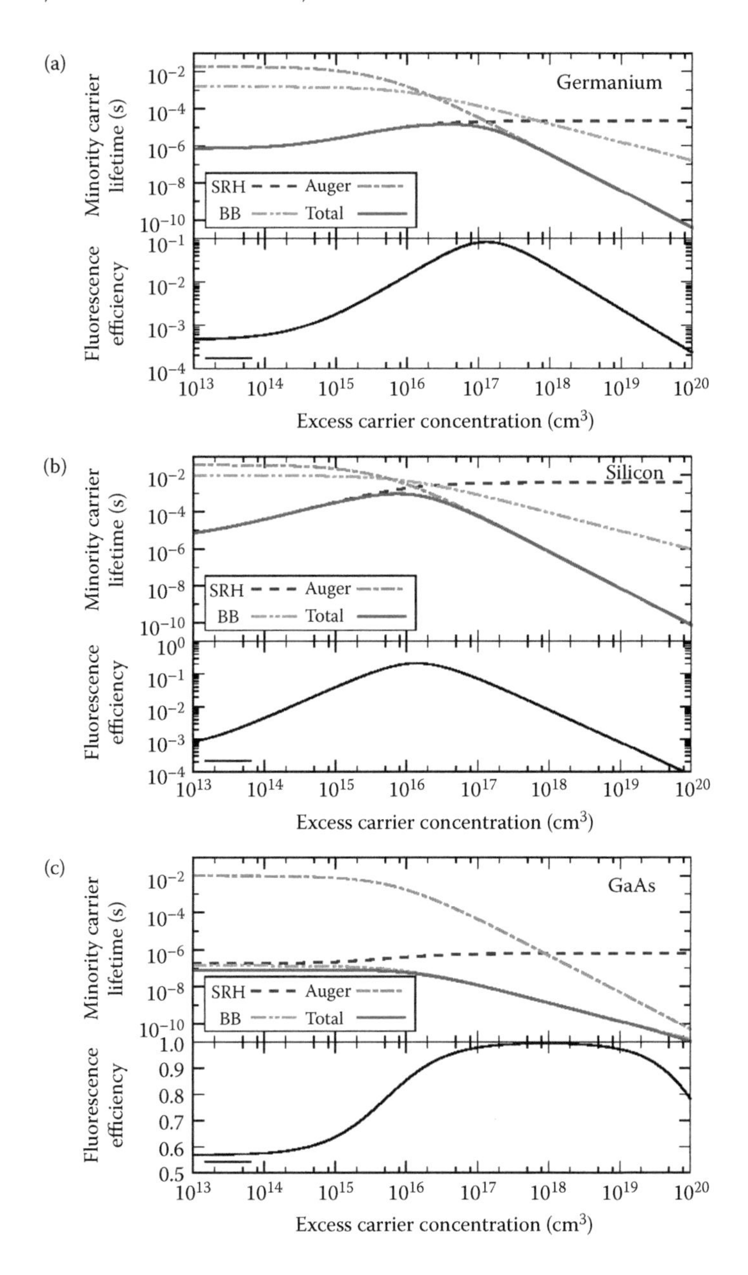

FIGURE 2.7 Minority carrier lifetime and internal fluorescence efficiency (luminescent IQE) of Ge, Si, and GaAs solar cells as a function of excess carrier concentration. (Reproduced with permission from Vossier, A. et al. *Journal of Applied Physics* 2015, 117, 015102.)

recombination under high injection, when the condition $n = p$ holds. The third term is trimolecular recombination that corresponds to Auger processes. An illustration is shown in Figure 2.8 for the lead halide perovskite solar cells, indicating first the PL intensity as a function of the excitation intensity, and then the IQE that allows to recognize the dominant recombination mechanism. At lower intensities of illumination, the PL is controlled by recombination via trap states. When the traps become saturated, the PL is dominated by band-to-band mechanism, and finally the Auger recombination sets in at very high-generation levels. The kinetic model based on Equation 2.41 allows to find the recombination lifetimes (Staub et al., 2016).

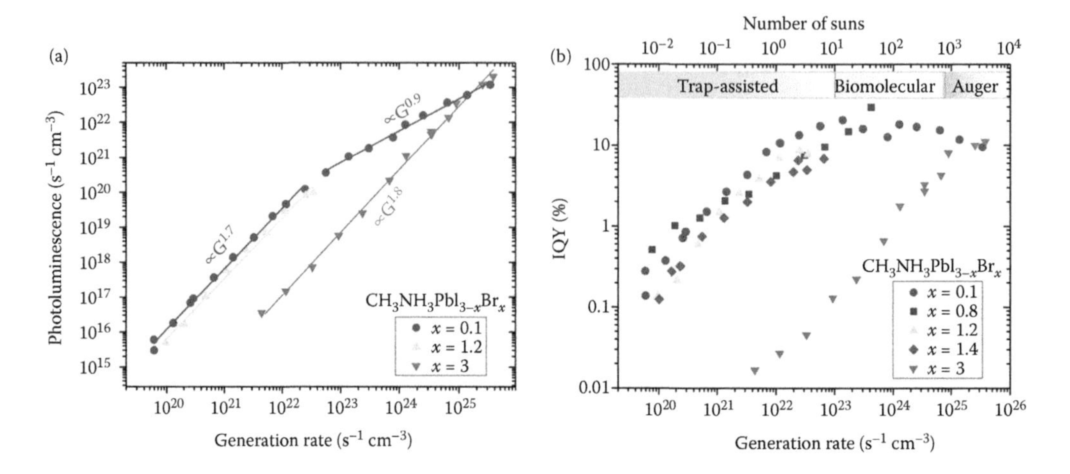

FIGURE 2.8 Steady-state PL of lead halide perovskite solar cells. (a) Pump-power dependence of the integrated PL signal for three Br concentrations. (b) Pump-power dependence of the IQE. (Reproduced with permission from Sutter-Fella, C. M. et al. *Nano Letters* 2016, 16, 800–806.)

In nanostructured semiconductors and organic blends, recombination occurs by the transference of a carrier between two different phases. In these systems, the charge transfer process is often influenced by disorder and a combination of surface states. The electron transfer rate can often be expressed using a phenomenological model as follows (Bisquert and Marcus, 2014):

$$U_n = k_{rec}\left(n^{\beta} - n_0^{\beta}\right) \tag{2.42}$$

where $0 < \beta \leq 1$ is a recombination order.

2.5 RECOMBINATION LIFETIME

Let us consider the question of the probability of survival of a quantity of excess electrons injected to a bulk or nanostructured semiconductor. We assume that the electrons are minority carriers and that electro-neutrality is maintained by fast displacement of the majorities. The decay of a population of electrons is governed by their rate of recombination according to the equation

$$\frac{dn}{dt} = -U_n(n) \tag{2.43}$$

We take first the simplest recombination model, which is that of linear recombination introduced in Equation 2.35. Excess electrons injected can be written as $\Delta n = n - n_0$, and their decay is determined by

$$\frac{d(\Delta n)}{dt} = -k_{rec}\Delta n \tag{2.44}$$

Therefore, the decay with time takes the form

$$\Delta n(t) = \Delta n(0)e^{-t/\tau_n} \tag{2.45}$$

where the lifetime, τ_n, is

$$\tau_n = k_{rec}^{-1} \tag{2.46}$$

In general, we define the lifetime as the constant in the denominator of the exponential decay law of Equation 2.45. However, we observe that such decay law depends critically on the fact that our starting recombination law in Equation 2.35 is linear, which is far from being the general case, as discussed in Section PSC.2.4. It is useful to develop a procedure whereby a lifetime can be established in any type of decay process, so that the results of experiments can be well defined. Here, we describe a method that is widely used in device characterization, in which the determination of the lifetime is based on small perturbation measurement. The general rationale for small perturbation method was already introduced in Section ECK.3.10. It has the advantage that the measured quantity is independent of the amplitude of the measurement signal, since all the equations become linear in the small perturbation domain. Small perturbation recombination lifetime has been amply adopted in the field of dye-sensitized and organic solar cells (Bisquert et al., 2009), and it has also been developed in amorphous silicon (Ritter et al., 1988) and crystalline silicon solar cells (Brendel, 1995; Schmidt, 1999).

Let us take a system that is determined by any general recombination law that has the form $U_n(n)$ in terms of the carrier concentration n. The decay of injected carriers by recombination is outlined in Figure 2.9, which emphasizes that the measurement of lifetime basically consists of a perturbation of the Fermi level that induces the recombination toward a certain equilibrium value. Experimentally, the shift of the Fermi level will be recorded; for instance, measuring the transient of photovoltage. We assume that the stable carrier density $\bar{n}$ (the steady state) is maintained by background photogeneration. Considering the balance of generation and recombination, we have

$$\frac{dn}{dt} = G_\Phi - U_n(n) \tag{2.47}$$

which in equilibrium sets the background carrier density by the equation

$$U_n(\bar{n}) = G_\Phi \tag{2.48}$$

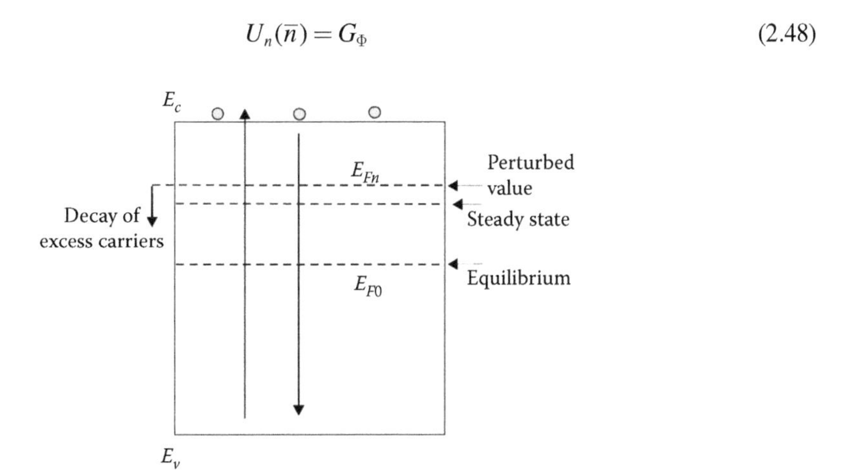

FIGURE 2.9 Scheme of the measurement of a lifetime. The semiconductor has an equilibrium density of electrons n_0 and a steady-state value $\bar{n}$, established by equilibrium of generation-recombination. The corresponding Fermi levels are indicated. An excess carrier density is injected, $\Delta n = n - n_0$, which decays in a characteristic time τ_n as the Fermi level E_{Fn} returns to the steady-state value.

Now a small perturbation $\hat{n}$ that is induced on top of the steady state provides the density dependence on time as

$$n(t) = \bar{n} + \hat{n}(t) \tag{2.49}$$

Let us determine the transient behavior of $\hat{n}$. With an expansion

$$U(n) = U(\bar{n}) + \frac{\partial U}{\partial n}\hat{n} \tag{2.50}$$

we obtain from Equation 2.47

$$\frac{d\hat{n}}{dt} = -\frac{\partial U}{\partial n}\hat{n} \tag{2.51}$$

The result we obtain in Equation 2.51 is that the linearization procedure always takes the evolution equation to a form of Equation 2.44, which will provide an exponential decay of the small perturbation excess density. Equation 2.51 defines a lifetime in terms of the recombination rate $U_n(n)$ as

$$\tau_n = \left(\frac{\partial U_n}{\partial n}\right)_{\bar{n}}^{-1} \tag{2.52}$$

As an example of the small perturbation procedure, consider the nonlinear recombination law introduced in Equation 2.42

$$U_n = k_{rec}\left(n^{\beta} - n_0^{\beta}\right) \tag{2.53}$$

Equation 2.47 can be written as

$$\frac{d\hat{n}}{dt} = G_{\Phi} - k_{rec}\left[(\bar{n} + \hat{n})^{\beta} - n_0^{\beta}\right] \tag{2.54}$$

Expanding the sum to first order in $\hat{n}$ and removing the steady-state terms (that cancel out), we have

$$\frac{d\hat{n}}{dt} = -k_{rec}\beta\bar{n}^{\beta-1}\hat{n} \tag{2.55}$$

Therefore, the lifetime is

$$\tau_n(\bar{n}) = (k_{rec}\beta\bar{n}^{\beta-1})^{-1} \tag{2.56}$$

Note that the lifetime is a function of the steady state; this is a general feature of nonlinear systems.

The decay of a small population that serves as a probe of the kinetics of the system has, therefore, been characterized for any recombination rate law and steady-state condition. More generally, the decay may require a number of sequential processes coupled to recombination, for example, when prior de-trapping of localized carriers is required. The linearizing method toward an exponential decay can easily be generalized and the characteristic decay time in general is denoted a *response time* (Rose, 1963).

In general, the response time τ of a carrier in a given type of electronic state with concentration c is determined by the decay of a small variation of the concentration $\hat{c}$ to equilibrium (Rose, 1963; Bisquert et al., 2009), according to the equation

$$\frac{\partial \hat{c}}{\partial t} = -\frac{1}{\tau}\hat{c} \tag{2.57}$$

Let us take Equation ECK.6.26 for capture and release of electrons in a trap. We recall that f is trap occupancy, determined by kinetic exchange with electrons from the conduction band with density n_c as indicated in Figure ECK.6.4. For a small perturbation of the trap occupancy (holding $\bar{n}_c$ fixed), we get the following equation that regulates the transient behavior:

$$\frac{\partial \hat{f}}{\partial t} = -(\beta_n \bar{n}_c + \varepsilon_n)\hat{f} \tag{2.58}$$

Comparing Equation 2.57, we observe that the response time for trapping is $\tau_c = (\beta_n \bar{n}_c)^{-1}$ and the response time for release is $\tau_r = \varepsilon_n^{-1}$. The time for decay is

$$\tau_t = \left(\tau_c^{-1} + \tau_r^{-1}\right)^{-1} \tag{2.59}$$

Using the definitions in Section ECK.6.3, Equation 2.59 can be written as

$$\tau_t = \frac{1}{\beta_n N_c e^{-E_c/k_BT}} \frac{1}{e^{E_t/k_BT} + e^{E_{Fn}/k_BT}} \tag{2.60}$$

If we assume that $E_{Fn} < E_t$ (so that the trap is unoccupied and available to capture electrons), we obtain

$$\tau_t = \frac{1}{\beta_n N_c} e^{(E_c - E_t)/k_BT} \tag{2.61}$$

Therefore, according to Equation 2.61, the release time from a trap to the conduction band becomes increasingly long the deeper the trap is with respect to E_c.

It is possible to adopt a different approach to define a recombination lifetime, which is to define an apparent lifetime as (Hornbeck and Haynes, 1955)

$$\tau_{app} = -\frac{n}{(dn/dt)} = -\frac{n}{U_n(n)} \tag{2.62}$$

It should be emphasized that Equation 2.62 uses total density and recombination rate instead of the small perturbation. Equation 2.62 is commonly used in silicon solar cells, as one can measure separately the recombination rate (equal to photogeneration) and the minority carrier density by the photovoltage transient or by other contactless methods, such as photoconductance (Sinton and Cuevas, 1996). In silicon solar cells, there are several mechanisms that lead to recombination of minority carriers, including radiative, SRH, Auger, and surface recombination (Richter et al., 2012), see Figure 2.7.

GENERAL REFERENCES

Optimization of light harvesting in solar cells: Redfield (1974), Yablonovitch (1982), Deckman et al. (1983), Campbell and Green (1987), Pettersson et al. (1999), Schueppel et al. (2010), and Wehrspohn et al. (2015).

Luminescence and phosphorescence: Pankove (1971), Zacks and Halperin (1972), Schmidt et al. (1992), Auzel (2003), Feldmann et al. (2003), de Wild et al. (2011), and Huang et al. (2013).

The quantum efficiency: Boroditsky et al. (2000), Schubert (2003), Heikkilä et al. (2009), Matioli and Weisbuch (2011), and Kivisaari et al. (2012).

Recombination in semiconductors: Hall (1952), Shockley and Read (1952), Blackmore (1962), Haug (1983), Wang et al. (2006), and Bisquert and Mora-Seró (2010).

Recombination lifetime: van Roosbroeck (1953), van Roosbroeck and Shockley (1954), Rose (1963), Schmidt (1999), Zaban et al. (2003), and Bisquert et al. (2009).

REFERENCES

Auzel, F. Upconversion and anti-Stokes processes with f and d ions in solids. *Chemical Reviews* 2003, 104, 139–174.

Wehrspohn, R. B.; Rau, U.; Gombert, A. *Photon Management in Solar Cells*; Wiley: Weinheim, 2015.

Bisquert, J.; Fabregat-Santiago, F.; Mora-Seró, I.; Garcia-Belmonte, G.; Giménez, S. Electron lifetime in dye-sensitized solar cells: Theory and interpretation of measurements. *The Journal of Physical Chemistry C* 2009, 113, 17278–17290.

Bisquert, J.; Marcus, R. A. Device modeling of dye-sensitized solar cells. *Topics in Current Chemistry* 2014, 352, 325–396.

Bisquert, J.; Mora-Seró, I. Simulation of steady-state characteristics of dye-sensitized solar cells and the interpretation of the diffusion length. *Journal of Physical Chemistry Letters* 2010, 1, 450–456.

Blackmore, J. S. *Semiconductor Statistics*; Dover Publications: New York, 1962.

Boroditsky, M.; Gontijo, I.; Jackson, M.; Vrijen, R.; Yablonovitch, E.; Krauss, T.; Chuan-Cheng, C.; Scherer, A.; Bhat, R.; Krames, M. Surface recombination measurements on III–V candidate materials for nanostructure light-emitting diodes. *Journal of Applied Physics* 2000, 87, 3497–3504.

Braslavsky, S. E.; Fron, E.; Rodriguez, H. B.; Roman, E. S.; Scholes, G. D.; Schweitzer, G.; Valeur, B.; Wirz, J. Pitfalls and limitations in the practical use of Forster's theory of resonance energy transfer. *Photochemical & Photobiological Sciences* 2008, 7, 1444–1448.

Brendel, R. Note on the interpretation of injection-level-dependent surface recombination velocities. *Applied Physics A* 1995, 60, 523.

Campbell, P.; Green, M. A. Light trapping properties of pyramidally textured surfaces. *Journal of Applied Physics* 1987, 62, 243–249.

de Wild, J.; Meijerink, A.; Rath, J. K.; van Sark, W. G. J. H. M.; Schropp, R. E. I. Upconverter solar cells: Materials and applications. *Energy & Environmental Science* 2011, 4, 4835–4848.

Deckman, H. W.; Wronski, C. R.; Witzke, H.; Yablonovitch, E. Optically enhanced amorphous silicon solar cells. *Applied Physics Letters* 1983, 42, 968–970.

Feldmann, C.; Jüstel, T.; Ronda, C. R.; Schmidt, P. J. Inorganic luminescent materials: 100 years of research and application. *Advanced Functional Materials* 2003, 13, 511–516.

Gerischer, H.; Michel-Beyerle, M. E.; Rebentrost, F.; Tributsch, H. Sensitization of charge injection into semiconductors with large band gap. *Electrochimica Acta* 1968, 13, 1509–1515.

Gerlach, W.; Schlangenotto, H.; Maeder, H. On the radiative recombination rate in silicon. *Physica Status Solidi (a)* 1972, 13, 277–283.

Green, M. A.; Zhao, J.; Wang, A.; Trupke, T. High-efficiency silicon light emitting diodes. *Physica E: Low-dimensional Systems and Nanostructures* 2003, 16, 351–358.

Hall, R. N. Electron-hole recombination in germanium. *Physical Review* 1952, 87, 387.

Haug, A. Auger recombination in direct-gap semiconductors: Band-structure effects. *Journal of Physics C: Solid State Physics* 1983, 16, 4159–4172.

Heikkilä, O.; Oksanen, J.; Tulkki, J. Ultimate efficiency limit and temperature dependency of light-emitting diode efficiency. *Journal of Applied Physics* 2009, 105, 093119.

Hornbeck, J. A.; Haynes, J. R. Trapping of minority carriers in silicon. I. p-Type silicon. *Physical Review* 1955, 97, 311.

Huang, X.; Han, S.; Huang, W.; Liu, X. Enhancing solar cell efficiency: The search for luminescent materials as spectral converters. *Chemical Society Reviews* 2013, 42, 173–201.
Kivisaari, P.; Riuttanen, L.; Oksanen, J.; Suihkonen, S.; Ali, M.; Lipsanen, H.; Tulkki, J. Electrical measurement of internal quantum efficiency and extraction efficiency of III-N light-emitting diodes. *Applied Physics Letters* 2012, 101, 021113.
Matioli, E.; Weisbuch, C. Direct measurement of internal quantum efficiency in light emitting diodes under electrical injection. *Journal of Applied Physics* 2011, 109, 073114.
Michaelis, W.; Pilkuhn, M. H. Radiative recombination in silicon p-n junctions. *Physica Status Solidi (b)* 1969, 36, 311–319.
Pankove, J. I. *Optical Processes in Semiconductors*; Prentice-Hall: Englewood Cliffs, NJ, 1971.
Pettersson, L. A. A.; Roman, L. S.; Inganas, O. Modeling photocurrent action spectra of photovoltaic devices based on organic thin films. *Journal of Applied Physics* 1999, 86, 487–496.
Redfield, D. Multiple-pass thin-film silicon solar cell. *Applied Physics Letters* 1974, 25, 647–648.
Richter, A.; Glunz, S. W.; Werner, F.; Schmidt, J.; Cuevas, A. Improved quantitative description of Auger recombination in crystalline silicon. *Physical Review B* 2012, 86, 165202.
Ritter, D.; Zeldov, E.; Weiser, K. Ambipolar transport in amorphous semiconductors in the lifetime and relaxation-time regimes investigated by the steady-state photocarrier grating technique. *Physical Review B* 1988, 38, 8296.
Rose, A. *Concepts in Photoconductivity and Allied Problems*; Interscience: New York, 1963.
Saxena, K.; Jain, V. K.; Mehta, D. S. A review on the light extraction techniques in organic electroluminescent devices. *Optical Materials* 2009, 32, 221–233.
Schmidt, J. Measurement of differential and actual recombination parameters in crystalline silicon wafers. *IEEE Transactions on Electron Devices* 1999, 46, 2018–2025.
Schmidt, T.; Lischka, K.; Zulehner, W. Excitation-power dependence of the near-band-edge photoluminescence of semiconductors. *Physical Review B* 1992, 45, 8989–8994.
Scholes, G. D. Long-range resonance energy transfer in molecular systems. *Annual Review of Physical Chemistry* 2003, 54, 57–87.
Schubert, E. F. *Light-Emitting Diodes*; Cambridge University Press: New York, 2003.
Schueppel, R.; Timmreck, R.; Allinger, N.; Mueller, T.; Furno, M.; Uhrich, C.; Leo, K.; Riede, M. Controlled current matching in small molecule organic tandem solar cells using doped spacer layers. *Journal of Applied Physics* 2010, 107, 044503.
Shockley, W.; Read, W. T. Statistics of the recombinations of holes and electrons. *Physical Review* 1952, 87, 835–842.
Sinton, R. A.; Cuevas, A. Contactless determination of current–voltage characteristics and minority-carrier lifetimes in semiconductors from quasi-steady-state photoconductance data. *Applied Physics Letters* 1996, 69, 2510.
Staub, F.; Hempel, H.; Hebig, J.-C.; Mock, J.; Paetzold, U. W.; Rau, U.; Unold, T.; Kirchartz, T. Beyond bulk lifetimes: Insights into lead halide perovskite films from time-resolved photoluminescence. *Physical Review Applied* 2016, 6, 044017.
Stephens, A. W.; Aberle, A. G.; Green, M. A. Surface recombination velocity measurements at the silicon–silicon dioxide interface by microwave-detected photoconductance decay. *Journal of Applied Physics* 1994, 76, 363–370.
Strickler, S. J.; Berg, R. A. Relationship between absorption intensity and fluorescence lifetime of molecules. *Journal of Chemical Physics* 1962, 37, 814–820.
Sutter-Fella, C. M.; Li, Y.; Amani, M.; Ager, J. W.; Toma, F. M.; Yablonovitch, E.; Sharp, I. D.; Javey, A. High photoluminescence quantum yield in band gap tunable bromide containing mixed halide perovskites. *Nano Letters* 2016, 16, 800–806.
Tiedje, T.; Yablonovitch, E.; Cody, G. D.; Brooks, B. G. Limiting efficiency of silicon solar cells. *IEEE Transactions on Electron Devices* 1984, 31, 711–716.
Trupke, T.; Daub, E.; Würfel, P. Absorptivity of silicon solar cells obtained from luminescence. *Solar Energy Materials and Solar Cells* 1998, 53, 103–114.
Trupke, T.; Green, M. A.; Wurfel, P.; Altermatt, P. P.; Wang, A.; Zhao, J.; Corkish, R. Temperature dependence of the radiative recombination coefficient of intrinsic crystalline silicon. *Journal of Applied Physics* 2003a, 94, 4930–4937.
Trupke, T.; Zhao, J.; Wang, A.; Corkish, R.; Green, M. A. Very efficient light emission from bulk crystalline silicon. *Applied Physics Letters* 2003b, 82, 2996–2998.
Usami, A. Theoretical study of application of multiple scattering of light to a dye-sensitized nanocrystalline photoelectrochemical cell. *Chemical Physics Letters* 1997, 277, 105–108.

van Roosbroeck, W. The transport of added current carriers in a homogeneous semiconductor. *Physical Review* 1953, 91, 282–289.

van Roosbroeck, W.; Shockley, W. Photon-radiative recombination of electrons and holes in germanium. *Physical Review* 1954, 94, 1558–1560.

Varshni, Y. P. Band-to-band radiative recombination in groups IV, VI, and III-V semiconductors (I). *Physica Status Solidi (b)* 1967a, 19, 459–514.

Varshni, Y. P. Band-to-band radiative recombination in groups IV, VI, and III–V semiconductors (II). *Physica Status Solidi (b)* 1967b, 20, 9–36.

Vossier, A.; Gualdi, F.; Dollet, A.; Ares, R.; Aimez, V. Approaching the Shockley-Queisser limit: General assessment of the main limiting mechanisms in photovoltaic cells. *Journal of Applied Physics* 2015, 117, 015102.

Vossier, A.; Hirsch, B.; Gordon, J. M. Is Auger recombination the ultimate performance limiter in concentrator solar cells? *Applied Physics Letters* 2010, 97, 193509.

Wang, Q.; Ito, S.; Grätzel, M.; Fabregat-Santiago, F.; Mora-Seró, I.; Bisquert, J.; Bessho, T.; Imai, H. Characteristics of high efficiency dye-sensitized solar cells. *The Journal of Physical Chemistry* 2006, 110, 19406–19411.

Yablonovitch, E. Statistical ray optics. *Journal of the Optical Society of America* 1982, 72, 899–907.

Zaban, A.; Greenshtein, M.; Bisquert, J. Determination of the electron lifetime in nanocrystalline dye solar cells by open-circuit voltage decay measurements. *ChemPhysChem* 2003, 4, 859–864.

Zacks, E.; Halperin, A. Dependence of the peak energy of the pair-photoluminescence band on excitation intensity. *Physical Review B* 1972, 6, 3072–3075.

3 Optical Transitions in Organic and Inorganic Semiconductors

This chapter provides a summary of the light absorption properties in a range of materials relevant to energy conversion including bulk inorganic semiconductors, semiconductor quantum dots, and organic molecules and materials. Emphasis is placed on the properties that determine the capabilities of solar energy harvesting when these materials are used as the light absorber in energy conversion devices, such as the spectral distribution of absorption features. A number of effects that are significant in the study of optical properties of semiconductors are revised, such as plasmonic absorption, the Burstein–Moss shift, and excitonic absorption. We also address the optical absorption features due to charge transfer complexes at heterojunctions and some of their applications.

3.1 LIGHT ABSORPTION IN INORGANIC SOLIDS

Let us consider the light absorption process in an inorganic solid material as indicated in Figure 3.1a. In the interband transition process, an electron jumps from a state of energy E_i in the lower band to a state E_f in the upper band by absorption of a photon. By the law of conservation of energy,

$$E_f = E_i + h\nu \tag{3.1}$$

Figure 3.2a shows a semiconductor or insulator material with conduction and valence bands separated by the gap energy E_g. Furthermore, the VB maximum and CB minimum occur at the same value of the crystal momentum axis, so that the optical transition occurs satisfying the conservation of momentum. This is called a *direct transition*. The *joint density of states* of the valence band and conduction band states, D_{vc}, for the transition of an electron by an energy hv, is the probability that a state in the valence band is occupied by an electron (unoccupied by a hole) combined with the probability that a state in the conduction band, separated by the energy of the photon, is unoccupied by an electron. Taking the DOS for the parabolic band model described by Equation ECK.2.10, the joint DOS is given by the expression

$$D_{vc}(h\nu) = 0 \quad \text{for} \quad h\nu < E_g \tag{3.2}$$

$$D_{vc}(h\nu)d(h\nu) = \frac{\left(2m^*_{eh}\right)^{3/2}}{2\pi^2\hbar^3}(h\nu - E)^{1/2}d(h\nu) \quad \text{for} \quad h\nu \geq E_g \tag{3.3}$$

where m^*_{eh} is the reduced mass for electrons and holes,

$$m^*_{eh} = \frac{m^*_e m^*_h}{m^*_e + m^*_h} \tag{3.4}$$

Obviously, photons of energy lower than E_g cannot be absorbed. At energies hv larger than the bandgap, the joint density of states increases with the square root of the energy as a result of the parabolic (effective mass) approximation for $E(k)$ when $k \to 0$. As a consequence of this,

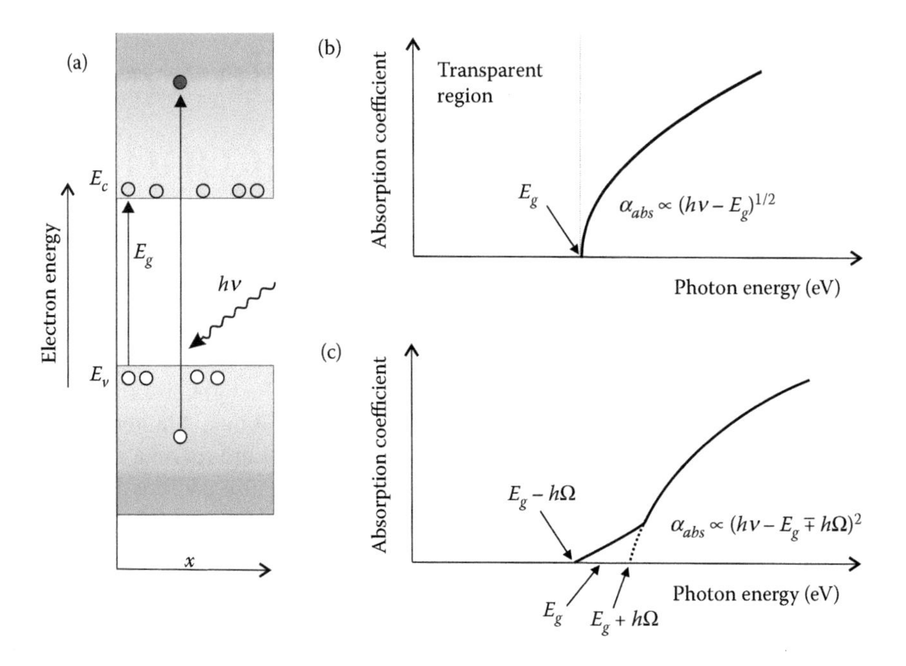

FIGURE 3.1 (a) Illustration of a photoinduced electronic transition across the semiconductor bandgap, (b) the absorption coefficient as a function of photon energy for a direct gap semiconductor, and (c) for an indirect bandgap semiconductor.

photoexcited electrons arrive in the conduction band with energies in excess of E_c, as indicated in Figure 3.1a. As described in Section ECK.5.5, an electron in a high-energy state in Figure 3.1a is termed a hot electron, as it is a very short-lived state (about 1 ps). In this time scale, the electron loses the kinetic energy and cools down to the bottom of the band by collisions with phonons.

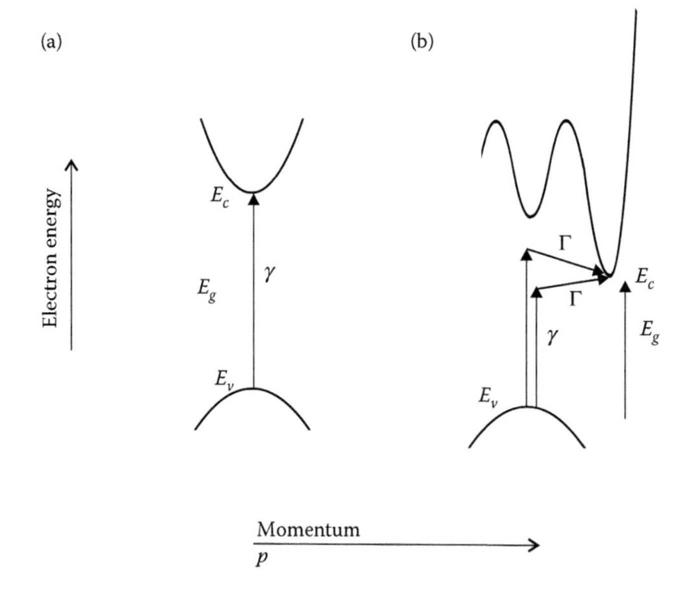

FIGURE 3.2 (a) Lowest energy electronic transition from the valence band to the conduction band for a direct bandgap semiconductor by absorption of a photon γ. (b) Electronic transitions in a semiconductor of indirect bandgap by the absorption of a photon γ and simultaneous absorption or release of a phonon Γ.

The bonding character of a semiconductor largely determines the molecular orbitals that are involved in optical electronic transitions. For ionic solids, the valence band is usually formed from the highest occupied p orbitals of the anions, with a mixture of d levels to a certain extent. The conduction band is formed by the s levels of the cation. In the case of covalently bound solids, the valence and conduction band come, respectively, from bonding and antibonding states of sp^3 hybrid orbitals.

The optical absorption coefficient α is determined by the quantum mechanical transition rate $W_{i \to f}$ from the initial to the final quantum state by absorption of a photon of frequency ν. The transition rate is given by Fermi's golden rule

$$W_{i \to f} = \frac{2\pi}{\hbar} |M|^2 D_{vc}(h\nu) \tag{3.5}$$

The matrix element M represents the electric dipole moment for the transition. Furthermore, the transition must be allowed by the selection rules that depend on the crystal symmetry (Klingshirn, 1995). According to Equation 3.5, the absorption rate is proportional to the joint density of states given by Equation 3.3. Therefore, we obtain the following spectral dependence:

$$\alpha(h\nu) \propto (h\nu - E)^{1/2} \tag{3.6}$$

Note that this relatively simple relation for band-to-band transitions in direct absorption materials is in many cases obeyed only approximately due to defects in the bandgap, Coulomb interaction (exciton formation), and other factors, as discussed below.

In semiconductors where the VB maximum and CB minimum do not have the same crystal momentum, the direct transition excited by only a photon cannot occur, since the photon has a very small linear momentum. However an *indirect transition* is possible that is assisted by phonons, which take the difference of momentum of the electron in the final and initial states. This process is indicated in Figure 3.2b. The spectral dependence of the absorption coefficient is

$$\alpha(h\nu) \propto (h\nu - E_g \mp h\Omega)^2 \tag{3.7}$$

Here, E_g is the indirect bandgap and $h\Omega$ is the phonon energy. The signs $\mp$ depend on whether the phonon is absorbed or emitted. The spectral dependence of the absorption coefficient is shown in Figure 3.1c. Since the absorption event is a two-particle quantum transition, the absorption coefficient for indirect bandgap material is usually much smaller than for direct bandgap semiconductors.

In summary, direct semiconductors are characterized by an absorption edge and their bandgap E_g. Photons with energies less than E_g are not absorbed. Photons with energies larger than E_g are absorbed with increasing absorbance as their energy increases. However, in practice, the absorption (and luminescence) spectral features of semiconductors can be influenced by different factors that modify the above-mentioned ideal features for an abrupt bandgap. Phonon-assisted transitions in indirect absorption produce a tail of absorption below the bandgap energies. In addition, any symmetry-breaking disorder in the lattice disrupts the perfect crystalline structure and causes the appearance of localized states close to the band edge that generate the Urbach tail at energies below the main absorption edge (Urbach, 1953). Spatial fluctuations of the fundamental bandgap also cause tailing absorption and emission (Mattheis et al., 2007). The precise shape of subbandgap absorption is an important feature for studies of PL and photovoltaic properties of a semiconductor. A general model for absorption coefficient including different physical effects for subbandgap tails is reported by Katahara and Hillhouse (2014). As an example, Figure 3.3a

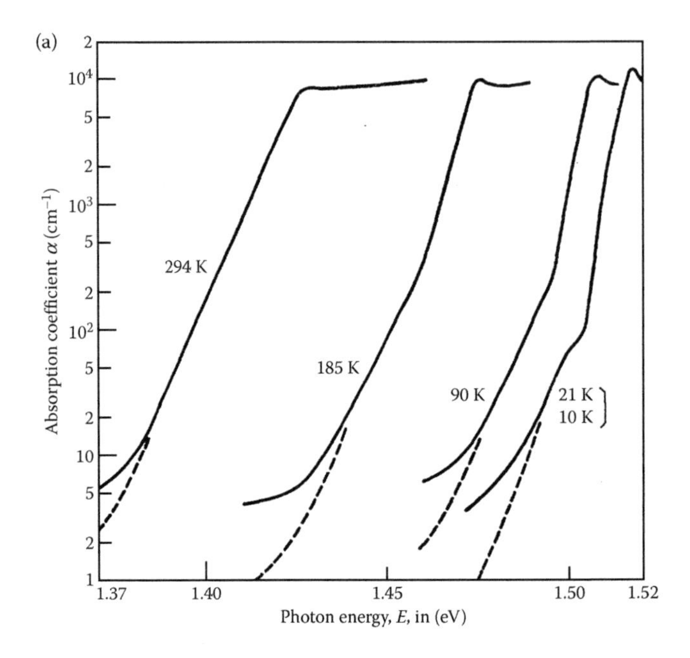

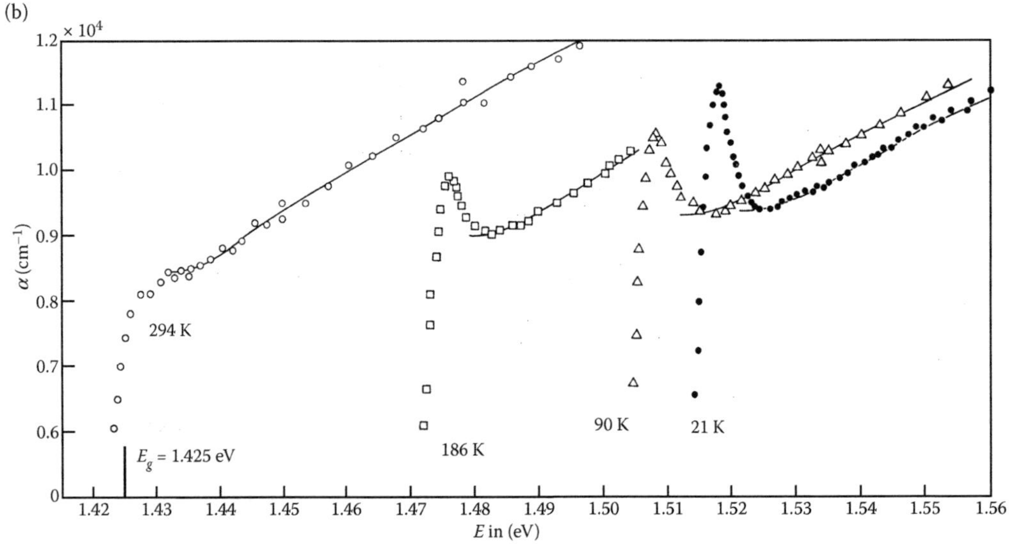

FIGURE 3.3 (a) Absorption of GaAs in a broad photon energy range and (b) in the exciton region. (Reproduced with permission from Sturge, M. D. *Physical Review* 1962, 127, 768–773.)

shows the main absorption edge of GaAs, which is a direct absorption semiconductor. The high absorption coefficient in combination with a bandgap of 1.42 eV constitutes excellent features for solar energy harvesting that make GaAs the highest efficiency solar cell material at this time. The absorption of GaAs is nearly constant at $E > E_g$, and it is well described by the expression (Miller et al., 2012)

$$\alpha = \alpha_0 \left(1 + \frac{E - E_g}{E_1}\right) \quad E > E_g \tag{3.8}$$

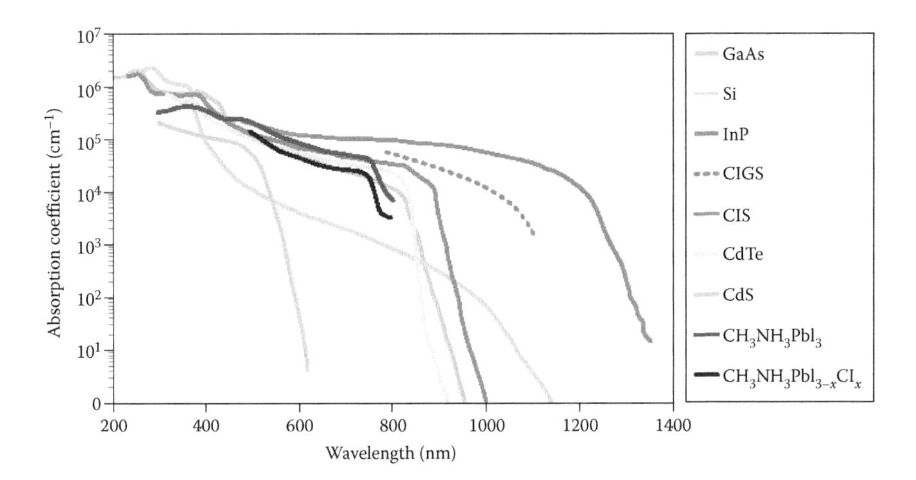

FIGURE 3.4 Effective absorption coefficient with respect to wavelength of several photovoltaic materials. (Reproduced with permission from Green, M. A. et al. *Nature Photonics* 2014, 8, 506–514.)

where E_1 is a constant. The optical absorption coefficient for subbandgap photons displays an Urbach tail and shows an exponential spectral dependence:

$$\alpha(E) = \alpha_0 \exp\left(\frac{E - E_g}{E_0}\right) \quad E \leq E_g \tag{3.9}$$

Here, E_0 is a tailing parameter. The hump at the absorption edge is the excitonic absorption, shown in Figure 3.3b and discussed in Section 3.3.

Figure 3.4 shows the absorption coefficient of several semiconductors with respect to wavelength. Figure 3.5 shows the absorption coefficient with respect to energy indicating the Urbach

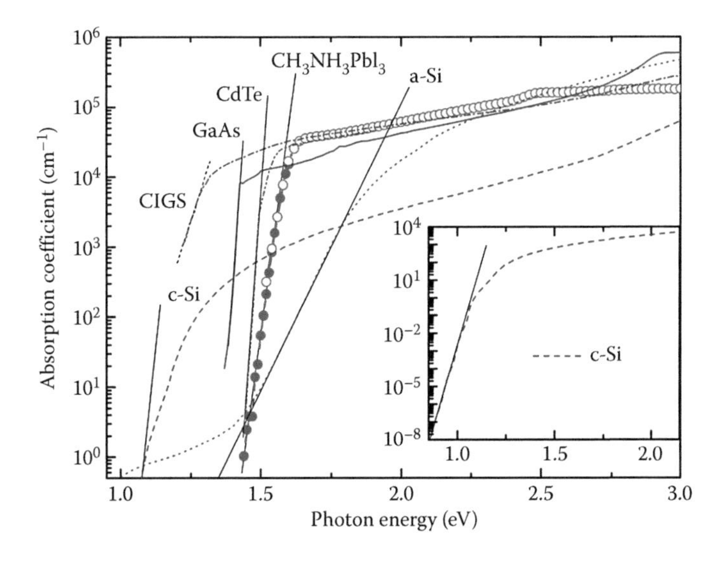

FIGURE 3.5 Effective absorption coefficient of several photovoltaic materials with respect to photon energy, all measured at room temperature. For each material, the slope of the Urbach tail is shown. The inset shows the data for c-Si down to low-absorption values. (Reproduced with permission from De Wolf, S. et al. *The Journal of Physical Chemistry Letters* 2014, 5, 1035–1039.)

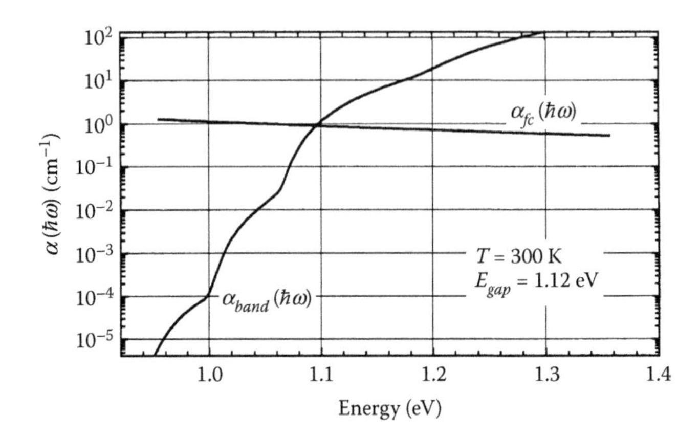

FIGURE 3.6 The band-to-band and free carrier absorption coefficients of silicon. (Reproduced with permission from Trupke, T. et al. *Solar Energy Materials and Solar Cells* 1998, 53, 103–114.)

tail. As noted earlier, direct semiconductors show a sharp increase of the absorption coefficient above the bandgap energy. In contrast to this, in indirect semiconductors, the absorption coefficient is considerably lower and rises very slowly. Silicon has a direct bandgap of 3.4 eV and an indirect bandgap of 1.1 eV at 300 K. The latter is relevant for the conversion of solar energy but the absorption coefficient at the band edge energy is very small (Green, 2008; Wang et al., 2013a). To measure the entire range of the absorption coefficient below E_g in silicon, many samples with increasing thickness would be needed. Alternatively, techniques based on reciprocity of luminescence and absorption that are developed in Chapter PSC.6 allow for extraordinary accuracy at very long wavelengths down to about 10^{-7} cm^{-1} (Daub and Würfel, 1995; Trupke et al., 1998; Barugkin et al., 2015) and provide absorption coefficient values shown in Figure 3.6. The staircase processes observed in this figure correspond to successive phonon absorptions of the indirect transition.

As mentioned in Section PSC.2.1, the absorption length L_α of the absorber material is a central characteristic determining the necessary solar cell thickness. If we take as a reference the absorption coefficient at 775 nm wavelength (1.6 eV photon energy) for silicon, we observe in Figure 3.4 that the absorption length is in the range of $L_\alpha = 10$ μm, while GaAs solar cells with $\alpha = 10^4$ cm^{-1} have values $L_\alpha = 1$ μm that require considerably thinner films. The $CH_3NH_3PbI_3$ perovskite requires an even thinner film of 300 nm to absorb all the incoming radiation at this wavelength.

3.2 FREE CARRIER PHENOMENA

The presence of a large density of free carriers in metals, or in the CB, and/or VB of semiconductors, produces significant effects in their optical properties. We discuss briefly several important effects: the plasma reflectivity, the Burstein shift, and the techniques based on the absorption of radiation by free carriers.

A *plasma* is a neutral gas of positive or negative carriers, as electrons or holes. In metals and heavily doped semiconductors, the plasma is hence formed by an electron gas electrically compensated by fixed ions. The plasma absorption is due to a collective excitation of free carriers by the electrical field of the electromagnetic waves. The quantization leads to quasiparticles termed *plasmons* that obey the Bose statistics. Nonetheless, the oscillations of the electrons in the metal can be well described as a classical effect by combining Maxwell's equations, the Drude model of free electron conductivity, and the Lorentz model of dipole oscillators. For an undamped displacement of the electrons, the relative dielectric constant depends on the angular frequency of the incoming light as follows:

$$\varepsilon_r(\omega) = 1 - \frac{\omega_p^{\,2}}{\omega^2} \tag{3.10}$$

where the *plasmon frequency* is given by

$$\omega_p = \left(\frac{nq^2}{\varepsilon_0 m_e} \right)^{1/2} \tag{3.11}$$

Here, n is the density of free electrons. Using the relationships (PSC.2.3) and (PSC.2.4), it is obtained that the reflectivity R is near unity at frequencies $\omega \leq \omega_p$, and decreases from $\omega = \omega_p$ approaching zero at $\omega \gg \omega_p$. Due to electrical field screening, all light of frequencies below the plasmon frequency is reflected. Equation 3.11 indicates that the plasmon frequency is proportional to the square root of the carrier concentration, n. In metals, the plasmon energy $\hbar\omega_p$ is around 5 eV. Thus, metals reflect all light below this energy and become transparent in the ultraviolet. The transparent conducting oxides (TCO, Section ECK.5.5) usually have a lower electron concentration than metals by 1–2 orders of magnitude, hence plasmon energies lie around 1 eV for the highest doped TCOs. Consequently, the TCOs do not transmit infrared radiation and are used as low-emissivity window coatings (Klein, 2013). Figure 3.7 shows the reduction of the transmittance at long wavelengths by the decrease of the plasmon wavelength λ_P with increasing carrier concentration as indicated by the conductivity of the oxide.

Considering a metal nanoparticle, the plasmon causes an enhancement of the local field intensity in the nearby space by redistribution of the optical field. A localized surface plasmon resonance induces sharp spectral absorption and scattering peaks so that chromophores situated in the region of enhanced field effectively absorb more light (Munechika et al., 2010). Plasmonic nanostructures can thus redistribute the optical field, increasing the fluorescence intensity. The shift of the plasmon resonance spectral peak allows the detection of molecular interactions near the metal nanoparticle surface (Mayer and Hafner, 2011). These effects have been utilized in applications such as biological labeling, sensing and imaging, LEDs, and single photon sources. The plasmonic nanostructures can maximize the light absorption in solar cells (Ming et al., 2012).

The semiconductor optical gap is defined as the energy of the lowest electronic transition accessible via absorption of a single photon (Bredas, 2014). The effective increase of the optical gap due

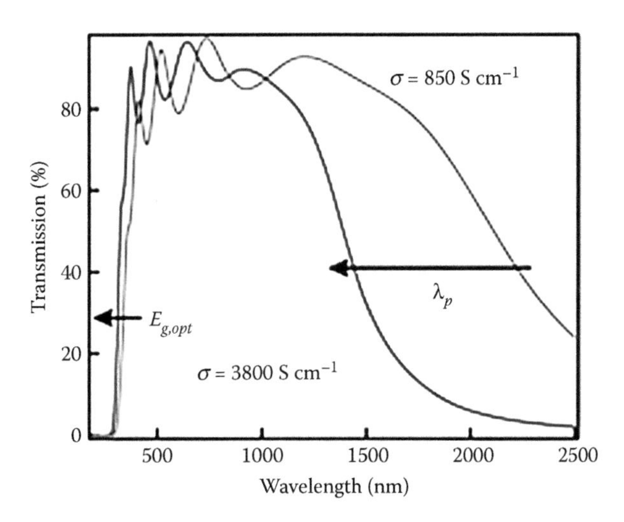

FIGURE 3.7 The optical transmission spectra of two ITO films with different electrical conductivity that indicate the free carrier density. The increase in the optical bandgap is due to the Burstein shift and the reduction in the plasmon wavelength λ_p with increasing carrier concentration changes the absorption at long wavelength. (Reproduced with permission from Klein, A. *Journal of the American Ceramic Society* 2013, 96, 331–345.)

to the creation of a very large density of electrons (holes) in the conduction (valence) band, which fills up the energy levels near the edge of the band, causes the blue shift of the optical absorption edge known as the *Burstein–Moss shift*. This effect is illustrated in Figure 3.8a. In the short wavelength region of Figure 3.7, the shift of optical absorption edge of differently doped TCOs should be noted. It is observed that the transmission onset shifts to shorter wavelength for the degenerately doped, high-conductivity material.

Changes in the occupation of the CB can be spectrally detected by the Burstein–Moss shift. The transient absorption spectra (TAS) shown in Figure 3.9a for a $CH_3NH_3PbI_3$ perovskite film consist of the difference of the semiconductor absorption at different, short times after photogeneration by a pump signal, with respect to the dark spectrum (Manser and Kamat, 2014). The negative parts are termed bleaches and indicate enhanced transient absorption due to specific photoinduced transitions. According to Equation PSC.1.4, the bleach at 760 nm is associated

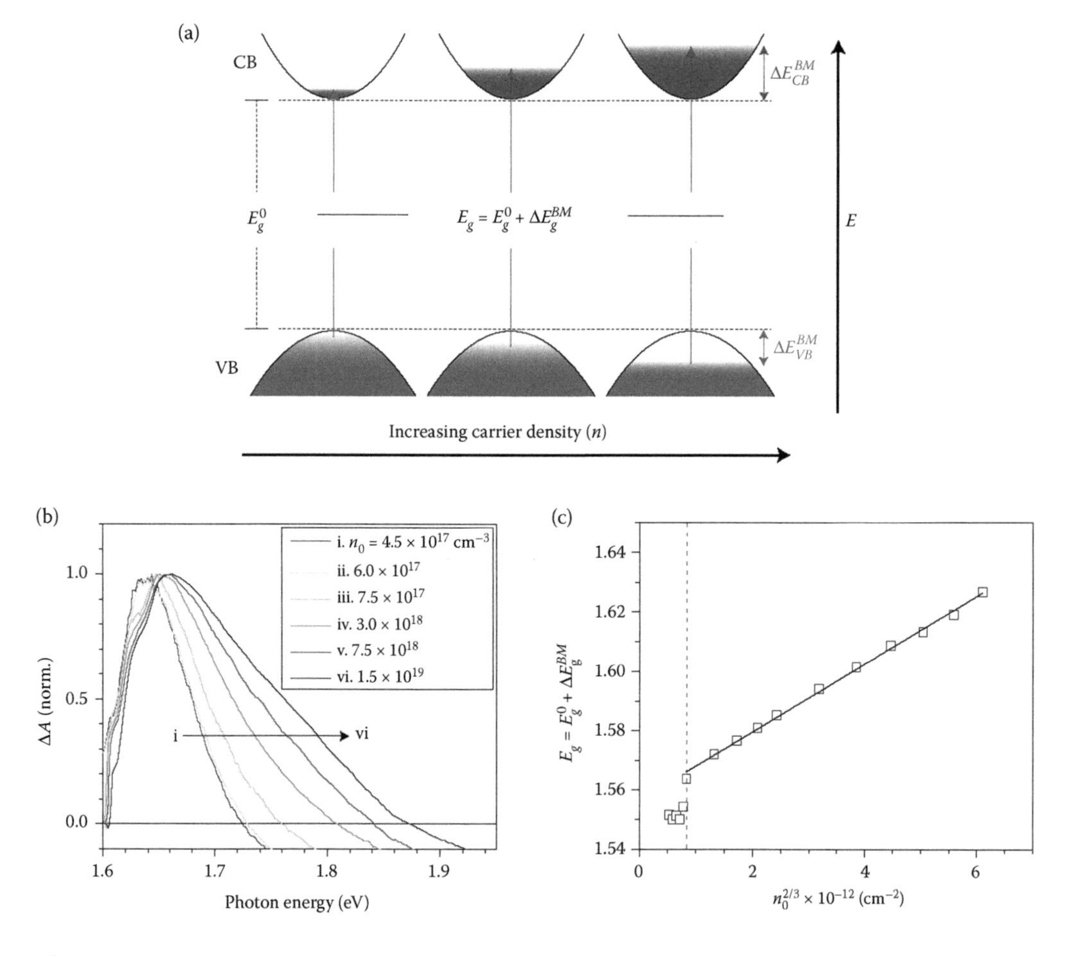

FIGURE 3.8 (a) Schematic representation of the Burstein–Moss effect showing the contribution from both electrons in the conduction band and holes in the valence band in the case of similar effective masses. (b) Normalized TAS of the band-edge transition in $CH_3NH_3PbI_3$ recorded at the maximum bleach signal (5 ps) after 387 nm pump excitation of varying intensity. The corresponding carrier densities are indicated in the legend. (c) Modulation of the intrinsic bandgap of $CH_3NH_3PbI_3$ according to the Burstein–Moss model. The vertical dashed line marks the onset of bandgap broadening. The solid line is a linear fit to the data after the onset threshold. The linear trend indicates agreement with band filling by free charge carriers. (Reproduced with permission from Manser, J. S.; Kamat, P. V. *Nature Photonics* 2014, 8, 737–743.)

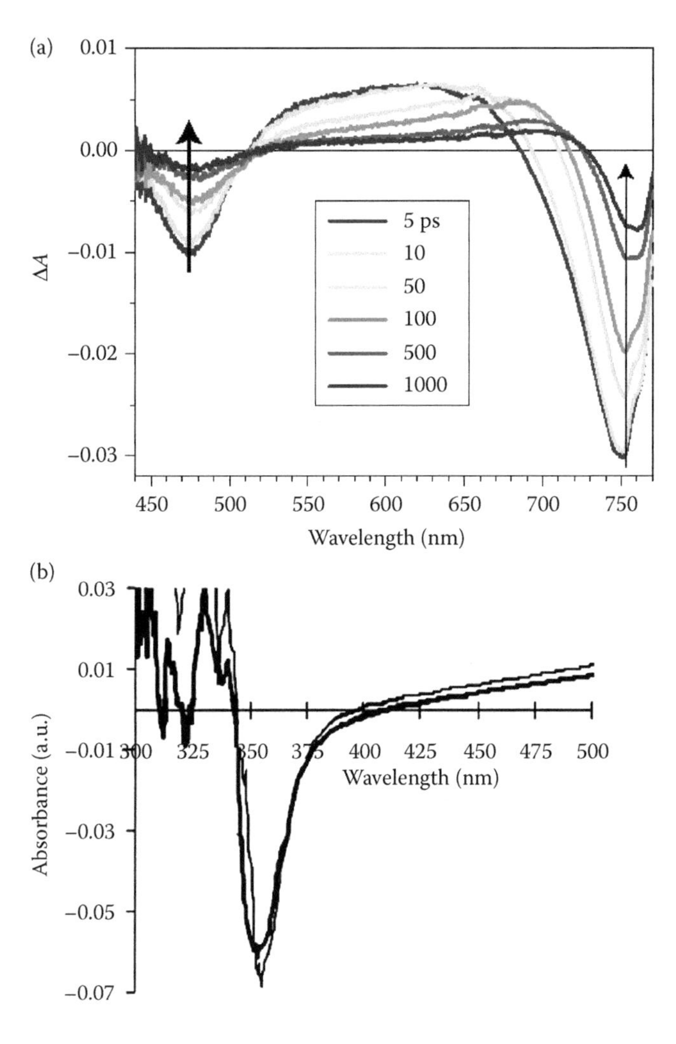

FIGURE 3.9 (a) Time-resolved TAS of $CH_3NH_3PbI_3$ at different probe delay times following 387 nm laser excitation with an energy density of 8 μJ cm^{-2}. Arrows indicate low- and high-energy bleach recovery. (Reproduced with permission from Manser and Kamat, 2014). (b) Spectral changes of the bare TiO_2 (thin line) and $SrTiO_3$-coated TiO_2 electrodes (thick line) at applied −0.9 V versus SCE in $HClO_4$ aqueous solution (pH 1.8). The short wavelength bleach that is related to the electron accumulation in TiO_2 is not affected by the coating. (Reproduced with permission from Diamant, Y. et al. *The Journal of Physical Chemistry B* 2003, 107, 1977–1981.)

with an optical transition at 1.63 eV that corresponds to the direct transition across the bandgap of the $CH_3NH_3PbI_3$ perovskite. The change of absorption is incremented when the number of photo-generated carriers is increased, as shown in Figure 3.8b. Using the absorbance of the film, the absorption coefficient of $CH_3NH_3PbI_3$, and the excitation energy density of the laser, it is possible to determine the initial photo-generated carrier density. The result of this calculation at various pump powers enables to fit the difference in absorbance using the expression that relates the shift of absorption to the carrier number density in parabolic-band theory (Muñoz et al., 2001):

$$\Delta E_g^{BM} = \frac{\hbar^2}{2m^*_{eh}}(3\pi^2 n)^{2/3} \tag{3.12}$$

This model provides a very good agreement with the results of photoexcitation of $CH_3NH_3PbI_3$ at 760 nm, as shown in Figure 3.8c. The results indicate that the perovskite becomes degenerate at relatively low carrier densities of order 5×10^{17} cm^{-3}, which points to a low-effective density of states in the conduction band, associated with a small effective mass (sharp $E(k)$ curvature), cf. Table ECK.2.1.

Free carrier absorption involves the absorption of a photon by excitation of an electron in the CB to a higher-energy level. The conservation of momentum is satisfied by phonons or by impurity scattering. Free carrier absorption usually decreases monotonically at higher photon energies as shown for silicon crystal in Figure 3.6. The Burstein–Moss shift reveals the change of electronic occupation of the CB of nanostructured semiconductors such as TiO_2 by large negative voltage that induces strong electron accumulation (Figure 3.9b). The spectra may be divided into two regions separated by the bandgap energy, which lies approximately at 400 nm. At long wavelengths, the intensity increases because of the absorbance of free electrons in the CB. The bleaching peak below 400 nm appears due to the increase of excitation energy across the gap.

Another technique to detect transient phenomena associated with free carriers is the time-resolved microwave conductivity (TRMC) method. This technique is based on the measurement of the relative change of the microwave power reflected from a semiconductor that is caused by a small increase in the conductivity (Savenije et al., 2013). However, TRMC operates in the GHz frequencies and the trap states in the bandgap of nanocrystalline semiconductors may contribute to the observed response. Carriers in bandgap localized states, discussed in Section ECK.8.4, also induce light absorption in the long wavelength region of the spectrum. Figure 3.10 illustrates different

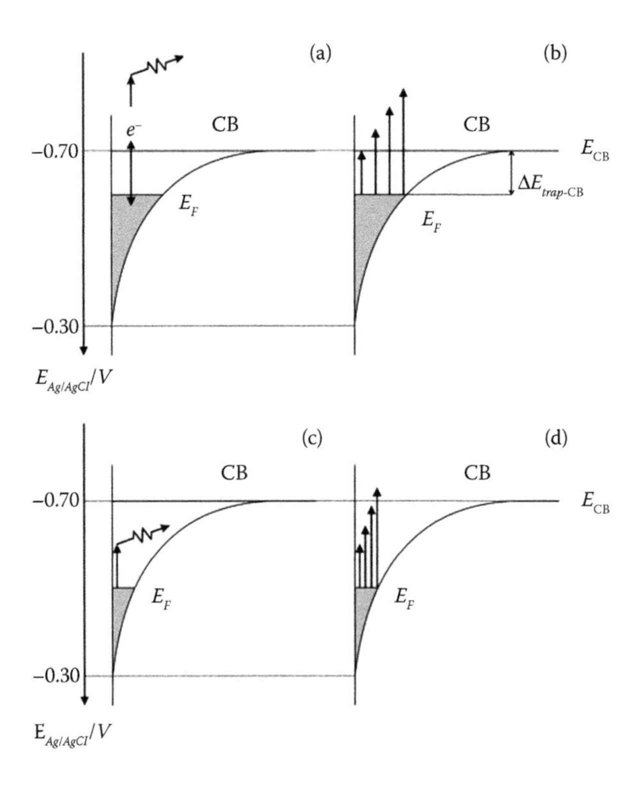

FIGURE 3.10 Possible transitions in the middle infrared of accumulated electrons in anatase TiO_2 nanocrystal electrodes. (a) Intra-conduction band transitions associated with phonon absorption. (b) Excitation of shallow trapped electrons to the conduction band. (c) Polaron excitations by coupling with two-dimensional surface phonons. (d) Intra-bandgap transitions. (Reproduced with permission from Berger, T. et al. *The Journal of Physical Chemistry C* 2012, 116, 11444–11455.)

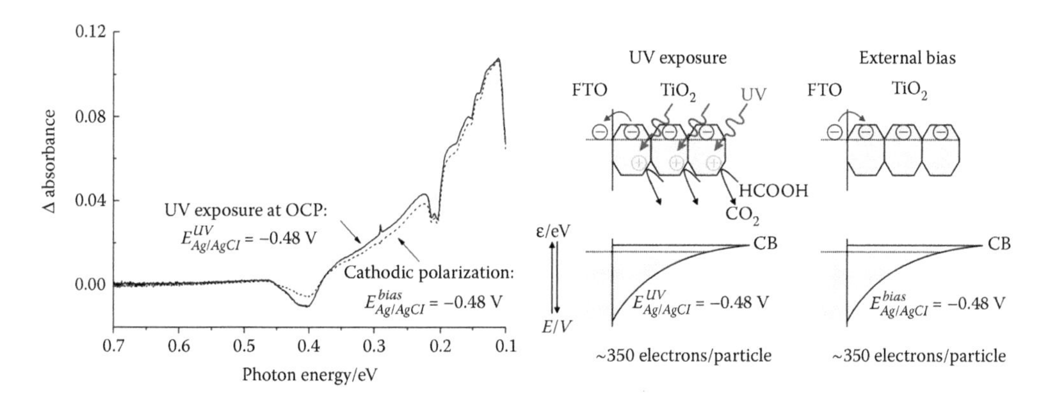

FIGURE 3.11 IR spectra of an anatase TiO_2 nanocrystal electrode during UV exposure at open circuit potential and after polarization to $V_{Ag/AgCl} = -0.48$ V. (Reproduced with permission from Berger, T.; Anta, J. A. *Analytical Chemistry* 2011, 84, 3053–3057.)

mechanisms whereby the electrons in traps in TiO_2 provide transitions that contribute to the optical absorption. These effects result in absorption features in the middle infrared at energies much lower than the bandgap, 0.2 eV, corresponding to a wavelength of 1 μm, as shown in Figure 3.11. The same difference in spectra can be obtained either from UV generation of carriers or from the electrons injected at negative bias. In the case of ZnO nanostructures, electrons accumulated within the conduction band cause a bleach of the excitonic band (Subramanian et al., 2003).

The method of time-resolved Terahertz spectroscopy (TRTS) makes use of sub-picosecond pulses of freely propagating electromagnetic radiation in the Terahertz range, that is characterized by sub-mm wavelengths (300 μm for 1 THz), and very low-photon energies (4.2 meV at 1 THz) (Ulbricht et al., 2011; Canovas et al., 2013). As the detection energy corresponds to less-than-thermal energies at room temperature (1 THz corresponds to 48 K), this method directly investigates the dynamics of free carriers and provides information about the complex dielectric function of the material, yielding the complex conductivity. It is a powerful tool for studying ultrafast charge carrier dynamics and carrier transfer processes in semiconductor nanostructures.

3.3 EXCITONS

Free electrons in the conduction band and free holes in the valence band of a semiconductor interact by the Coulomb attraction and their relative motion is quantized, provided that this interaction is stronger than the random thermal fluctuation (caused by phonon collisions). The stable quasi-particle state associated with this electron–hole interaction is an *exciton*, Figure 3.12a. In a first approximation, the exciton can be treated as a hydrogenic atom of reduced mass m_{eh} in a medium of dielectric constant ε_r. Using the Bohr model, the binding energy of the n_l level relative to the ionization limit is

$$E_B(n_l) = -\frac{m^*_{eh}}{m_0\varepsilon_r^2}\frac{R_H}{n_l^2} = -\frac{R_X}{n_l^2} \tag{3.13}$$

Here, m_0 is the mass of the free, noninteracting electron; $R_H = 13.6$ eV is the Rydberg constant of the hydrogen atom; and

$$R_X = \frac{m^*_{eh}}{m_0\varepsilon_r^2} R_H \tag{3.14}$$

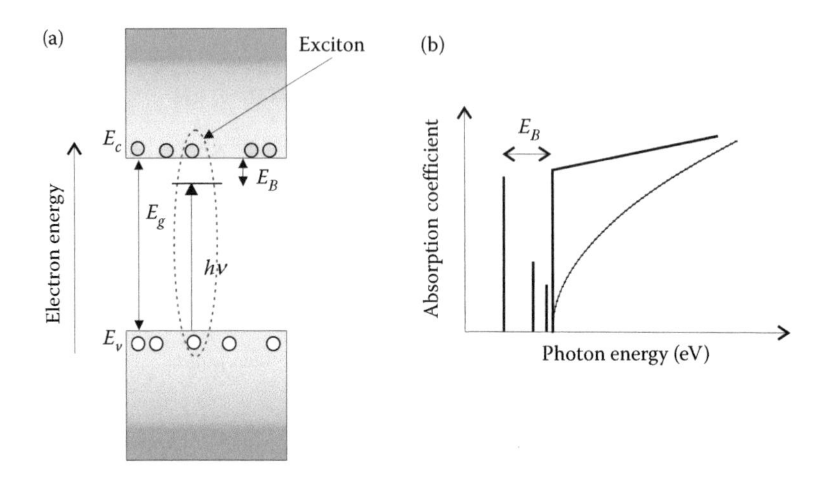

FIGURE 3.12 (a) Illustration of a photoinduced excitonic transition indicating the binding energy of the exciton. (b) Characteristic absorption associated with the excitonic electron–hole correlations. The smooth spectrum of the direct bandgap semiconductor is also shown in thin line.

is known as the effective Rydberg of the exciton. The average separation between an electron and a hole in the $n_l = 1$ exciton or effective Bohr radius is hence defined as

$$a_X = \frac{m^*_{eh}}{m_0} \frac{1}{\varepsilon_r} a_B \tag{3.15}$$

where $a_B = 0.053$ nm is the Bohr radius of the hydrogen atom.

Here, we describe briefly the excitonic characteristics of inorganic solids with direct transitions. The strong optical coupling of the exciton with the radiation field exerts a great influence on the optical absorption features as outlined in Figure 3.12b. The photon energy required to create the electron–hole pair is the bandgap energy minus the binding energy due to the Coulomb interaction, given in Equation 3.13. Therefore, a strong optical absorption is expected below the band edge energy, so that the optical gap is lower than the transport gap. In addition, higher-excitonic levels produce a flat shape at $E = E_g$ instead of the smooth rise of the parabolic band-to-band absorption, as predicted by the theory of Elliott (1957). For GaAs, the binding energy of the $n_l = 1$ exciton calculated by values in Table ECK.2.1 is $E_B = 5.6$ meV, which is very weak with respect to $k_B T$ at room temperature. Therefore, the exciton, with an effective Bohr radius $a_X = 10.6$ nm, can remain bound only at very low temperatures. However, the *Sommerfeld enhancement* due to the exciton continuum of the absorption edge is observed from low-to-high temperatures (as explained by the theory of Elliot) instead of the parabolic absorption. The excitonic absorption features of GaAs are shown in Figure 3.3 and for InP in Figure 3.13. GaN is a large bandgap semiconductor, $E_g = 3.4$ eV, where $m^*_e = 0.067\, m_e$, $m^*_h = 0.51\, m_e$, and $\varepsilon_r = 12$. The exciton-binding energy and radius for this material are $E_B = 26$ meV and $a_X = 2.8$ nm, respectively, so that the exciton effects are more significant than in the former case. In general, for semiconductors with a small bandgap, the dielectric constant is large and the effective masses are small. Hence, binding energy tends to increase and the radius decreases as the bandgap E_g of the semiconductor increases.

Self-consistency of Equation 3.14 requires the exciton radius to be much larger than the lattice spacing, so that the dielectric screening applies. Excitons with small-binding energy and a large radius are called the *Wannier–Mott excitons*. These are delocalized states that can move freely through the crystal (represented by the exciton center of mass). In organic materials, the dielectric constant tends to be low and the exciton-binding energy E_0 is larger than in inorganic semiconductors, so that excitons are often stable in organic materials. In insulators and some organic crystals,

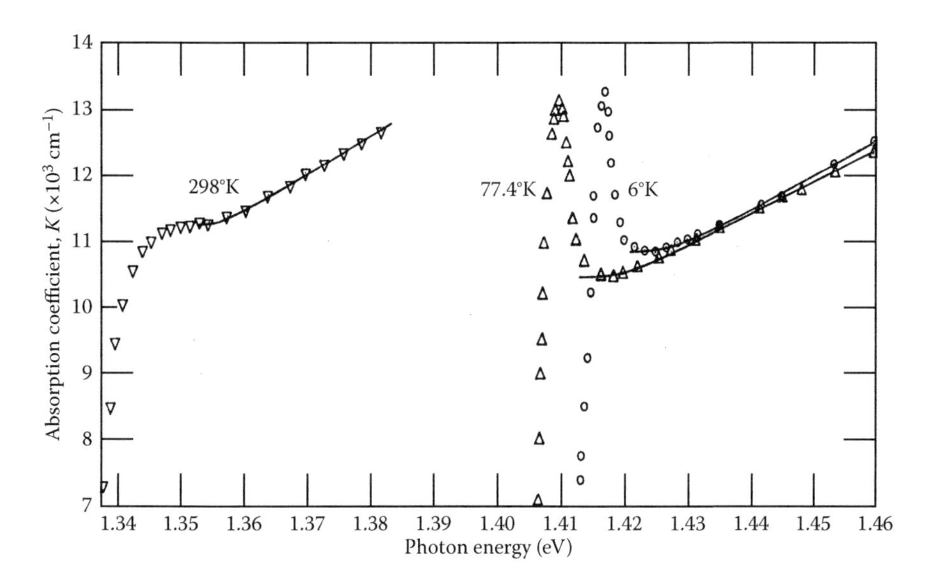

FIGURE 3.13 Absorption of InP in the exciton region. (Reproduced with permission from Turner, W. J. et al. *Physical Review* 1964, 136, A1467.)

the tightly bound electron–hole pair wave function is confined to one crystal unit and then it is called a *Frenkel exciton*.

Free carriers present in heavily doped samples shield the Coulomb interaction and reduce the binding forces by screening, impeding the formation of excitons (Mahan, 1967). Simple arguments based on the Debye–Hückel theory show that the exciton-binding energy decreases strongly at concentrations of $n \approx 10^{17}\,\text{cm}^{-3}$ (Gay, 1971). If the Debye length is λ_D, the exciton-binding energy E_X is reduced from the ground state energy E_0 as (Sachenko and Kryuchenko, 2000)

$$E_X \approx E_0\left(1-\frac{a_X}{\lambda_D}\right)^2 \tag{3.16}$$

This expression is plotted in Figure 3.14 with respect to the free carrier concentration. It is observed that when the Debye length becomes of similar size as the exciton radius, $\lambda_D \approx 4a_X/3$, the exciton-binding energy is reduced to half the ground state value. If $a_X = 3$–5 nm and $\varepsilon = 3.5$, the reduction of binding energy happens at $n_0 \approx 5 \times 10^{16}\,\text{cm}^{-3}$ with the consequence of increased dissociation probability. In addition, at $a_X \approx \lambda_D$, the exciton Mott transition occurs, and the bound electron–hole pairs do not occur anymore.

3.4 QUANTUM DOTS

Quantum dots (QDs) are regular fragments of a material whose size is smaller than the exciton radius, usually in the range of a few nm. QDs have fully developed crystalline structure but the conduction and valence bands are replaced by discrete levels due to the confinement effect, while some of the semiconductor bulk properties such as high-extinction coefficient are preserved. In semiconductor QDs, the bandgap can be easily tuned by the control of their size and shape, providing an excellent tool for nanoscale design of light absorber materials. The induced dipole moment by the electromagnetic field, as represented by the oscillator strength, is superior to their bulk counterparts, so that the absorption coefficient is larger. The increased specific surface area also plays an important role in the photophysical properties of QDs.

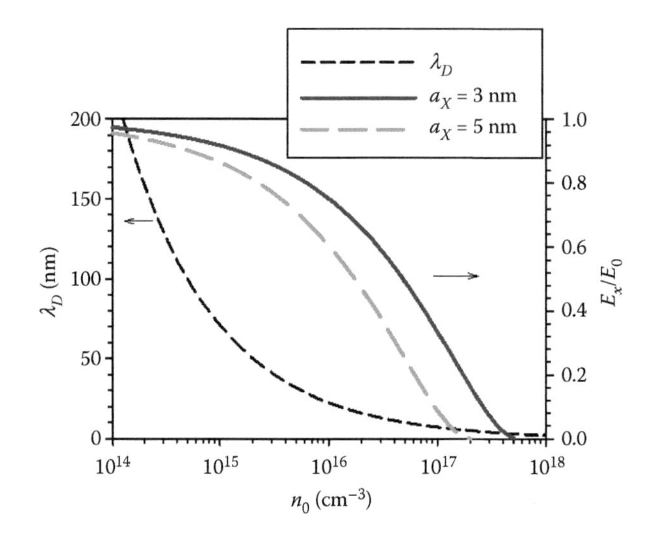

FIGURE 3.14 Reduction of exciton-binding energy E_X/E_0 by screening for a semiconductor with $\varepsilon = 3.5$ and two values of exciton Bohr radius as indicated. The Debye screening length λ_D is also shown.

Colloidal QDs are prepared by low-temperature wet chemical routes that provide a very precise control of shape and size and also high monodispersity that is especially important for light emission applications, see Figure PSC.1.6. The resulting QDs are freestanding in colloidal solutions, which is made possible by their surrounding organic ligands that provide solubility in many nonpolar solvents, thus further facilitating the formation of films for application in devices. Figures 3.15 and 3.16 illustrate main absorption and PL properties of semiconductor QDs. Figure 3.15 shows the room temperature absorption and emission spectra of InP QDs with a mean diameter of 3.2 nm. The absorption spectrum shows a broad excitonic peak at about 590 nm due to the inhomogeneous size distribution of the QDs. The PL spectrum shows two emission bands. The weak emission near the band edge with a peak at 655 nm is attributed to the ground exciton recombination. The larger band above 850 nm is due to radiative surface states. It has been widely observed that recombination at surface states produces broadband and very slow PL in QDs (Harruff and Bunker, 2003;

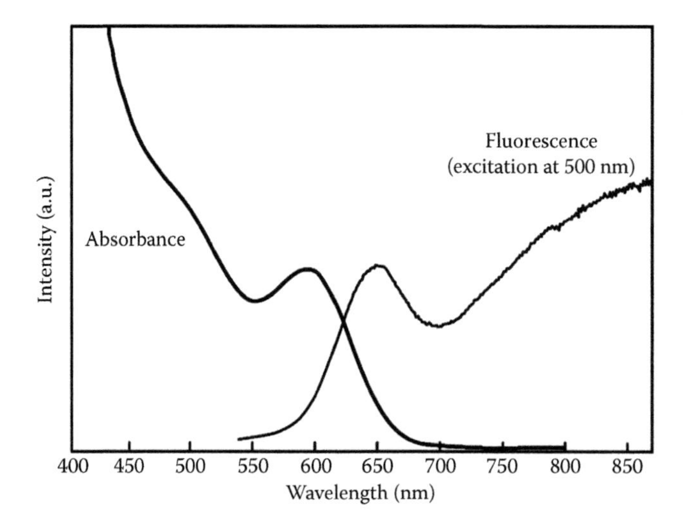

FIGURE 3.15 Absorption and emission spectra at 298 K of untreated 32 Å InP QDs. (Reprinted with permission from Micic, O. I. et al. *Applied Physics Letters* 1996, 68, 3150–3152.)

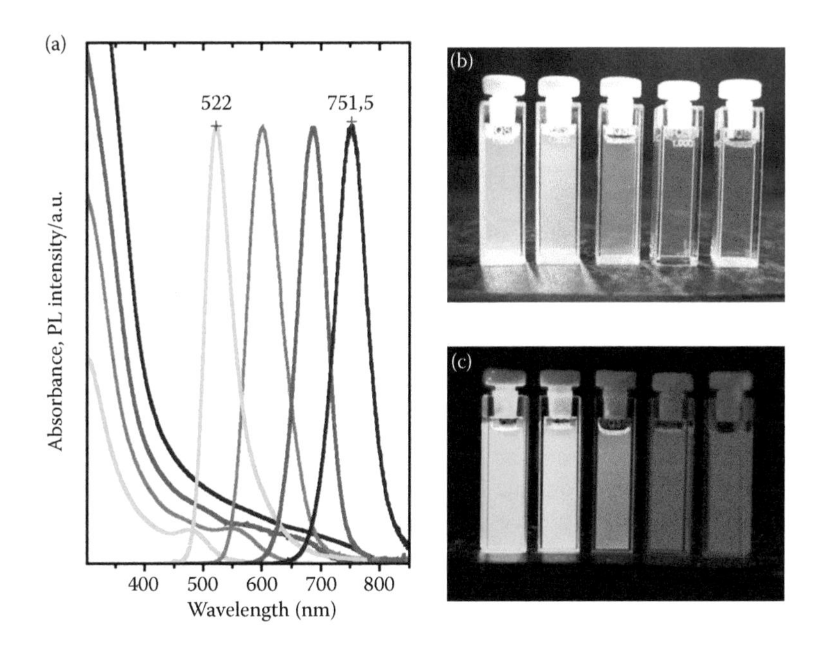

FIGURE 3.16 (a) Absorption and PL spectra of HF-photoetched InP nanocrystals of different size. (b, c) Visualization of size-dependent change of the PL color of HF-photoetched InP nanocrystals. The smallest (1.7 nm) particles emit green, whereas 4 nm particles emit deep red. Larger InP nanocrystals emit in near-IR (not shown). The high PL quantum yield makes the colloidal solutions "glow" in room light (b). Photo (c) shows the luminescence of the etched InP nanocrystals placed under UV-lamp. (Reprinted with permission from Talapin, D. V. et al. *The Journal of Physical Chemistry B* 2002, 106, 12659–12663.)

Shea-Rohwer and Martin, 2007). The room temperature absorption spectra of InP nanocrystals as a function of size between 1.7 and 4 nm are shown in Figure 3.16. The spectra shift to higher energy as the QD size decreases due to the change of the bandgap by the quantum size effect.

The properties of semiconductor nanocrystals can be modified by the formation of nanoheterostructures consisting of QDs with a combination of core and shell materials, or embedding a single material QD in a solid framework. The combination of sizes, composition, and shapes provides a large versatility for the engineering of specific properties and physical effects. In core–shell QDs, the electronic and optical properties of the combined material are largely determined by the band alignment type, as indicated in Figure 3.17. All these structures produce a modification of the original core

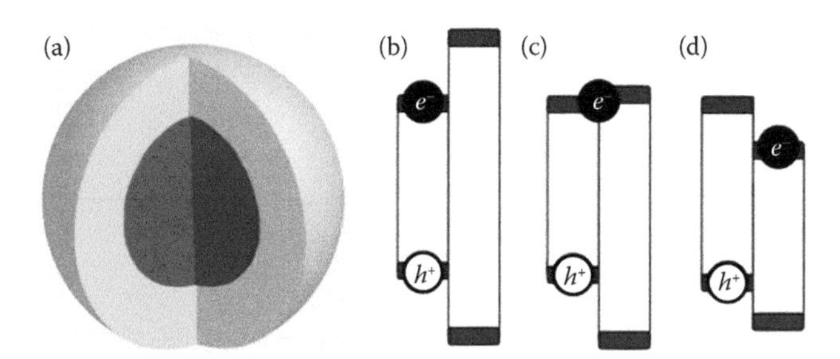

FIGURE 3.17 (a) Schematic of a core/shell nanocrystal. Differing alignments of core (*left* in each case) and shell (*right* in each case) conduction (*red*) and valence (*blue*) bands afford distinct electronic structures: (b) Type I, (c) Quasi-type II, and (d) Type II. Location of an excited-state electron or hole in the core–shell structure is indicated by e^- or h^+, respectively, and determined by the particular band alignments. (Reprinted with permission from Hollingsworth, J. A. *Coordination Chemistry Reviews* 2014, 263–264, 197–216.)

QDs. The type I heterojunction occurs when the bandgap of the shell layer encloses that of the central core. This structure causes the confinement of both the electron and the hole in the core, away from surface traps, which reduces surface recombination and improves the radiative quantum yield (Hines and Guyot-Sionnest, 1996). In type II, the band alignment is staggered, which causes light absorption and emission at lower energies than both core and shell original materials (Kim et al., 2003), as further discussed in Section PSC.3.6. Finally, in the intermediate case, either conduction or valence band is aligned producing a larger delocalization of either the electron or the hole.

The assembly of semiconductor QDs into QD films is significantly cheaper compared to their bulk semiconductor counterparts since their synthesis takes place at lower temperatures and with solution-based approaches (Vanmaekelbergh, 2011; Guyot-Sionnest, 2012). Unlike sintered nanoparticulate networks studied in Chapters ECK.8 and ECK.9, one may form films in which QDs are held together by organic linkers so that each QD maintains its electronic identity and the conductivity is governed by transitions between dots, Figure FCT.1.5. Solar cells based on QD layers have been developed based on PbS, which has a bulk bandgap in the IR. High-power conversion efficiencies have been obtained employing the bidentate organic linkers, which bring the nanoparticles into close packing while achieving the best-available surface passivation (Luther et al., 2008; Tang et al., 2011; Ip et al., 2012). On the other hand, for LED applications, the proximity of QDs causes a quenching of the luminescence by FRET, and the EQEs are greatly decreased with respect to the colloidal suspension in solution (Shirasaki et al., 2012).

3.5 ORGANIC MOLECULES AND MATERIALS

Organic materials and molecules consist of carbon and hydrogen with a few heteroatoms such as sulfur or nitrogen, see Section ECK.8.5. The light absorption features of organic materials and chromophores are of high importance for artificial photosynthesis schemes as well as for all solar energy conversion devices based on organic molecules and materials.

Light absorption in natural organic pigments provides the basic fuel for life on the earth. In the process of oxygenic photosynthesis, pigment molecules absorb sunlight and use its energy to synthesize carbohydrates from CO_2 and water. The absorption of the main pigments such as chlorophylls and carotenoids is shown in Figure PSC.1.2, in comparison with the solar spectral photon flux at the top of the earth's atmosphere and on the earth's surface. We note that oxygenic photosynthesis utilizes photons in a restricted range of wavelengths between 400 nm and 700 nm. A photon having a wavelength of 700 nm has an energy of 1.8 eV. This energy is required to drive the required reactions in the photosynthetic apparatus (Milo, 2009).

As discussed in the preceding sections of this chapter, in inorganic materials at room temperature the excited electron–hole pair is formed by very weakly bound carriers. Therefore, the excited carriers are delocalized in their bands with small exchange energy. Furthermore, there is a continuum of electronic states in the CB and VB over a broad energy range. Consequently, absorption bands extend all the way to very short wavelengths in the energy range above the bandgap.

In contrast to these characteristic features of inorganic crystalline semiconductors, the main factors determining the interaction of photons with electrons in organic molecules are caused by the strong influence of the correlation of spin states of the electrons in the ground and excited states and by the fact that molecules have many vibrational degrees of freedom. The absorption features of organic molecules show specific bands centered at the wavelengths that correspond to transitions between quantum states having a large spatial overlap. The exchange energy corresponding to the excited and unexcited electrons can have a large value 0.7–1.0 eV (Köhler and Bässler, 2009).

In organic materials, the highest molecular level filled with electrons is the HOMO (highest occupied molecular state), which for a conjugated molecule will be a π orbital and the first available level above the HOMO is the LUMO (lowest unoccupied molecular state), which is an excited configuration of the π orbital termed a π^* state. The main transition for optical absorption in organic molecules is the π–π^* transition due to its high-molar absorptivity. The π–π^* transition involves

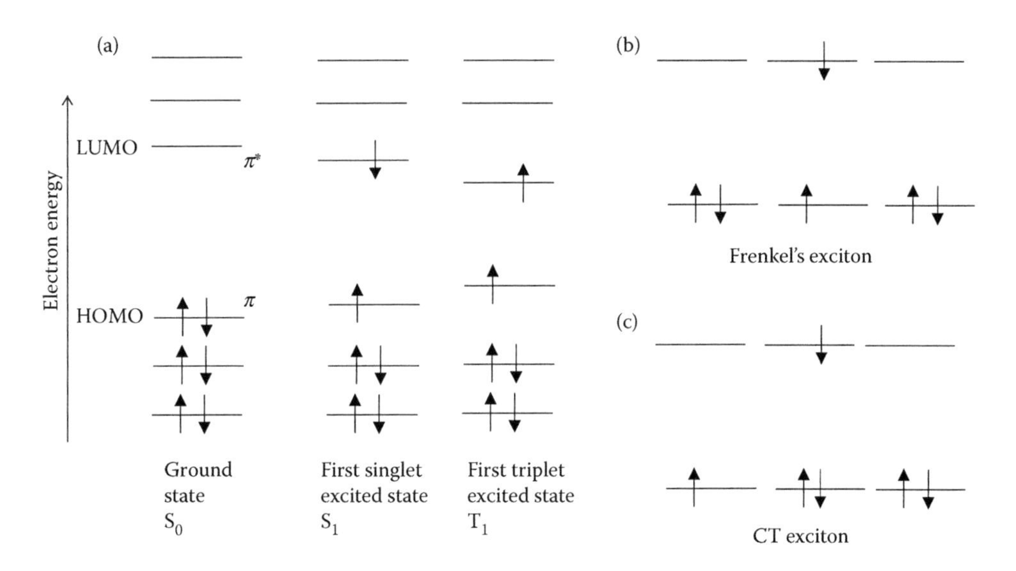

FIGURE 3.18 (a) The unpaired electrons in the excited state of a molecule can have their spins parallel or antiparallel, forming, respectively, a singlet or triplet state. The ground state is a singlet. The diagram also indicates the effect of the Coulomb and exchange energy in the energy levels of the frontier orbitals involved in the first excited singlet and triplet states. (b) Frenkel exciton. (c) CT exciton.

electrons that are strongly correlated and it is important to consider in detail the possible pairing of the spin of the electron in the excited state with that remaining in the ground state (Figure 3.18). Electrons in the ground state are paired off with their spins antiparallel and the net spin angular momentum is zero, according to the Pauli exclusion principle. This is called a *singlet state*, S_0. In the excited state, the electrons can have their spin states either antiparallel or parallel. The first case is the excited singlet state S_1. In the second case, the total spin angular momentum is 1 and there are three possible spin wave functions according to the z-component of the angular momentum; hence these are called *triplet states*, T_1. Since photons carry no spin, a photon-induced transition $S_0 \rightarrow T_1$ is not allowed and the main optical absorption route corresponds to the transition $S_0 \rightarrow S_1$.

In an organic crystal or chain, the promotion of an electron occurs predominantly in one molecule or to a neighboring molecule in a charge transfer (CT) event. Accordingly, distinct classes of excitons exist in organic crystals depending on the extent of charge separation (Bardeen, 2014). In the Frenkel exciton, the electron and hole associated with the excited state reside in the same molecule (Figure 3.18b). Two electrons are localized on the LUMO and HOMO level of the same site, implying that the exchange interaction between them can be large. This effect separates the Frenkel exciton into two singlet and triplet bands. In a CT event, the electron is transferred to the LUMO of another molecule, leaving a hole in the HOMO, see Figure 3.18c, and the resulting CT exciton has ionic character. Long-range transport of excitons in the spin singlet state by hopping occurs by the Forster energy transfer. Triplet exciton hopping is dominated by the Dexter energy transfer, since excitons in the triplet state are spin forbidden from emitting.

Because of the vibrational degrees of freedom of the molecule or polymer, there are many singlet excited states, above the ground state S_1, that are termed S_2, S_3, etc. and similarly for the triplet states T_2, T_3, etc. In Figure 3.19, the ground state of the vibronic ensemble of a class of spin states is indicated with a thick line while the upper energy states are indicated by thinner lines. The vibronic levels in a given type of spin-correlated system correspond to the quantum number of the harmonic oscillator $v = 0, 1, 2\ldots$. The transitions involving photon absorption or emission will start from the lowest energy state $v = 0$ of the ground or excited singlet ensemble, respectively. The excited state will rapidly relax to the lowest energy state, in a process similar to that explained in Figure ECK.6.9 and

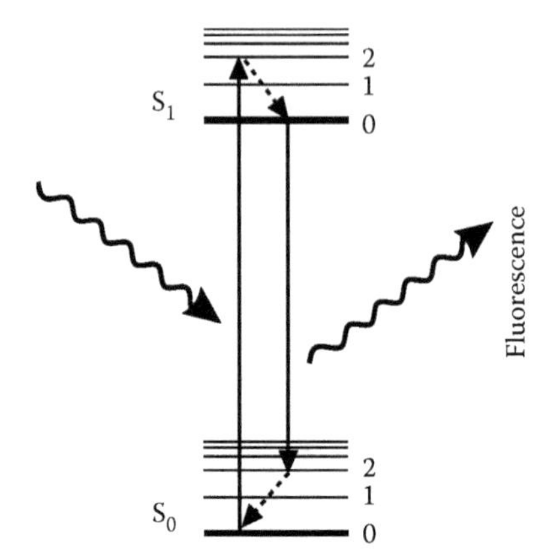

FIGURE 3.19 The photon-induced transition from the ground singlet to the first excited singlet states. The excited state relaxes rapidly in the manifold of vibronic states and decays back to the ground singlet state emitting a photon in the process of fluorescence.

shown in Figure 3.20 for the case of a vibronic manifold. Then, the excited state S_1 may decay back to one of the energy states in the ground singlet, emitting a photon in the process termed *fluorescence.*

While the photon absorption and emission process starts from the lowest energy state of the respective singlet vibronic manifold, the intensity of the different transitions will determine the lineshape of absorption and fluorescence spectra. In Figure 3.20a, the vibronic manifolds of the singlet states are represented by the quantum states of oscillators. The oscillators are displaced in the horizontal axis that indicate the nuclear coordinates of the molecule, due to the fact that equilibrium position of excited and ground states is not the same. The nuclear coordinates can be taken as normal coordinates of the vibration, Q, since rotations and translations can be neglected. The process of the electronic transition is so fast that the nuclear coordinates do not change their values during the transition. This general rule is termed the *Franck–Condon principle*, as discussed in Chapter ECK.6.

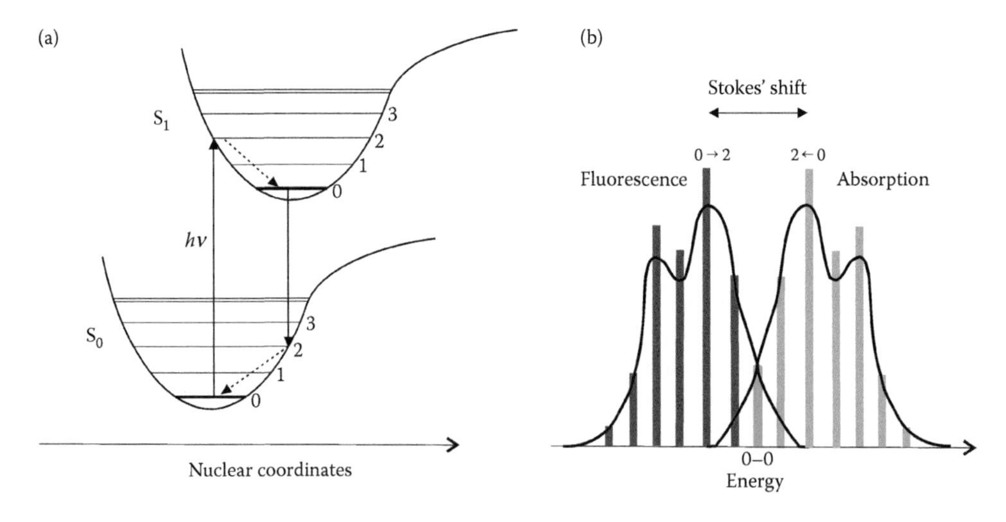

FIGURE 3.20 (a) The photon-induced transition from the 0-ground singlet to the 2-level of the excited singlet state in an oscillator model. The initial excitation in 2-level excited state relaxes rapidly to the 0-level of the excited state and decays by fluorescence to the 2-level of the ground singlet state, which then relaxes to 0-level of the ground singlet state. (b) The intensity of individual transitions is determined by the spatial overlap of the electronic states involved in the transition. The smoother lines correspond to the actual lineshape expected in the observation of absorption and fluorescence.

Therefore, the probability of a specific transition is largely determined by the spatial overlap of the initial and final states of the electronic vibrational wave functions.

If the ground and excited states have a similar distribution of oscillator quantum states, then the upward transitions from the 0 level of the ground singlet state will have the same probabilities as the downward transitions from the 0 level of the excited singlet state. However, the fluorescent transitions will occur at lower energies (larger wavelengths) than the absorption transitions, thus causing Stokes shift. The fluorescent peak is often a mirror image of the absorption peak and the intercept occurs at the 0–0 (zero–zero) transition that has the same energy in both absorption and fluorescence. Some examples are shown in Figure 3.21. The fluorescent luminescence shows the

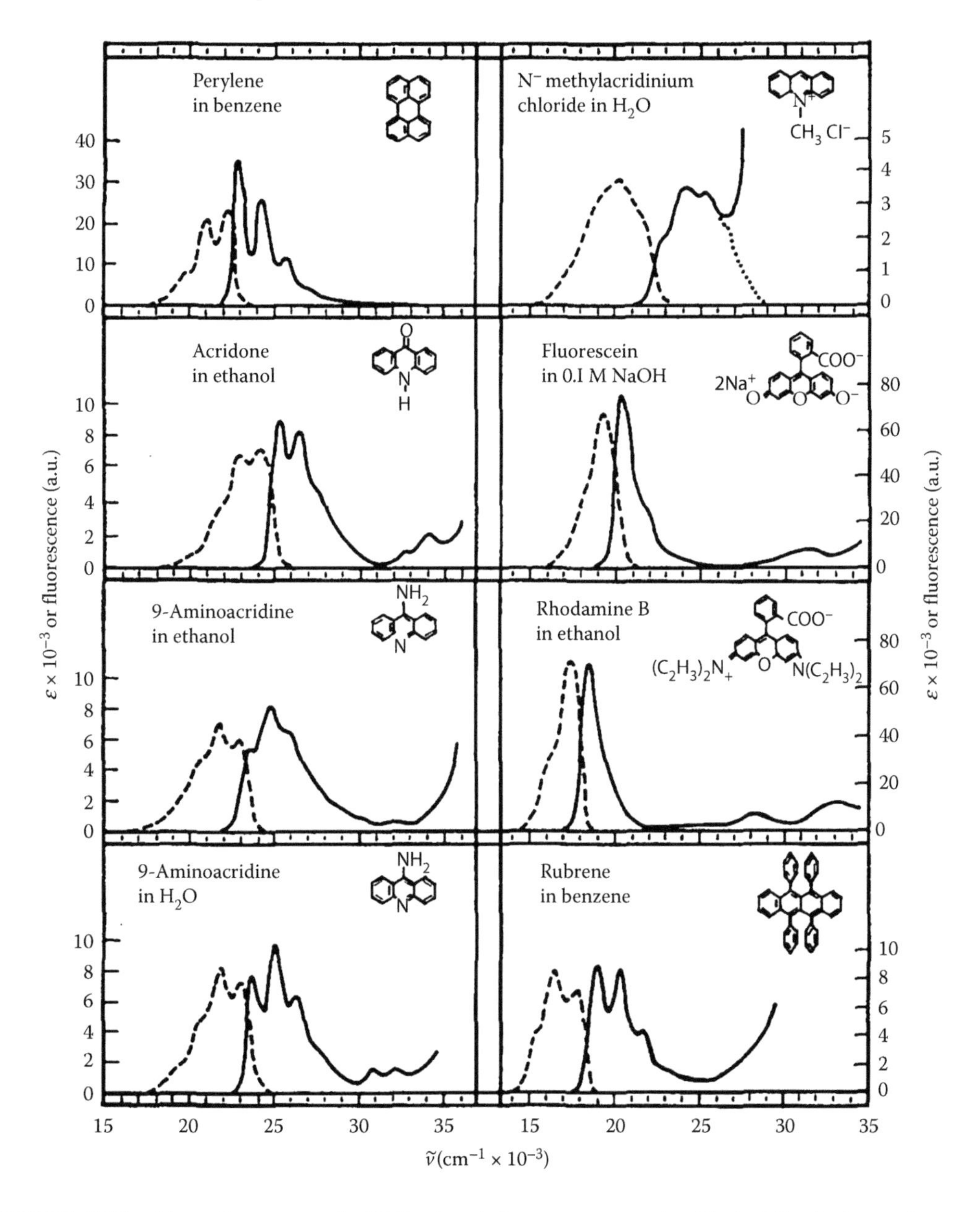

FIGURE 3.21 Absorption spectra (*solid lines*) and fluorescence spectra (*dashed lines*) of organic compounds. (Reproduced with permission from Strickler, S. J.; Berg, R. A. *Journal of Chemical Physics* 1962, 37, 814–820.)

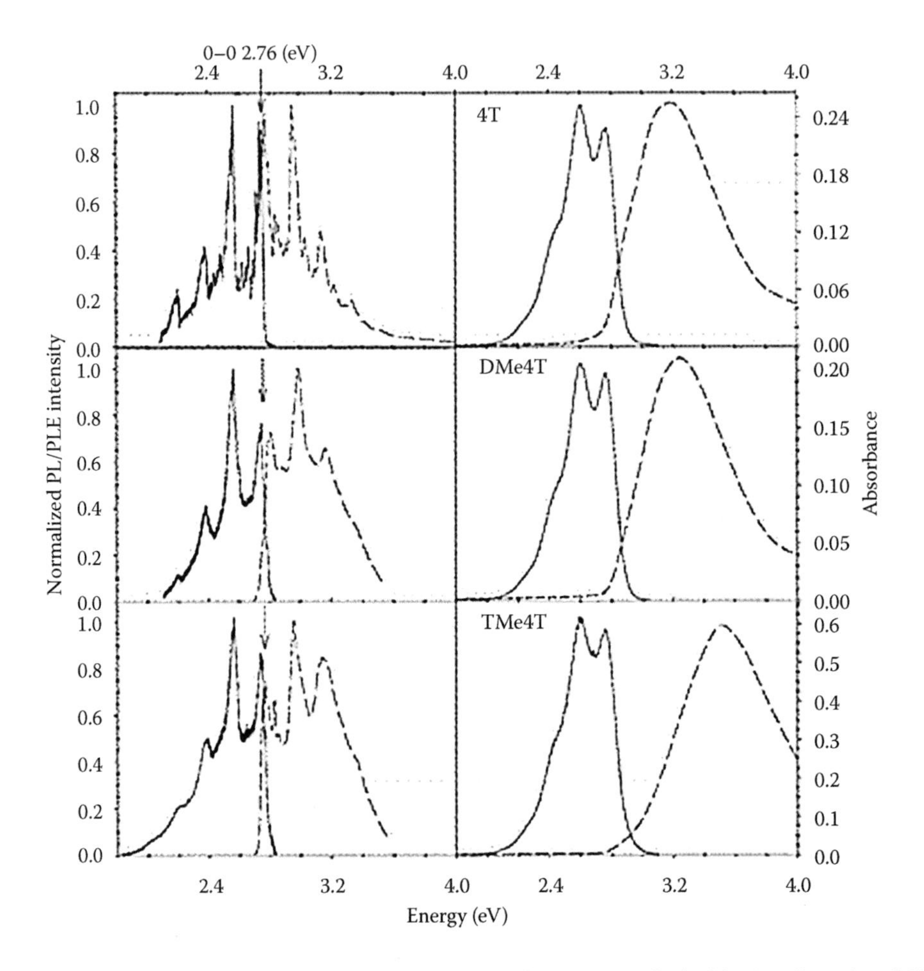

FIGURE 3.22 (b) Normalized PL (*solid line*) and absorption spectra (*dashed line*) collected at 4 K (*left*) and 300 K (*right*) of several oligothiophenes. (Reproduced with permission from Macchi, G. et al. *Physical Chemistry Chemical Physics* 2009, 11, 984–990.)

vibronic splitting. The vibrational states can be closely spaced and are coupled with normal thermal motion. Therefore, the emission and absorption spectra occur over a band of wavelengths rather than in sharp lines. The vibronic features of oligothiophenes are shown in several peaks that are clearly resolved at low temperature in the left column of Figure 3.22, while at higher temperatures, those features are smoothed out in the right column of Figure 3.22.

The direct transition from the ground singlet state to a triplet state requires a spin flip that is forbidden on the basis of the conservation of the spin angular momentum alone. However, the transition in which an electron is moved between orbitals of differing symmetry becomes possible if assisted by the spin–orbit coupling, since the change in spin angular momentum compensates for the change in orbital angular momentum. The most effective way to increase the efficiency of these transitions is the presence of heavy atoms incorporated in the molecule that provoke a very large, metal-induced spin–orbit coupling. The excited triplet state is usually more stable than the excited singlet manifold due to larger stabilization by the exchange interaction between the electrons. Therefore, T_1 is lower in energy than S_1. The transition $S_1 \rightarrow T_1$ is termed *intersystem crossing* (ISC) and depends on spin–orbit coupling and on the vibrational overlap between the singlet and triplet states, see Figure 3.23. From the relaxed state, a luminescent

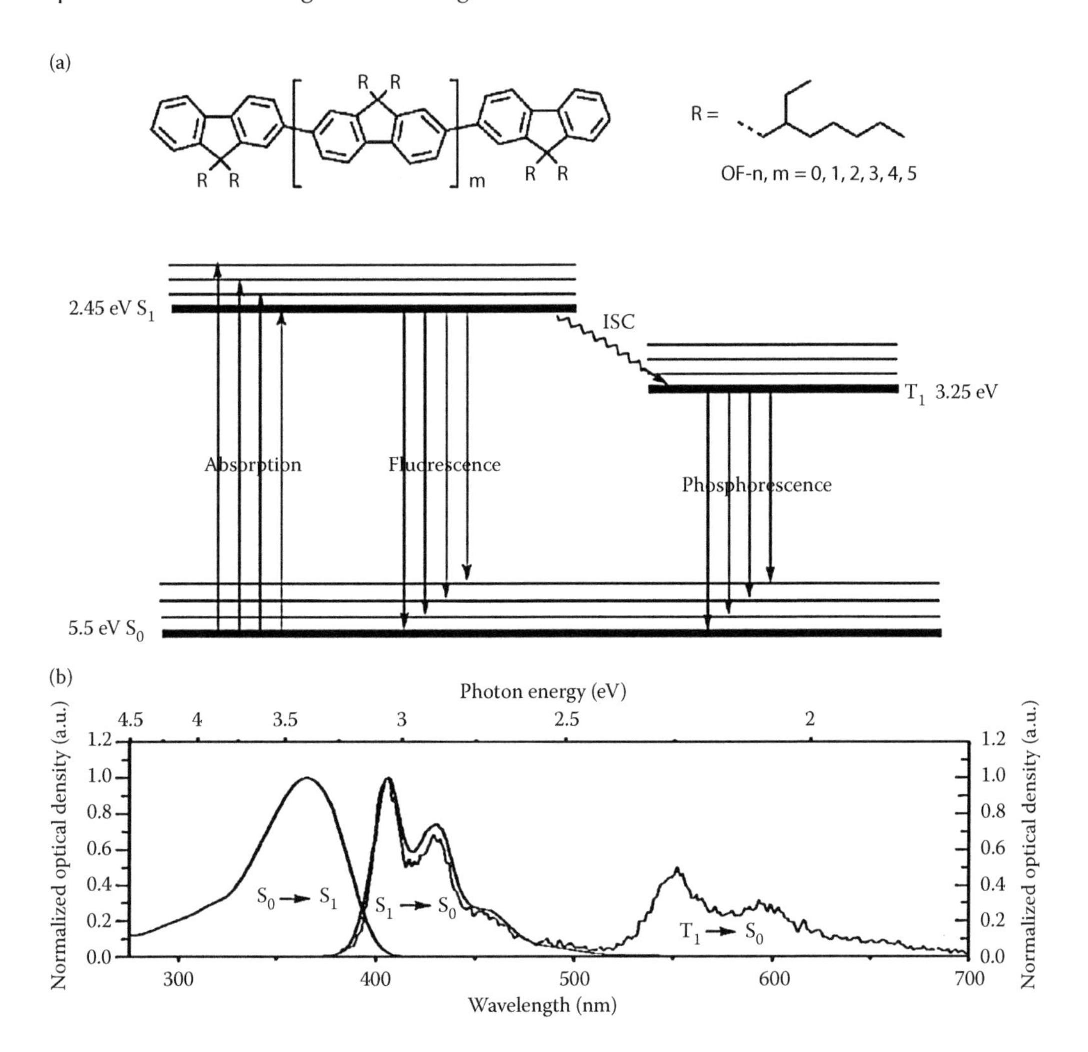

FIGURE 3.23 Structure of oligofluorenes with 2-ethylhexyl side chains (OFn). (a) Energy diagram and (b) absorption and luminescence of OF5 in a matrix of methyltetrahydrofurane (m-THF). (Reproduced with permission from Wegner, G. et al. *Macromolecular Symposia* 2008, 268, 1–8.)

transition $T_1 \rightarrow S_0$ is weakly allowed. This process occurs at lower energy and longer wavelengths than fluorescence, and with much longer radiative lifetime (microseconds to miliseconds), and is called *phosphorescence*. Since ISC is facilitated if the states have similar energy, the transition from S_1 may preferentially occur to a higher-lying triplet state T_n that subsequently relaxes vibronically to T_1.

The metal-to-ligand charge transfer (MLCT) in excited states of $d\pi^6$ coordination compounds has been widely applied in solar energy harvesting using dye-sensitized solar cells (Ardo and Meyer, 2009). In this type of transition, light absorption promotes an electron from the metal d orbitals to the ligand π^* orbitals. The processes following the vertical excitation of [Ru(bpy)3]2+ to the Franck–Condon state are shown in Figure 3.24. Absorption of visible light (450 nm) by $Ru(bpy)_3^{2+}$ leads to a $d - \pi^*$ MLCT transition that proceeds with unit quantum efficiency. The lifetime of the excited state, $Ru(bpy)_3^{2+*}$, is quite long (~600 ns) (Henry et al., 2008). The excited electron undergoes ISC to the lowest triplet states of $Ru(bpy)_3^{\bullet 2+}$ from where phosphorescent emission of a photon of energy $E = 2.1$ eV (610 nm) occurs.

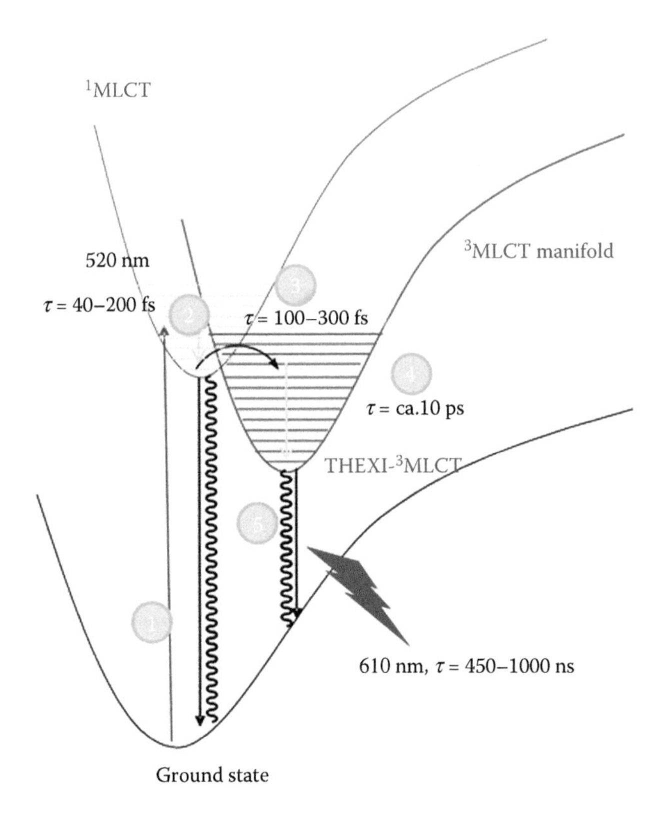

FIGURE 3.24 Overview of processes following the excitation of [Ru(bpy)3]2+ to the Franck–Condon (FC) state. (1) Excitation from the groundstate to the FC state, (2) relaxation of the FC state and fluorescence (520 nm), the lifetime of which is determined by the rate of ISC, ca. <300 fs, (3) to the vibrationally hot ^{3}MLCT state followed by (4) vibrational cooling to the THEXI–^{3}MLCT state (complete by 20 ps), which itself undergoes both (5) nonradiative and radiative relaxation with a lifetime of 400–1000 ns. (Reproduced with permission from Henry, W. et al. *The Journal of Physical Chemistry A* 2008, 112, 4537–4544.)

The size of a conjugated system greatly influences all the photophysical quantities because the confinement effect increases when the number of repetition units decreases. The energy gap, ionization potential, and electron affinity of oligomers show a reciprocal dependence on the number of repeat unit (or degree of polymerization) (Wegner et al., 2008). When the molecules are condensed to form an organic solid, the absorption properties of the solid may either remain similar to those of the molecule in solution or else may be strongly modified. An example is shown in Figure 3.25. In a random molecular distributed solid, the molecular absorption still predominates. However, if the solid is a crystal or forms ordered aggregates, the electronic transition obtains a predominant CT character, which induces efficient coupling of intermolecular vibrational modes (Gierschner et al., 2005; Oksana et al., 2005).

3.6 THE CT BAND IN ORGANIC BLENDS AND HETEROJUNCTIONS

The combination of a donor and acceptor pair of molecules, or the formation of an intimate heterojunction between semiconductors opens new optical transitions with respect to those of the separate materials by excitation of an electron from the VB (HOMO) of the acceptor to the CB (LUMO) of the donor, as shown in Figure 3.26. The heterojunction state formed in this type of transition is termed the *exciplex* (shorthand for excited states complex formation) or *charge transfer complex* (CTC). The CTC can be strongly excitonic in which case the opposite carriers remain localized at

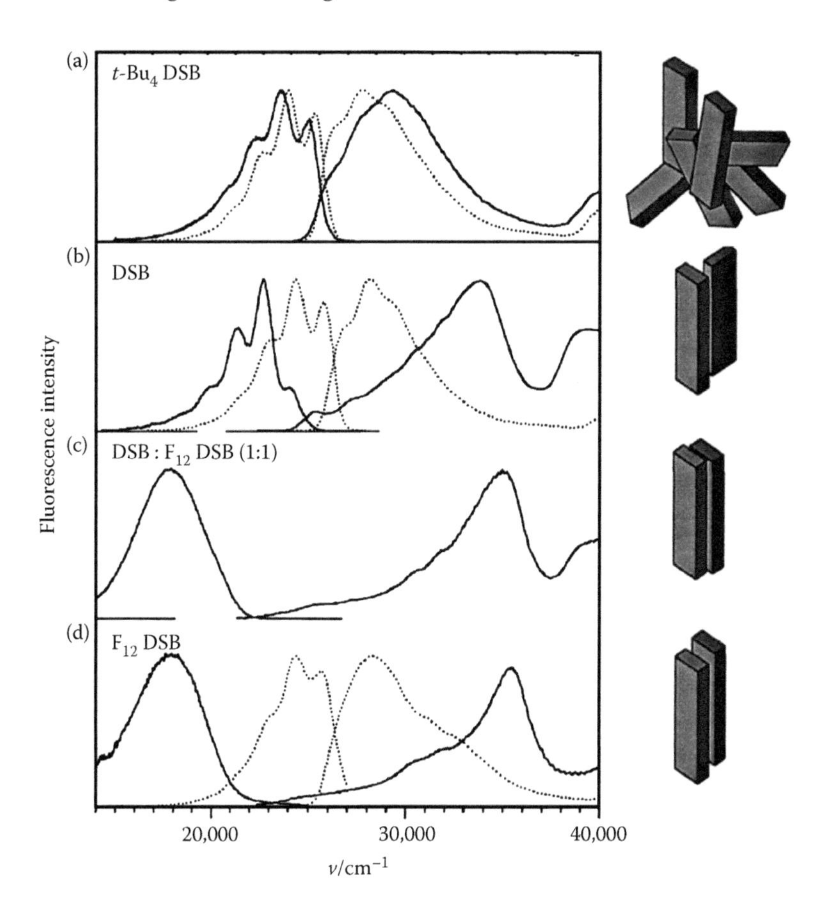

FIGURE 3.25 Fluorescence (*left*) and absorption spectra (*right*) of distyrylbenzene nanoparticles: (a) t-Bu_4DSB, (b) DSB, (c) cocrystallized DSB/F_{12}DSB, and (d) F_{12}DSB. Spectra in hexane solution (*dashed lines*) are shown for comparison. A schematic representation of the respective condensed phase structures is given on the right. (Reproduced with permission from Gierschner, J. et al. *The Journal of Chemical Physics* 2005, 123, 144914.)

the interface as indicated in Figure 3.26, or it may give rise to fast charge separation, and the optical gap will be smaller than the transport gap of the organic blend (Bredas, 2014).

CT absorption or emission processes in donor–acceptor molecular systems have been amply investigated comparing the absorption properties of the blend with respect to the parent materials. CT absorption bands of π-conjugated polymers and oligomers combined with electron acceptors correspond to the transition from the ground state to the charge-separated state. The peak absorbance energy is seen to correlate with the difference between oxidation potential of the donor and reduction potential of the acceptor, as shown in Figure 3.27. The CTC usually provides featureless, redshifted emission spectra and long-radiative decay times (Gebler et al., 1997; Morteani et al., 2004). The CTC can also be realized in core–shell quantum dots with type II band alignment, indicated in Figure 3.17d. Figure 3.28 shows that the absorptivity of CdTe/CdSe and CdSe/ZnTe core–shell QDs is extended to the infrared and the PL occurs at energies that are smaller than the bandgap of both materials that form the QD (Kim et al., 2003). The absorptivity in the infrared is weak since the QD behaves as an indirect semiconductor, nonetheless it provides a significant increase of light harvesting for photovoltaic applications (Wang et al., 2013a,b). On the other hand, the application of the exciplex emission for LEDs requires donor–acceptor blends or bilayer systems where the CTC emission is highly efficient (Cocchi et al., 2002).

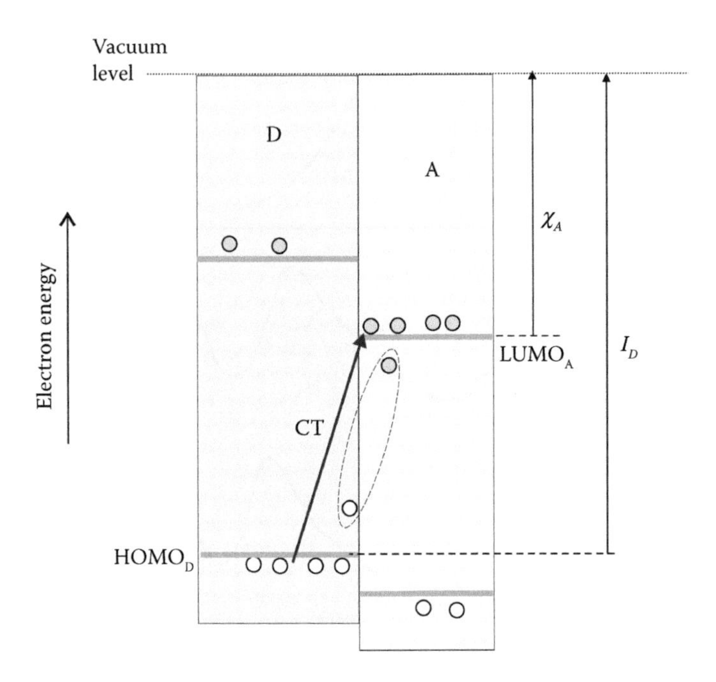

FIGURE 3.26 Interface between an electron donor (D) and electron acceptor (A) material showing optical excitation across the interface known as exciplex or charge-transfer complex. The energy of the CT state may be less than $I_D - \chi_A$ due to Coulomb-binding energy.

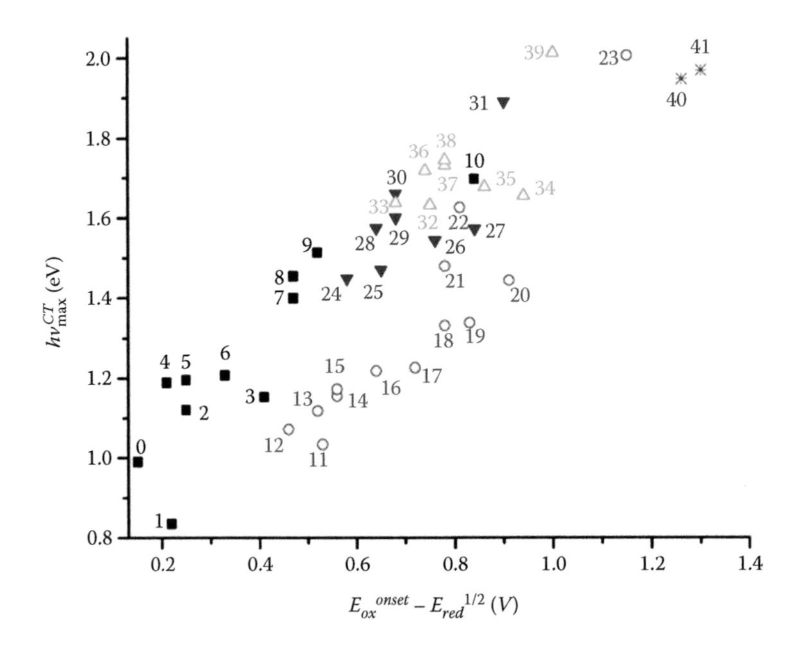

FIGURE 3.27 Correlation of the photon energy of maximum absorbance for the first CT band ($h\nu_{max}^{CT}$) with the difference between oxidation potential of the donor and reduction potential of the acceptor for various donors mixed with different acceptors (see Table 2 of the original publication). (Reproduced with permission from Panda, P. et al. *The Journal of Physical Chemistry B* 2007, 111, 5076–5081.)

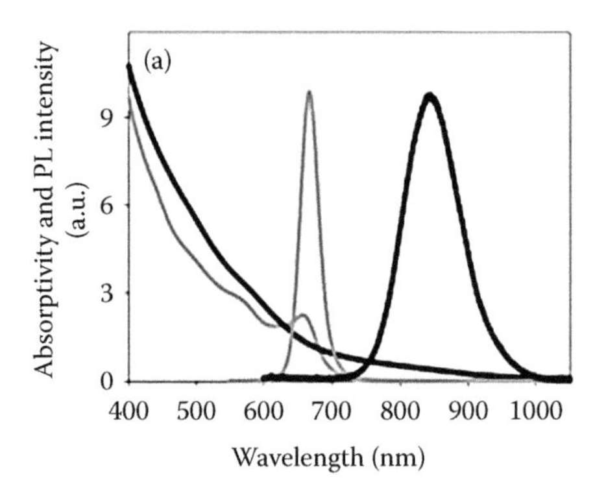

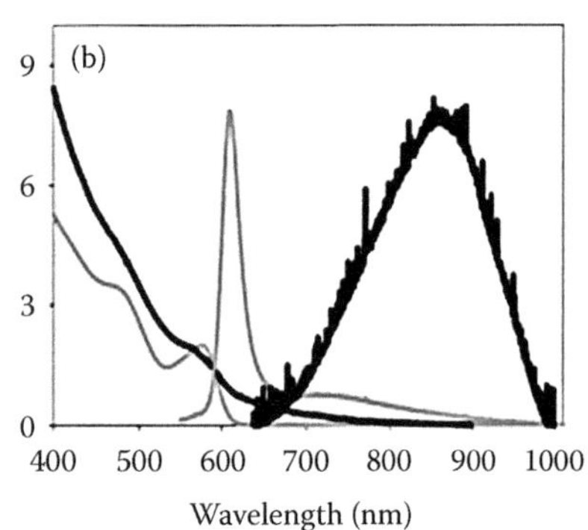

FIGURE 3.28 Absorptivity and normalized PL spectra of 3.2 nm radius CdTe QD (gray lines on (a)), CdTe/CdSe (3.2 nm radius core/thickness of 1.1 nm shell) QD (black lines on (a)), 2.2 nm radius CdSe QD (gray lines on (b)), and CdSe/ZnTe (2.2 nm radius core/thickness of 1.8 nm shell) QD (black lines on (b)). (Reproduced with permission from Kim, S. et al. *Journal of the American Chemical Society* 2003, 125, 11466–11467.)

GENERAL REFERENCES

Light absorption in inorganic solids: Pankove (1971), Klingshirn (1995), and Fox (2001).

Free carrier absorption: Burstein (1952), Liu and Bard (1989), and Fox (2001).

Burstein shift: Burstein (1952), Moss (1954), Kamat et al. (1989), Muñoz et al. (2001), Kawamura et al. (2003), and Manser and Kamat (2014).

Time resolved microwave conductivity: Friedrich and Kunst (2011), Dunn et al. (2012), Fravventura et al. (2013), and Savenije et al. (2013).

Excitons: Pankove (1971), Fox (2001), Rosencher and Vinter (2002), and Bardeen (2014).

Exciton screening: Gay (1971) and Amo et al. (2006).

Quantum dots: Alivisatos (1996), Yu et al. (2003), Kamat (2008), and de Mello (2010).

REFERENCES

Alivisatos, A. P. Perspectives on the physical chemistry of semiconductor nanocrystals. *The Journal of Physical Chemistry* 1996, 100, 13226–13239.

Amo, A.; Martin, M.; Viña, L.; Toropov, A.; Zhuravlev, K. Interplay of exciton and electron-hole plasma recombination on the photoluminescence dynamics in bulk GaAs. *Physical Review B* 2006, 73, 035205.

Ardo, S.; Meyer, G. J. Photodriven heterogeneous charge transfer with transition-metal compounds anchored to TiO_2 semiconductor surfaces. *Chemical Society Reviews* 2009, 38, 115–164.

Bardeen, C. J. The structure and dynamics of molecular excitons. *Annual Review of Physical Chemistry* 2014, 65, 127–148.

Barugkin, C.; Cong, J.; Duong, T.; Rahman, S.; Nguyen, H. T.; Macdonald, D.; White, T. P.; Catchpole, K. R. Ultralow absorption coefficient and temperature dependence of radiative recombination of $CH_3NH_3PbI_3$ perovskite from photoluminescence. *The Journal of Physical Chemistry Letters* 2015, 6, 767–772.

Berger, T.; Anta, J. A. IR and spectrophotoelectrochemical characterization of mesoporous semiconductor films. *Analytical Chemistry* 2011, 84, 3053–3057.

Berger, T.; Anta, J. A.; Morales-Florez, V. Electrons in the band gap: Spectroscopic characterization of anatase TiO_2 nanocrystal electrodes under Fermi level control. *The Journal of Physical Chemistry C* 2012, 116, 11444–11455.

Bredas, J.-L. Mind the gap! *Materials Horizons* 2014, 1, 17–19.

Burstein, E. Anomalous optical absorption limit in InSb. *Physical Review* 1952, 93, 632–633.

Canovas, E.; Pijpers, J.; Ulbricht, R.; Bonn, M. Carrier Dynamics in Photovoltaic Structures and Materials Studied by Time-Resolved Terahertz Spectroscopy, Chapter 11. In *Solar Energy Conversion: Dynamics of Interfacial Electron and Excitation Transfer*. The Royal Society of Chemistry: London, 2013, pp 301–336.

Cocchi, M.; Virgili, D.; Giro, G.; Fattori, V.; Marco, P. D.; Kalinowski, J.; Shirota, Y. Efficient exciplex emitting organic electroluminescent devices. *Applied Physics Letters* 2002, 80, 2401–2403.
Daub, E.; Würfel, P. Ultralow values of the absorption coefficient of Si obtained from luminescence. *Physical Review Letters* 1995, 74, 1020–1023.
de Mello, C. Synthesis and properties of colloidal heteronanocrystals. *Chemical Society Reviews* 2010, 40, 1512–1546.
De Wolf, S.; Holovsky, J.; Moon, S.-J.; Löper, P.; Niesen, B.; Ledinsky, M.; Haug, F.-J.; Yum, J.-H.; Ballif, C. Organometallic halide perovskites: Sharp optical absorption edge and its relation to photovoltaic performance. *The Journal of Physical Chemistry Letters* 2014, 5, 1035–1039.
Diamant, Y.; Chen, S. G.; Melamed, O.; Zaban, A. Core-shell nanoporous electrode for dye sensitized solar cells: The effect of the SrTiO3 shell on the electronic properties of the TiO_2 core. *The Journal of Physical Chemistry B* 2003, 107, 1977–1981.
Dunn, H. K.; Peter, L. M.; Bingham, S. J.; Maluta, E.; Walker, A. B. In situ detection of free and trapped electrons in dye-sensitized solar cells by photo-induced microwave reflectance measurement. *The Journal of Physical Chemistry C* 2012, 116, 22063–22072.
Elliott, R. J. Intensity of optical absorption by excitons. *Physical Review* 1957, 108, 1384–1389.
Fox, M. *Optical Properties of Solids*; Oxford University Press: Oxford, 2001.
Fravventura, M. C.; Deligiannis, D.; Schins, J. M.; Siebbeles, L. D. A.; Savenije, T. J. What limits photoconductance in anatase TiO_2 nanostructures? A real and imaginary microwave conductance study. *The Journal of Physical Chemistry C* 2013, 117, 8032–8040.
Friedrich, D.; Kunst, M. Analysis of charge carrier kinetics in nanoporous systems by time resolved photoconductance measurements. *The Journal of Physical Chemistry C* 2011, 115, 16657–16663.
Gay, J. G. Screening of excitons in semiconductors. *Physical Review B* 1971, 4, 2567–2575.
Gebler, D. D.; Wang, Y. Z.; Blatchford, J. W.; Jessen, S. W.; Fu, D. K.; Swager, T. M.; MacDiarmid, A. G.; Epstein, A. J. Exciplex emission in bilayer polymer light-emitting devices. *Applied Physics Letters* 1997, 70, 1644–1646.
Gierschner, J.; Ehni, M.; Egelhaaf, H.-J.; Medina, B. M.; Beljonne, D.; Benmansour, H.; Bazan, G. C. Solid-state optical properties of linear polyconjugated molecules: Pi-Stack contra herringbone. *The Journal of Chemical Physics* 2005, 123, 144914.
Green, M. A. Self-consistent optical parameters of intrinsic silicon at 300 K including temperature coefficients. *Solar Energy Materials and Solar Cells* 2008, 92, 1305–1310.
Green, M. A.; Ho-Baillie, A.; Snaith, H. J. The emergence of perovskite solar cells. *Nature Photonics* 2014, 8, 506–514.
Guyot-Sionnest, P. Electrical transport in colloidal quantum dot films. *The Journal of Physical Chemistry Letters* 2012, 3, 1169–1175.
Harruff, B. A.; Bunker, C. E. Spectral properties of AOT-protected CdS nanoparticles: Quantum yield enhancement by photolysis. *Langmuir* 2003, 19, 893–897.
Henry, W.; Coates, C. G.; Brady, C.; Ronayne, K. L.; Matousek, P.; Towrie, M.; Botchway, S. W. et al. The early picosecond photophysics of ru(ii) polypyridyl complexes: A tale of two timescales. *The Journal of Physical Chemistry A* 2008, 112, 4537–4544.
Hines, M. A.; Guyot-Sionnest, P. Synthesis and characterization of strongly luminescing ZnS-Capped CdSe nanocrystals. *The Journal of Physical Chemistry* 1996, 100, 468–471.
Hollingsworth, J. A. Nanoscale engineering facilitated by controlled synthesis: From structure to function. *Coordination Chemistry Reviews* 2014, 263–264, 197–216.
Ip, A. H.; Thon, S. M.; Sjoerd Hoogland; Voznyy, O.; Zhitomirsky, D.; Debnath, R.; Levina, L. et al. Hybrid passivated colloidal quantum dot solids. *Nature Nanotechnology* 2012, 7, 577–582.
Kamat, P. V. Quantum dot solar cells. Semiconductor nanocrystals as light harvesters. *The Journal of Physical Chemistry C* 2008, 112, 18737–18753.
Kamat, P. V.; Dimitrijevic, N. M.; Nozik, A. J. Dynamic Burstein-Moss shift in semiconductor colloids. *The Journal of Physical Chemistry* 1989, 93, 2873–2875.
Katahara, J. K.; Hillhouse, H. W. Quasi-Fermi level splitting and sub-bandgap absorptivity from semiconductor photoluminescence. *Journal of Applied Physics* 2014, 116, 173504.
Kawamura, K.-i.; Maekawa, K.; Yanagi, H.; Hirano, M.; Hosono, H. Observation of carrier dynamics in CdO thin films by excitation with femtosecond laser pulse. *Thin Solid Films* 2003, 445, 182–185.
Kim, S.; Fisher, B.; Eisler, H.-J.; Bawendi, M. Type-II quantum dots: CdTe/CdSe(core/shell) and CdSe/ZnTe(core/shell) heterostructures. *Journal of the American Chemical Society* 2003, 125, 11466–11467.
Klein, A. Transparent conducting oxides: Electronic structure–property relationship from photoelectron spectroscopy with in situ sample preparation. *Journal of the American Ceramic Society* 2013, 96, 331–345.

Klingshirn, C. F. *Semiconductor Optics*; Springer-Verlag: Berlin, 1995.

Köhler, A.; Bässler, H. Triplet states in organic semiconductors. *Materials Science and Engineering: R: Reports* 2009, 66, 71–109.

Liu, C. Y.; Bard, A. J. Effect of excess charge on band energetics (optical absorption edge and carrier redox potentials) in small semiconductor particles. *The Journal of Physical Chemistry* 1989, 93, 3232–3237.

Luther, J. M.; Law, M.; Beard, M. C.; Song, Q.; Reese, M. O.; Ellingson, R. J.; Nozik, A. J. Schottky solar cells based on colloidal nanocrystal films. *Nano Letters* 2008, 8, 3488–3492.

Macchi, G.; Medina, B. M.; Zambianchi, M.; Tubino, R.; Cornil, J.; Barbarella, G.; Gierschner, J.; Meinardi, F. Spectroscopic signatures for planar equilibrium geometries in methyl-substituted oligothiophenes. *Physical Chemistry Chemical Physics* 2009, 11, 984–990.

Mahan, G. D. Excitons in degenerate semiconductors. *Physical Review* 1967, 153, 882–889.

Manser, J. S.; Kamat, P. V. Band filling with free charge carriers in organometal halide perovskites. *Nature Photonics* 2014, 8, 737–743.

Mattheis, J.; Rau, U.; Werner, J. H. Light absorption and emission in semiconductors with band gap fluctuations—A study on Cu(In,Ga)Se2 thin films. *Journal of Applied Physics* 2007, 101, 113519.

Mayer, K. M.; Hafner, J. H. Localized surface plasmon resonance sensors. *Chemical Reviews* 2011, 111, 3828–3857.

Micic, O. I.; Sprague, J.; Lu, Z.; Nozik, A. J. Highly efficient band-edge emission from InP quantum dots. *Applied Physics Letters* 1996, 68, 3150–3152.

Miller, O. D.; Yablonovitch, E.; Kurtz, S. R. Strong internal and external fluorescence as solar cells approach the Shockley-Queisser limit. *IEEE Journal of Photovoltaics* 2012, 2, 303–311.

Milo, R. What governs the reaction center excitation wavelength of photosystems I and II? *Photosynthesis Research* 2009, 101, 59–67.

Ming, T.; Chen, H.; Jiang, R.; Li, Q.; Wang, J. Plasmon-controlled fluorescence: Beyond the intensity enhancement. *The Journal of Physical Chemistry Letters* 2012, 3, 191–202.

Morteani, A. C.; Sreearunothai, P.; Herz, L. M.; Friend, R. H.; Silva, C. Exciton regeneration at polymeric semiconductor heterojunctions. *Physical Review Letters* 2004, 92, 247402.

Moss, T. S. The interpretation of the properties of indium antimonide. *Proceedings of the Physical Society* 1954, 76, 775.

Munechika, K.; Chen, Y.; Tillack, A. F.; Kulkarni, A. P.; Plante, I. J.-L.; Munro, A. M.; Ginger, D. S. Spectral control of plasmonic emission enhancement from quantum dots near single silver nanoprisms. *Nano Letters* 2010, 10, 2598–2603.

Muñoz, M.; Pollak, F.; Kahn, M.; Ritter, D.; Kronik, L.; Cohen, G. Burstein-Moss shift of n-doped In0.53Ga0.47As/InP. *Physical Review B* 2001, 63, 233302.

Oksana, O.; Svitlana, S.; David, G. C.; Ray, F. E.; Frank, A. H.; Rik, R. T.; Sean, R. P.; John, E. A. Optical and transient photoconductive properties of pentacene and functionalized pentacene thin films: Dependence on film morphology. *Journal of Applied Physics* 2005, 98, 033701.

Panda, P.; Veldman, D.; Sweelssen, J.; Bastiaansen, J. J. A. M.; Langeveld-Voss, B. M. W.; Meskers, S. C. J. Charge transfer absorption for pi-conjugated polymers and oligomers mixed with electron acceptors. *The Journal of Physical Chemistry B* 2007, 111, 5076–5081.

Pankove, J. I. *Optical Processes in Semiconductors*; Prentice-Hall: Englewood Cliffs, NJ, 1971.

Rosencher, E.; Vinter, B. *Optoelectronics*; Cambridge University Press: Cambridge, 2002.

Sachenko, A. V.; Kryuchenko, Y. V. Excitonic effects in band-edge luminescence of semiconductors at room temperatures. *Semiconductor Physics, Quantum Electronics & Optoelectronics* 2000, 3, 150–156.

Savenije, T. J.; Ferguson, A. J.; Kopidakis, N.; Rumbles, G. Revealing the dynamics of charge carriers in polymer: Fullerene blends using photoinduced time-resolved microwave conductivity. *The Journal of Physical Chemistry C* 2013, 117, 24085–24103.

Shea-Rohwer, L. E.; Martin, J. E. Luminescence decay of broadband emission from CdS quantum dots. *Journal of Luminescence* 2007, 127, 499–507.

Shirasaki, Y.; Supran, G. J.; Bawendi, M. G.; Bulovic, V. Emergence of colloidal quantum-dot light-emitting technologies. *Nature Photonics* 2012, 7, 13–23.

Strickler, S. J.; Berg, R. A. Relationship between absorption intensity and fluorescence lifetime of molecules. *Journal of Chemical Physics* 1962, 37, 814–820.

Sturge, M. D. Optical absorption of Gallium Arsenide between 0.6 and 2.75 eV. *Physical Review* 1962, 127, 768–773.

Subramanian, V.; Wolf, E. E.; Kamat, P. V. Green emission to probe photoinduced charging events in ZnO/Au nanoparticles. Charge distribution and Fermi-level equilibration. *The Journal of Physical Chemistry B* 2003, 107, 7479–7485.

Talapin, D. V.; Gaponik, N.; Borchert, H.; Rogach, A. L.; Haase, M.; Weller, H. Etching of colloidal InP nanocrystals with fluorides: Photochemical nature of the process resulting in high photoluminescence efficiency. *The Journal of Physical Chemistry B* 2002, 106, 12659–12663.

Tang, J.; Wang, X.; Brzozowski, L.; Barkhouse, D. A. R.; Debnath, R.; Levina, L.; Sargent, E. H. Schottky quantum dot solar cells stable in air under solar illumination. *Advanced Materials* 2011, 22, 1398–1402.

Trupke, T.; Daub, E.; Würfel, P. Absorptivity of silicon solar cells obtained from luminescence. *Solar Energy Materials and Solar Cells* 1998, 53, 103–114.

Turner, W. J.; Reese, W. E.; Pettit, G. D. Exciton absorption and emission in InP. *Physical Review* 1964, 136, A1467.

Ulbricht, R.; Hendry, E.; Shan, J.; Heinz, T. F.; Bonn, M. Carrier dynamics in semiconductors studied with time-resolved terahertz spectroscopy. *Reviews of Modern Physics* 2011, 83, 543–586.

Urbach, F. The long-wavelength edge of photographic sensitivity and of the electronic absorption of solids. *Physical Review* 1953, 92, 1324–1324.

Vanmaekelbergh, D. Self-assembly of colloidal nanocrystals as route to novel classes of nanostructured materials. *Nano Today* 2011, 6, 419–437.

Wang, H.; Liu, X.; Zhang, Z. Absorption coefficients of crystalline silicon at wavelengths from 500 nm to 1000 nm. *International Journal of Thermophysics* 2013a, 34, 213–225.

Wang, J.; Mora-Sero, I.; Pan, Z.; Zhao, K.; Zhang, H.; Feng, Y.; Yang, G.; Zhong, X.; Bisquert, J. Core/shell colloidal quantum dot exciplex states for the development of highly efficient quantum dot sensitized solar cells. *Journal of the American Chemical Society* 2013b, 135, 15913–15922.

Wegner, G.; Baluschev, S.; Laquai, F.; Chi, C. Managing photoexcited states in conjugated polymers. *Macromolecular Symposia* 2008, 268, 1–8.

Yu, W.; Qu, L. H.; Guo, W. Z.; Peng, X. G. Experimental determination of the extinction coefficient of CdTe, CdSe, and CdS nanocrystals. *Chemistry of Materials* 2003, 15, 2854–2860.

4 Fundamental Model of a Solar Cell

In this chapter, we study the models and physical features that explain those general properties of diodes that are central to their application for energy conversion. We first examine the operation of realistic semiconductor device models, starting with majority carrier diodes controlled by charge transference at the contact. Then we discuss in detail the recombination diode formed by a semiconductor layer of long diffusion length with asymmetric contacts to either electrons or holes. The latter model is the basis for formulating a fundamental model of a solar cell, which consists of a semiconductor light absorber with selective contacts. We discuss the conversion of light-generated carriers to voltage, and we describe the general function of the selective contacts as well as their specific materials and interfacial properties.

4.1 MAJORITY CARRIER INJECTION MECHANISMS

In Section ECK.5.8, we introduced the basic properties of a diode. We defined forward and reverse voltage, and we remarked the current–voltage characteristic associated with rectification:

$$j = j_0(e^{qV/mk_BT} - 1) \tag{4.1}$$

In the case $m = 1$, Equation 4.1 is the *Shockley ideal diode equation* or the *diode law*. The parameter m is the *ideality factor*, also known as the *quality factor*, and it describes the specific exponential voltage dependence of the current. Often, the current density–voltage curve depends on a combination of processes and m describes the best exponential approximation. For example, ideality factors approaching $m = 2$ occur by the generation/recombination in the space charge region (SCR) shown later in Figure PSC.10.16. j_0 is the *reverse saturation current*. In this chapter, we aim to describe the basic models for solar cells based on operation of semiconductor devices; therefore, we start with a detailed view of semiconductor devices, which lead to the diode structure that realizes Equation 4.1. There are two main kinds of diodes: based on either majority carrier injection at one interface, or those controlled by minority carrier injection and recombination. We begin with the diodes of the first class.

The properties of a Schottky barrier (SB) have been commented on in Chapter ECK.9. We examine the types of diodes that can be formed with an SB. Figure 4.1a shows an SB at the contact of an n-type semiconductor with a metal, see also Figure ECK.9.3. The semiconductor has an *ohmic contact* to majority carriers at the right side. This is a convenient condition in order to focus our attention on only one junction, as it means that the polarization by applied voltage V translates into a difference of Fermi levels across the SB junction. As described in Section ECK.5.8, the forward bias occurs when the voltage is made negative at the ohmic contact of the semiconductor, which means that positive voltage applied to the SB side decreases the size of the barrier (Figure 4.1b and Figure ECK.9.2). The interfacial barriers for injection of electrons and holes, $\Phi_{B,n}$ and $\Phi_{B,p}$, are defined as the energy levels at the contact measured from the Fermi level, as discussed in Section ECK.4.5. Figure 4.2 shows the rectification at semiconductor–electrolyte interface discussed in Section ECK.6.6, where the voltage governs the surface concentration of electronic carriers, and the rate of transfer is determined by the interfacial charge transfer rates.

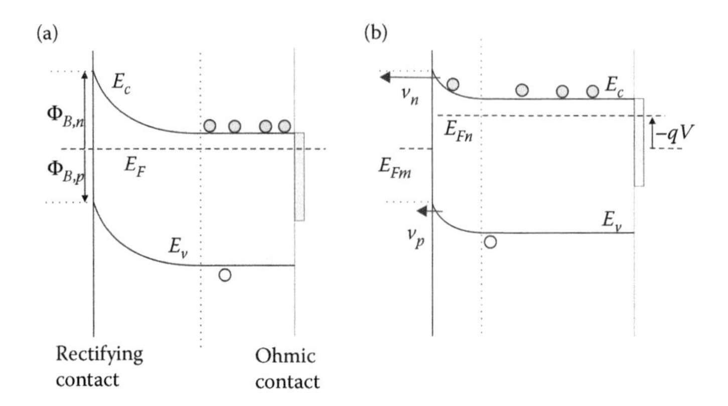

FIGURE 4.1 Model for an SB with a metal contact at (a) equilibrium and (b) at forward bias, showing the majority and minority carrier injection velocities from semiconductor to metal. The injection of majorities predominates at the contact, and the diode is a majority carrier device.

As shown in Figure 4.1b, a change of voltage modifies the Fermi level of the majority carrier with respect to the conduction band edge level E_c at the contact. Therefore, the voltage produces a change in the concentration of electrons at the semiconductor surface, n_s. This property allows to govern the rate of electron transfer by the voltage applied to the barrier as further discussed below, or alternatively the hole injection as shown in Figure 4.2. The rates will be increased or decreased according to the number of electrons available at the surface.

In the equilibrium situation of Figure 4.1a, the same current j_0 flows in both directions, from the metal to the semiconductor and vice versa. The electron transfer rate is quantified by the electron current density *in equilibrium*, j_0, which plays a similar role to the *exchange current density* in electrochemistry.

The fundamental assumption about a biased SB is that the Fermi levels in the two phases in contact (the metal and the semiconductor) are homogeneous up to the interface where the step of the Fermi level occurs as discussed in Section ECK.9.2. The details of the transition of the Fermi level across the interface have important influence on the majority carrier rate of transfer. The location of the Fermi level step actually determines whether a disturbance with respect to equilibrium affects

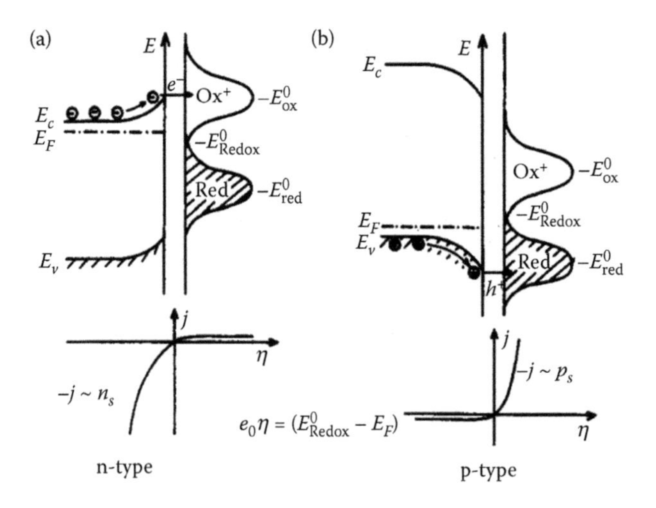

FIGURE 4.2 Redox reactions at n-type (a) and p-type (b) semiconductor electrodes, showing the energy diagrams and the current-voltage curves. (Reproduced with permission from Gerischer, H. *Electrochimica Acta* 1990, 35, 1677–1699.)

either the concentration of electrons in the semiconductor or in the metal, which sets the mechanism of electron transfer. These assumptions correspond respectively to the *diffusion theory*, which sets the gradient of the Fermi level in the semiconductor side, and the *Bethe thermionic-emission theory*, which states the drop at the metal side. A synthesis of these mechanisms is formulated in the combined *thermionic emission-diffusion theory*, which considers the two Fermi level drops in series (Crowell and Sze, 1966). These questions have been amply discussed in the literature of crystalline semiconductors and are summarized by Rhoderick and Williams (1988). It has been concluded that in the Schottky diodes with high mobility semiconductors, the forward current is limited by thermionic emission provided the forward bias is not too large. In the following, we apply the generalized thermionic emission theory discussed in Section ECK.6.7 to derive the current across the SB.

4.2 MAJORITY CARRIER DEVICES

The surface concentration of majority carriers is a crucial quantity to determine the current flowing in an SB biased by a voltage V. In the derivation of the properties of the SB in Chapter ECK.9, we have assumed that the only significant electric charge in the barrier is the donor density implying $n \ll n_0 = N_D$, Equation ECK.9.9. As we move from the quasi-neutral region toward the surface, the electron density decreases to a value n_s related to the concentration in the bulk by the expression $n_s = n_0 e^{qV_{sc}/k_BT}$, where V_{sc} is the potential difference across the barrier in the semiconductor surface, Equation ECK.9.4. Therefore, the surface concentration dependence on voltage is

$$n_s = n_{s0} e^{qV/k_BT} \tag{4.2}$$

where n_{s0} is the carrier concentration at the surface in the unbiased SB. In the equilibrium situation, electron flow across the barrier, between the metal and the semiconductor conduction band, is balanced in both directions. When a forward bias is applied to the diode, as in Figure 4.1b, the surface concentration of electrons n_s increases as indicated in Equation 4.2; hence, the thermionic flow of electrons from the semiconductor to the metal increases, while the thermal emission from the metal remains the same. Therefore, the resulting current is given by the diode law Equation 4.1 as follows:

$$\begin{aligned} j &= qv_n(n_s - n_{s0}) \\ &= qv_n n_{s0}(e^{qV/k_BT} - 1) \end{aligned} \tag{4.3}$$

where v_n is a transfer velocity (Crowell and Sze, 1966), further discussed below, see Figure 4.1b. Since the rectification is due to the injection or suppression of majority carriers, the diode is called a *Schottky diode* or more generally a *majority carrier device* (Shannon, 1979). The reverse saturation current j_0 is given by Equation ECK.6.92

$$j_0 = qn_{s0}v_n \tag{4.4}$$

The barrier height in equilibrium, $\Phi_{B,n}$, is drawn in Figure 4.1 and defined in Equation ECK.4.24. We can write Equation 4.4 as

$$j_0 = qv_n N_c e^{-\Phi_{B,n}/k_BT} \tag{4.5}$$

where N_c is the effective density of states of the conduction band. Therefore, a fit of log j_0 with respect to reciprocal temperature provides the barrier height (Sze, 1981).

Figure 4.3 compares the parameters of majority carrier diodes formed on n-type Si by Au and PEDOT:PSS contacts; two materials that have the same work function of 5.1 eV and hence similar

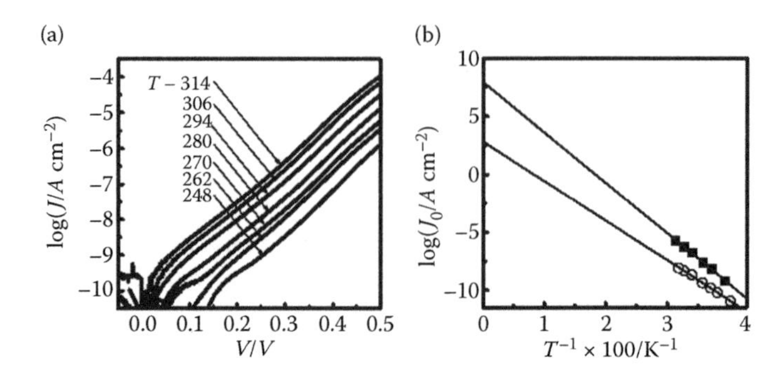

FIGURE 4.3 (a) Forward bias current density–voltage response for an n-Si/PEDOT:PSS device measured at several temperatures (K). (b) Temperature dependence of j_0 for (■) n-Si/Au and (O) n-Si/PEDOT:PSS contacts. (Reproduced with permission from Price, M. J. et al. *Applied Physics Letters* 2010, 97, 083503.)

$\Phi_{B,n}$ (Price et al., 2010). The current–voltage characteristic of n-Si/PEDOT:PSS at different temperatures is shown in Figure 4.3a, and it follows well the model of Equation 4.3. Figure 4.3b shows the temperature dependence of j_0. The slopes are well described by the exponential dependence of Equation 4.5 with a large offset, which indicates that the carrier charge transfer velocity v_n is much larger for n-Si/Au contact than for n-Si/PEDOT:PSS contact. This result shows that v_n is dramatically influenced by carrier transference properties at the interface between the two materials. Figure 4.4 shows the jV characteristics and the derived parameters of a Cr/n-GaAs/In Schottky contact.

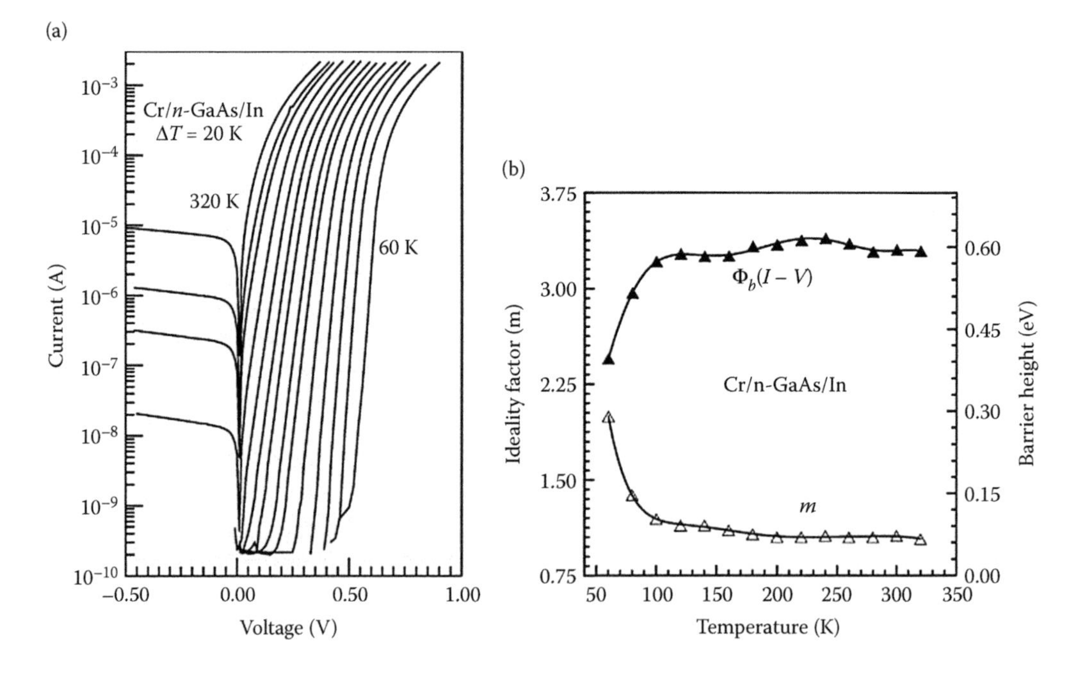

FIGURE 4.4 (a) Experimental forward bias current–voltage characteristics of Cr/n-GaAs/In Schottky contact at various temperatures. (b) Temperature dependence of the ideality factor (*the open triangles*) and barrier height (*filled triangles*). (Reproduced with permission from Korkut, H. et al. *Microelectronic Engineering* 2009, 86, 111–116.)

4.3 MINORITY CARRIER DEVICES

Based on the SB with an n-type semiconductor, we discuss a different operation mechanism of a semiconductor diode, governed by injection and recombination of minority carriers as shown in Figure 4.5. As explained in Section 4.2, the voltage modulates the size of the depletion layer, so that the forward voltage makes the Fermi level of majority carrier electrons, E_{Fn}, move toward the conduction band in the depletion zone and consequently reduces the barrier width. In the device shown in Figure 4.5, the contact facilitates the exchange of *minority carrier* (holes) over that of majority carrier (electrons), in contrast to the majority carrier exchange, seen in Figure 4.1. Minority carrier injection predominates because of higher hole transfer velocity at the interface $v_p \gg v_n$, or the barrier for injection of holes $\Phi_{B,p}$ is small (Green and Shewchun, 1973). In Figure 4.5b, the dominant current flows by hole transfer from the metal to the valence band. The injected holes travel toward the interior of the semiconductor and eventually recombine in the quasi-neutral region. This SB diode is a *minority carrier device.*

Another example is shown in Figure 4.6. In these two figures, we observe that the Fermi levels across the space charge region (SCR) are flat. This is due to the common assumption that neglects recombination in the SCR, as discussed in Chapter PSC.10.

In general, in the SB diode, we must take into account two effects due to the modification of surface concentrations by the forward voltage:

1. Increased flux of majorities to the metal
2. Decreased injection of minorities to the metal that results in an increased minority flux from the metal to the semiconductor

The dominant condition, according to the interfacial kinetics, will determine the type of diode, either majority or minority carrier type.

Now we go back to Figures ECK.5.11 and ECK.5.12, and we observe that the devices in these figures are minority carrier diodes operating in the same mechanism as that of Figure 4.5. One obvious difference is that Figure ECK.5.12 shows minority electrons device and Figure 4.5 a minority holes device. Another difference is the minority carrier distribution inside the diode at forward bias. In Figure ECK.5.12b, the Fermi level of minorities remains homogeneous across the whole bulk semiconductor. Meanwhile, in Figures 4.5b and 4.6, the Fermi level of minorities progressively decays to the equilibrium value when the injected holes travel further away from the selective contact for holes. These differences are controlled by the minority carrier diffusion length, as shown in Section PSC.10.2. Another common characteristic is illustrated in Figure 4.6: the Fermi level of holes remains in equilibrium with a metal and across a dielectric layer that forms an interfacial dipole at

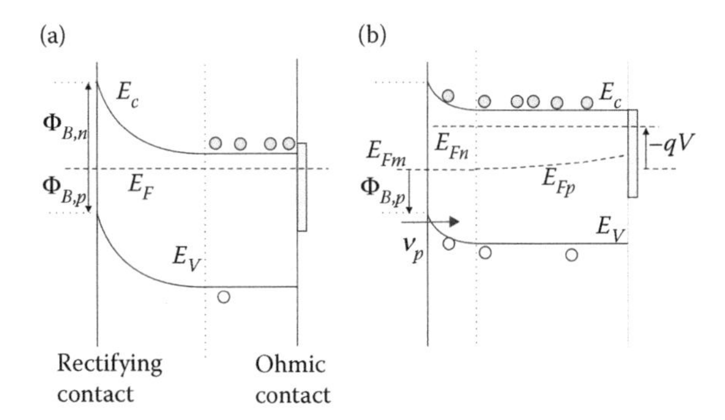

FIGURE 4.5 Model for an SB with a metal contact at (a) equilibrium and (b) at forward bias. The injection of minorities (holes) predominates at the contact and the diode is a minority carrier device.

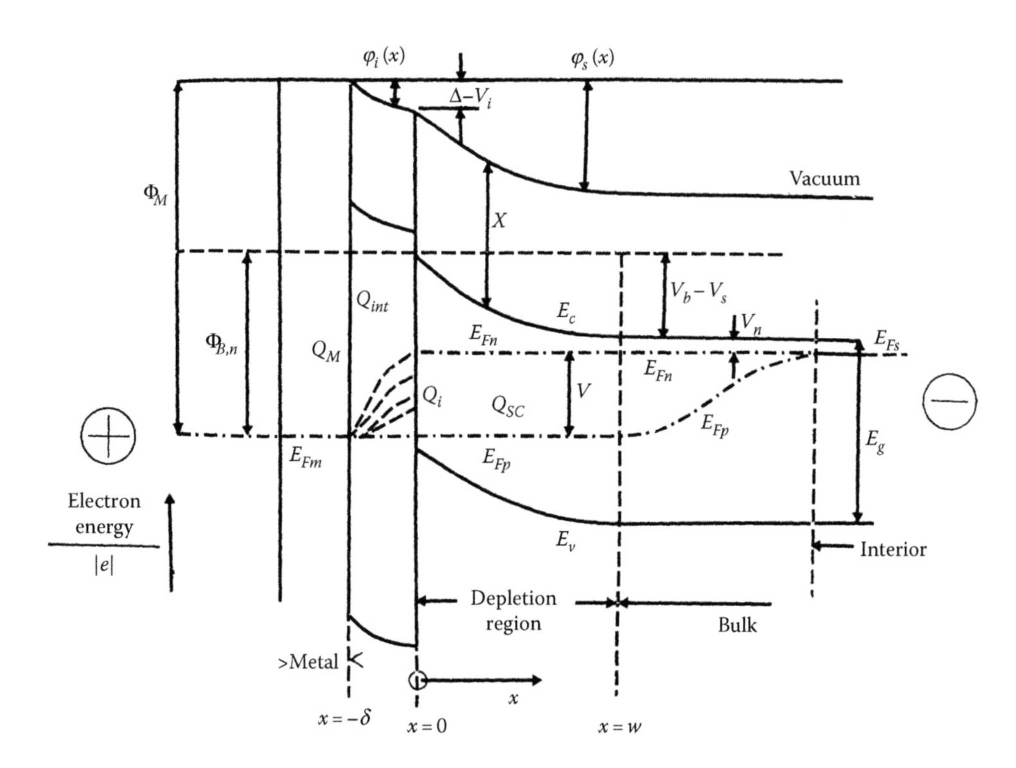

FIGURE 4.6 Model for an SB with a contact to the minority carrier (holes) at forward bias. There is a barrier formed by a thin metal oxide at the metal–semiconductor contact. This is a model for an SB solar cell. (Reproduced with permission from Landsberg, P. T.; Klimpke, C. *Proceedings Royal Society (London) A* 1977, 354, 101–118.)

the contact. In the bulk, there is full recombination of injected minorities, before they reach the back contact. This is a diode with a short diffusion length as extensively analyzed in Chapter PSC.10.

4.4 FUNDAMENTAL PROPERTIES OF A SOLAR CELL

The solar cell is a device capable of absorbing photons of the sunlight, as a result of which free electron–hole pairs are produced. These pairs can be extracted and made to circulate through an external load that supports a positive voltage, thereby resulting in net delivery of electrical power to the outer circuit just from the consumption of photons and with no other change in the device (Araújo and Martí, 1994).

This process of conversion of light to electricity is called the *photovoltaic effect* and it will be progressively explained in the following chapters. The interaction of radiation with the electron–hole system in the semiconductor light absorber results in the transference of energy from the incoming photons to the electron–hole gas, causing the modification of the Fermi levels that can be used to produce useful work on an external system. And vice versa, the electrons and holes system may lose energy by creating photons that are emitted outward.

These processes are outlined in Figure 4.7, where a model semiconductor without contacts is indicated. Previously in Chapter ECK.5, we described the properties of an electron–hole gas in a semiconductor, consisting of the set of electrons in the extended states of the conduction band and set of holes in the valence band that thermalize to the temperature of the absorber, T_A. In equilibrium, the two sets of carriers share a common electrochemical potential that determines the equilibrium Fermi level E_{F0}, as indicated in Figure 4.7a. When the semiconductor film is illuminated in excess of the ambient radiation, it is displaced to a different steady state of recombination-generation

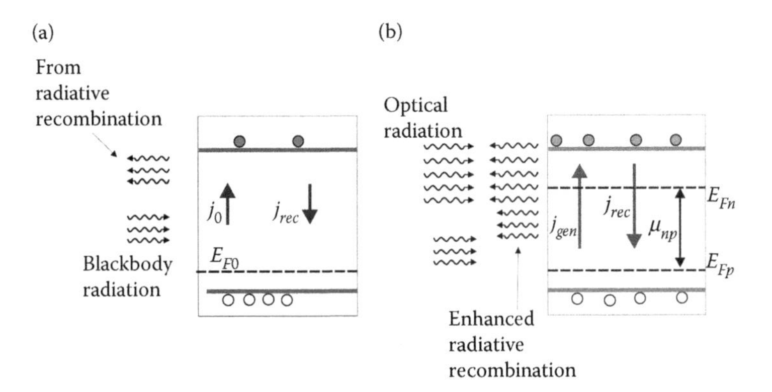

FIGURE 4.7 (a) Scheme of p-type semiconductor material in dark equilibrium, showing the balance of incoming and outgoing photons absorbed or produced in the material, and the equilibrium Fermi level E_{F0}. The full dots at the conduction band energy level and the empty dots at the valence band energy level represent electrons and holes, respectively. The generation by incoming thermal radiation produces a generation flux j_0/q and radiative recombination internal flux j_{rec}/q. (b) Under excess optical radiation, the rate of generation increases to j_{gen}/q, the minority carrier density increases producing the larger internal recombination flux and the splitting of the Fermi levels of electrons and holes, expressed as an internal chemical potential μ_{np}. The increased radiative recombination produces excess radiation so that external fluxes of photons can be equilibrated. Eventually, the radiation field spectral flux becomes that of photons at nonzero chemical potential with $\eta_{ph} = \mu_{np}$, as shown in Section PSC.6.4.

equilibrium with excess populations of generated carriers (Equation PSC.9.5) that cause a separation of the Fermi levels, as shown in Figure 4.7b. The difference of electrochemical potentials of electrons and holes μ_{np} is stated in Equation ECK.2.46:

$$\mu_{np} = \eta_n + \eta_p = \mu_n + \mu_p = E_{Fn} - E_{Fp} \tag{4.6}$$

When the free energy of the photons has been captured in the form of a difference of electrochemical potentials of electronic carriers, additional steps are still necessary for the conversion into useful electrical energy. Electronic devices for energy and light production require a transformation of the electronic carrier concentration to a voltage. This functionality is realized by the diode structure, which imparts directionality to the carrier flow with respect to the external contacts. When trying to exit the absorber, electrons move preferentially in one direction and are impeded to move in the opposite one, as in a valve, while holes experience a similar effect in the contrary direction.

We have discussed so far two classes of diodes. In the majority carrier device, the diode operation is controlled by the transference rate across the rectifying contact. This device is called *unipolar*, and it is regulated by injection. However, the minority carrier device is called *bipolar*. It is controlled by the flow of minority carriers whose distribution is established in the active layer by a competition between transport and recombination. We will use the shorthand *recombination diode* for this type of diode. The minority carrier device requires good selective contacts (as discussed below), especially if the device is thin and the minorities arrive to the selective contact of the majority carrier.

In the analysis of realistic solar cell technologies, there are a wide variety of detailed considerations concerning the material properties and operation mechanisms. At this stage, we are interested in a fundamental model that allows us to establish the dominant mechanisms that set the limits to efficiency and a benchmark for evaluation of actual solar cells. The basic structure of a solar cell that we focus our attention on in the following chapters consists of a semiconductor diode composed of a good light absorber material with selective contacts to electrons and holes, forming a recombination diode.

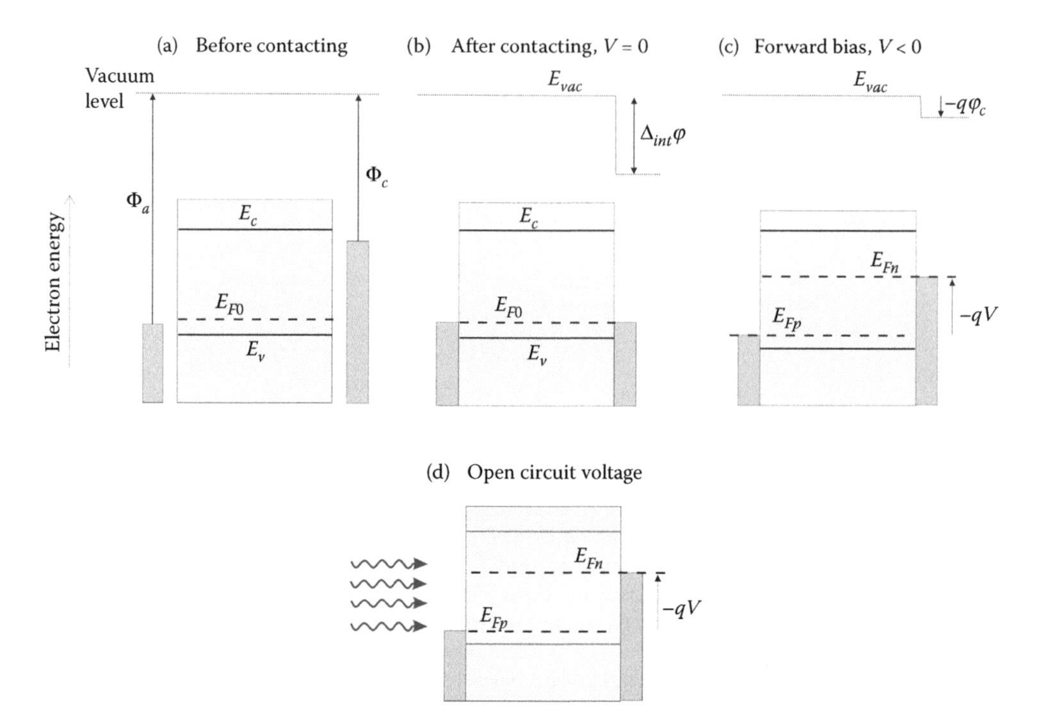

FIGURE 4.8 Schematic representation energy diagram of a p-type layer with conduction band for electrons E_c and valence band for holes E_v and two contacts: an electron selective contact (ESC) with work function Φ_c and hole selective contact (HSC) with work function Φ_a. The contacts are considered as metals in which the carrier energy level is at the Fermi level. The voltage produces a modification of the Fermi level of minority carriers (electrons) with respect to E_c at the right contact, while holes remain at equilibrium at the left contact. (a) Energies of the separate materials. (b, c) Different situations of bias voltage V indicating the Fermi levels of electrons (E_{Fn}) and holes (E_{Fp}). In (d), the semiconductor is illuminated and minority carrier generation raises the Fermi level of electrons consequently producing a photovoltage.

In order to discuss the properties of this model, we show in Figure 4.8 the formation of a recombination diode from the separate materials by application of contacts with different work function to the semiconductor light absorber layer. The process of equilibration of the separate materials to a common Fermi level has been shown in Figure FCT.1.9, and it was noted that the main question is the resulting distribution of the original difference of work functions. In the previous study in Chapter FCT.1 we noted two fundamentally different possibilities. One is the distribution of the band bending in the absorber layer, like in the SB model of Figure 4.6. In the other possibility, the modification of the vacuum level (VL) when the materials achieve equilibrium is absorbed by the minority carrier selective contact as a dipole layer, described in Section ECK.4.6. Any potential drop (change of VL) thereafter due to nonzero voltage appears just at this interface, as in the case of forward voltage (Figure 4.8c). This is just one possibility among many types of device behavior, as discussed in chapter PSC.10. However, Figure 4.8 is probably the simplest diode model in that the bands remain flat in the absorber and transport occurs entirely by diffusion. We also assume low-device thickness, high mobilities, and neglect surface recombination. These properties imply that the Fermi levels are flat as shown in Figure 4.8. The internal chemical potential μ_{np} is extracted and a measurable voltage V occurs, of value given by the expression (Bisquert et al., 2004; Luque et al., 2012)

$$qV = E_{Fn} - E_{Fp} \tag{4.7}$$

Instead of providing a voltage, the outer contacts of the cell can be adapted to produce the synthesis of chemical substances that store the energy in the form of chemical bonds, and then it is called a solar fuel converter, see Section PSC.11.3.

4.5 PHYSICAL PROPERTIES OF SELECTIVE CONTACTS IN SOLAR CELLS

To form a solar cell, it is necessary to obtain selective contacts with electrons and holes at each side of the semiconductor absorber layer (Bisquert et al., 2004). The use of contacts for facile injection of one specific carrier has been previously discussed in Section FCT.1.2. When the contact barrier to one specific carrier is reduced, it becomes progressively more *ohmic* and hence better suited for injection, as shown in Figure FCT.1.4 (Abkowitz et al., 1998). Now we introduce the idea of a selective contact, which is more specific and demanding than the injection ohmic contact, especially in the case of photovoltaic devices. An ideally selective contact is one that is transparent to one carrier type and blocks completely the other. The selective contact must tend to ohmic for one carrier, or at least have a negligible impedance, and reject the opposite charge carrier type. The ideal selective contact may operate reversibly for injection and extraction, which is a central property of the contact in solar cells type devices.

In a solar cell, the selective contacts perform two important functions. They introduce the asymmetry of extraction and injection of electrons and holes at the two contacts which is required to obtain a diode structure. Thereafter, the device is directional by this construction. Each side of the device equilibrates to a separate carrier, as outlined in Figure 4.8, see also Figure PSC.10.6. The electrons can be extracted from one contact and retrieved at the other one at a lower electrochemical potential, having delivered the difference of free energy to the external load. In addition, the selective contacts have the function of taking the carriers from a state in which the Fermi level in the absorber is displaced from thermal dark equilibrium, to a material (normally a metal) in which the carrier is at the equilibrium Fermi level (Honsberg et al., 2002). The light only creates carriers in a nonequilibrium state, and the selective contact converts such nonequilibrium Fermi levels into stable Fermi levels in the wires, and their difference is properly a voltage, as discussed in Chapter ECK.3.

There is a variety of ways to realize the asymmetry of the contacts usually required in devices for energy production, storage, or lighting. Selective contacts may be formed by materials that have the required kinetic properties to extract only one kind of carrier at the interface with the absorber. In electrochemical systems, the contact selectivity can be obtained by the kinetic asymmetry (for electrons and holes) of the interfacial reaction at the semiconductor–electrolyte junction, previously discussed in Section ECK.6.8 and also in Figure 4.2. A kinetic preference for extraction of holes at the semiconductor electrochemical contact, caused by the match of energy levels at both sides of the interface, is shown in Figure 4.9a. An extensive analysis of electrochemical diodes is presented in Chapter PSC.5. In a solution that contains different redox species formed by photochemical reaction, a contact that allows one reaction and blocks another species allows us to form a photogalvanic

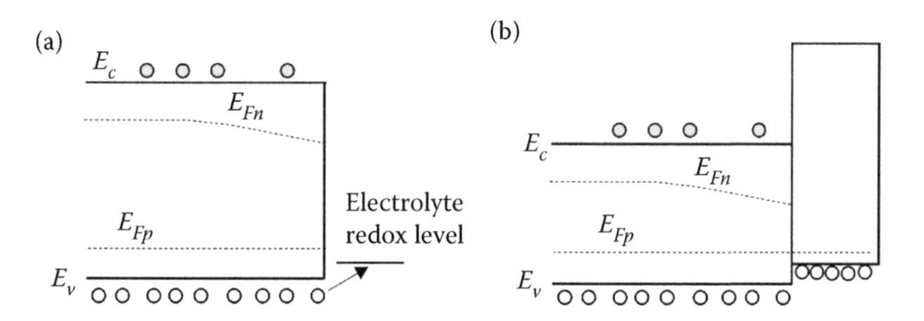

FIGURE 4.9 Mechanisms of selective contacts. (a) Preferential kinetic exchange at one semiconductor energy level at the semiconductor–electrolyte contact. (b) Equilibration of Fermi levels for holes at the semiconductor–hole transport layer contact.

solar cell (Albery, 1982). The selectivity to chemical reaction by biological molecules is used to build fuel cells that provide energy from a glucose-rich environment without the need to separate anode and cathode compartments (Heller, 2004).

Quantum dot films can be regulated for selective charge exchange at the contact as shown in Figure FCT.1.5, which indicates the preferential injection of electrons or holes by contact engineering. Hodes et al. (1992) first observed that a nanostructured solar cell, formed by an array of quantum dots in contact with electrolyte, operates by kinetic selectivity. Figure 4.10 shows a quantum dot film immersed in solution with only one oxide or organic contact. The sign of photovoltage can be inverted by the type of contact, which indicates that the contact layer may adapt itself either to the extraction of electrons or holes, as shown in the scheme in Figure 4.10c. Solar cells composed of a layer of quantum dots have provided promising conversion efficiency (Luther et al., 2008; Barkhouse et al., 2011; Ning et al., 2012).

In inorganic semiconductors, the selectivity is often formed by properties of the junction, involving SBs and space-charge regions, as discussed earlier in this chapter. A very important method to address one specific carrier selectively, by regulating the injection barrier at the interface, employs a semiconductor with adequate energy level for alignment to that of the absorber layer. In addition, the contact material has preferential doping of one type, and it is called a transport layer. This method to build a selective contact by match of the energy levels is indicated in the scheme of Figure 4.9b, and it has been applied in the models in Figure FCT.1.9 and Figure 4.8.

In order to impart selectivity to electrons and holes, the transport layers at the two sides of the absorber need to have a substantial difference of work function of the order of the absorber bandgap.

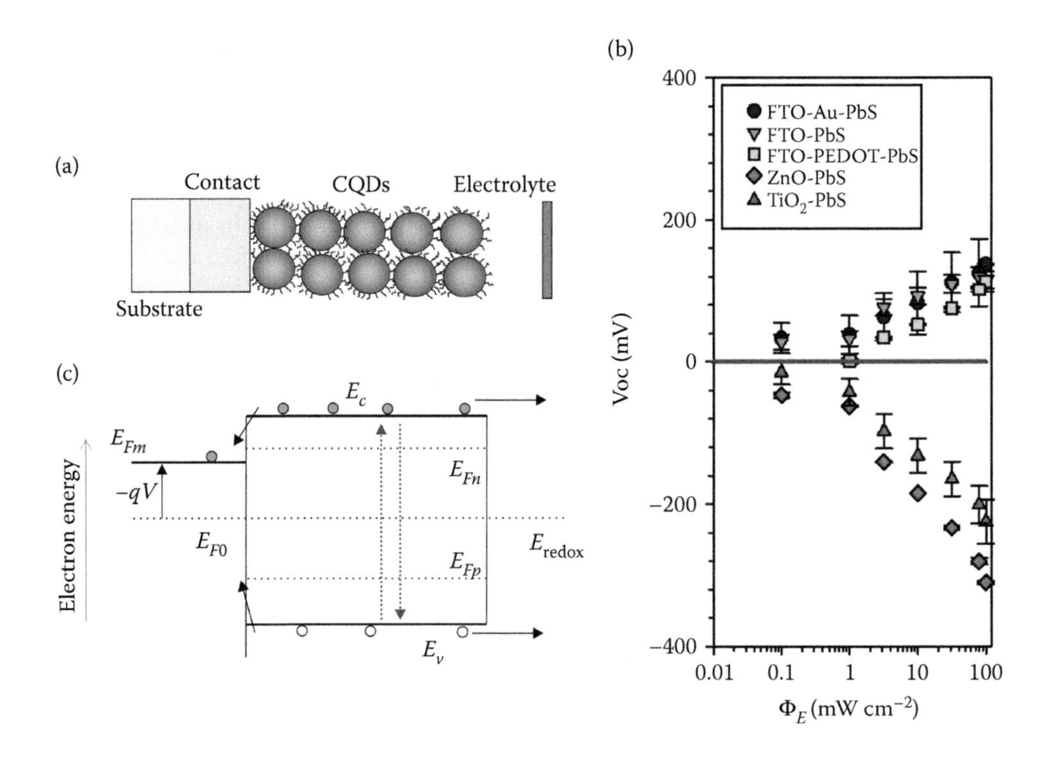

FIGURE 4.10 (a) Scheme of a QD film immersed in electrolyte and measured versus a counterelectrode. (b) Evolution of the open-circuit voltage V_{oc} with illumination intensity obtained experimentally for different metal or semiconductor–QD contacts. (c) Model for an SB with a contact to the minority carrier (holes) at forward bias. There is a barrier formed by a thin metal oxide at the metal–semiconductor contact. This is a model for an SB solar cell. (Adapted with permission from Mora-Sero, I. et al. *Nature Communications* 2013, 2, 2272.)

From the difference of work functions occurs another function of the contacts, which is to create an initial built-in potential in the device. This topic has been amply described in Section FCT.1.3 for the metal–insulator–metal (MIM) model (see Figure FCT.1.10), consisting of an insulator layer contacted by two metals of low and high-work function. Based on this previously discussed fundamental device, we can use as in Figure 4.11 an absorber "intrinsic" layer sandwiched between an n-doped layer that functions as the electron selective contact of the solar cell and a p-doped layer as the hole selective contact. Here, there are two main properties of the contacts. First, the difference of their work functions produces the slant of the bands at zero bias as in the MIM device. The second property is the conductivity to only one carrier that supports the selectivity function. This structure is usually termed a p–i–n solar cell. In contrast to Figure 4.8, the presence of a majority carrier, or charge accumulation in the absorber layer leading to band bending, is ignored. The external voltage affects the inclination of the straight bands (Figure FCT.1.10), so that charge collection is strongly influenced by an electrical field. In this class of models, the mechanism of selectivity is not associated with the properties of the contacts, rather the contacts have the function to produce the starting slant of the bands that impart directionality to carrier flow. The current–voltage curve is generated by the modification of charge collection induced by the electrical field, while recombination is neglected (Schilinsky et al., 2004). The open-circuit condition under illumination is achieved when the bands are flat (horizontal) and charge extraction ceases, as shown in Figure 4.11b. This approach was used in early analysis of polymer solar cells and also in ferroelectric solar cells, formed by materials that are very poor conductors (Lopez-Varo et al., 2016). In general, this type of model treats the device as a capacitor, that is it consists of an insulator that extracts photogenerated

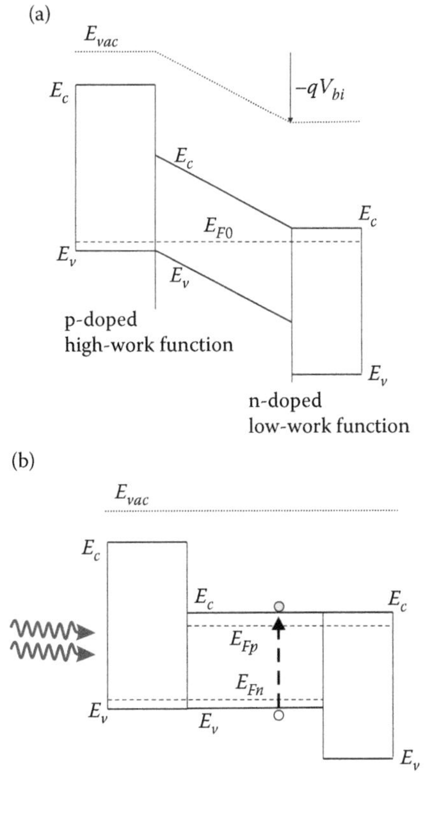

FIGURE 4.11 The model of a p–i–n solar cell, showing the slanted bands when the system is in equilibrium (a) and the flat bands when charge carrier extraction compensates the internal electrical field, in open-circuit condition (b).

charges as in early models of photoconductors (Goodman and Rose, 1971; Sokel and Hughes, 1982). Although this solar cell model is conceptually useful, the application to realistic high-quality photovoltaic devices is rather limited.

As mentioned, the main method to form a selective contact for holes is to introduce a hole transport layer that readily conducts hole carriers and blocks electrons. Devices based on organic materials exploit specifically doped layers with adequate thermodynamic and kinetic properties of charge transfer at the interface. Often, an electron layer for good injection at the cathode cannot function as effective hole blocking layer to confine the holes in the absorber layer region, hence different layers are used to perform the separate functions of ohmic carrier extraction of one type and blocking the complementary carrier. These *blocking layers* suppress injection or extraction of the undesired carrier at each contact and in addition, they conduct the carrier that is extracted with low impedance to avoid losses by the transport resistance. This structure has been amply developed in organic light emitted diodes (OLEDS), and it is described in more detail in Section PSC.5.4. The selective contact in solid solar cells should provide low rate of recombination resulting from collection of the undesired carrier and annihilation of the desired carrier. Buffer layers often improve the operation of inorganic solar cells by passivation of surface states and reducing recombination of carriers (Green et al., 2003). In silicon solar cell technology, the treatment of the majority carrier contact is very important for obtaining high efficiency because the back surface of the crystalline material produces a large recombination rate. To prevent the carrier loss, a back surface field is formed, which reflects the majority carrier or a SiO_2 passivation layer is introduced. These treatments facilitate large diffusion lengths that yield significant gains of sunlight conversion efficiency (Fossum, 1977; Bai et al., 1998).

GENERAL REFERENCES

Rectification at semiconductor/electrolyte junction: Memming (2001).

The Schottky diode: Sharma (1984) and Rhoderick and Williams (1988).

Fundamental model of the solar cell with selective contacts: Green (2002), Bisquert et al. (2004), and Würfel (2009).

Transition metal oxide contacts: Meyer et al. (2007), Toshinori et al. (2007), Meyer and Kahn (2011), Greiner et al. (2012), and Jasieniak et al. (2012).

REFERENCES

Abkowitz, M.; Facci, J. S.; Rehm, J. Direct evaluation of contact injection efficiency into small molecule based transport layers: Influence of extrinsic factors. *Journal of Applied Physics* 1998, 83, 2670–2676.

Albery, W. J. Development of photogalvanic cells for solar energy conservation. *Accounts of Chemical Research* 1982, 15, 142–148.

Araújo, G. L.; Martí, A. Absolute limiting efficiencies for photovoltaic energy conversion. *Solar Energy Materials and Solar Cells* 1994, 33, 213–240.

Bai, Y.; Phillips, J. E.; Barnett, A. M. The roles of electric fields and illumination levels in passivating the surface of silicon solar cells. *Electron Devices, IEEE Transactions on* 1998, 45, 1784–1790.

Barkhouse, D. A. R.; Debnath, R.; Kramer, I. J.; Zhitomirsky, D.; Pattantyus-Abraham, A. G.; Levina, L.; Etgar, L.; Grätzel, M.; Sargent, E. H. Depleted bulk heterojunction colloidal quantum dot photovoltaics. *Advanced Materials* 2011, 23, 3134–3138.

Bisquert, J.; Cahen, D.; Rühle, S.; Hodes, G.; Zaban, A. Physical chemical principles of photovoltaic conversion with nanoparticulate, mesoporous dye-sensitized solar cells. *The Journal of Physical Chemistry B* 2004, 108, 8106–8118.

Crowell, C. R.; Sze, S. M. Current transport in metal semiconductor barriers. *Solid-State Electronics* 1966, 9, 1035–1048.

Fossum, J. G. Physical operation of back-surface-field silicon solar cells. *Electron Devices, IEEE Transactions on* 1977, 24, 322–325.

Gerischer, H. The impact of semiconductors on the concepts of electrochemistry. *Electrochimica Acta* 1990, 35, 1677–1699.

Goodman, A. M.; Rose, A. Double extraction of uniformly generated electron-hole pairs from insulators with noninjecting contacts. *Journal of Applied Physics* 1971, 52, 2823–2830.
Green, M. A. Photovoltaic principles. *Physica E* 2002, 14, 11–17.
Green, M. A.; Shewchun, J. Minority carrier effects upon the small signal and steady state properties of Schottky diodes. *Solid-State Electronics* 1973, 16, 1141–1150.
Green, M. A.; Zhao, J.; Wang, A.; Trupke, T. High-efficiency silicon light emitting diodes. *Physica E: Low-dimensional Systems and Nanostructures* 2003, 16, 351–358.
Greiner, M. T.; Chai, L.; Helander, M. G.; Tang, W.-M.; Lu, Z.-H. Metal/metal-oxide interfaces: How metal contacts affect the work function and band structure of MoO_3. *Advanced Functional Materials* 2012, 23, 215–226.
Heller, A. Miniature biofuel cells. *Physical Chemistry Chemical Physics* 2004, 6, 209–216.
Hodes, G.; Howell, I. D. J.; Peter, L. M. Nanocristallyne photoelectrochemical cells. A new concept in photovoltaic cells. *Journal of the Electrochemical Society* 1992, 139, 3136–3140.
Honsberg, C. B.; Bremmer, S. P.; Corkish, R. Design trade-offs and rules for multiple energy levels solar cells. *Physica E* 2002, 14, 136.
Jasieniak, J. J.; Seifter, J.; Jo, J.; Mates, T.; Heeger, A. J. A solution-processed MoOx anode interlayer for use within organic photovoltaic devices. *Advanced Functional Materials* 2012, 22, 2594–2605.
Korkut, H.; Yildirim, N.; Turut, A. Temperature-dependent current–voltage characteristics of Cr/n-GaAs Schottky diodes. *Microelectronic Engineering* 2009, 86, 111–116.
Landsberg, P. T.; Klimpke, C. Theory of the Schottky barrier solar cell. *Proceedings Royal Society (London) A* 1977, 354, 101–118.
Lopez-Varo, P.; Bertoluzzi, L.; Bisquert, J.; Alexe, M.; Coll, M.; Huang, J.; Jimenez-Tejada, J. A. et al. Physical aspects of ferroelectric semiconductors for photovoltaic solar energy conversion. *Physics Reports* 2016, 653, 1–40.
Luque, A.; Marti, A.; Stanley, C. Understanding intermediate-band solar cells. *Nature Photonics* 2012, 6, 146–152.
Luther, J. M.; Law, M.; Beard, M. C.; Song, Q.; Reese, M. O.; Ellingson, R. J.; Nozik, A. J. Schottky solar cells based on colloidal nanocrystal films. *Nano Letters* 2008, 8, 3488–3492.
Memming, R. *Semiconductor Electrochemistry*; Wiley-VCH: Weinheim, 2001.
Meyer, J.; Hamwi, S.; Bulow, T.; Johannes, H. H.; Riedl, T.; Kowalsky, W. Highly efficient simplified organic light emitting diodes. *Applied Physics Letters* 2007, 91, 113506.
Meyer, J.; Kahn, A. Electronic structure of molybdenum-oxide films and associated charge injection mechanisms in organic devices. *Journal of Photonics for Energy* 2011, 1, 011109.
Mora-Sero, I.; Bertoluzzi, L.; Gonzalez-Pedro, V.; Gimenez, S.; Fabregat-Santiago, F.; Kemp, K. W.; Sargent, E. H.; Bisquert, J. Selective contacts drive charge extraction in quantum dot solids via asymmetry in carrier transfer kinetics. *Nature Communications* 2013, 2, 2272.
Ning, Z.; Ren, Y.; Hoogland, S.; Voznyy, O.; Levina, L.; Stadler, P.; Lan, X.; Zhitomirsky, D.; Sargent, E. H. All-inorganic colloidal quantum dot photovoltaics employing solution-phase halide passivation. *Advanced Materials* 2012, 24, 6295–6299.
Price, M. J.; Foley, J. M.; May, R. A.; Maldonado, S. Comparison of majority carrier charge transfer velocities at Si/polymer and Si/metal photovoltaic heterojunctions. *Applied Physics Letters* 2010, 97, 083503.
Rhoderick, E. H.; Williams, R. H. *Metal-Semiconductor Contacts*, 2nd edition; Clarendon Press: Oxford, 1988.
Schilinsky, P.; Waldauf, C.; Hauch, J.; Brabec, C. J. Simulation of light intensity dependent current characteristics of polymer solar cells. *Journal of Applied Physics* 2004, 95, 2816–2819.
Shannon, J. M. A majority-carrier camel diode. *Applied Physics Letters* 1979, 35, 63–65.
Sharma, B. L. *Metal-Semiconductor Schottky Barrier Junctions and Their Applications*; Plenum: New York, 1984.
Sokel, R.; Hughes, R. C. Numerical analysis of transient photoconductivity in insulators. *Journal of Applied Physics* 1982, 53, 7414–7424.
Sze, S. M. *Physics of Semiconductor Devizes*, 2nd edition; John Wiley and Sons: New York, 1981.
Toshinori, M.; Yoshiki, K.; Hideyuki, M. Formation of ohmic hole injection by inserting an ultrathin layer of molybdenum trioxide between indium tin oxide and organic hole-transporting layers. *Applied Physics Letters* 2007, 91, 253504.
Würfel, P. *Physics of Solar Cells. From Principles to New Concepts*, 2nd edition; Wiley: Weinheim, 2009.

5 Recombination Current in the Semiconductor Diode

We continue with the study of the fundamental model of the solar cell, based on the assumption of radiative recombination that produces the maximal theoretical efficiency. We analyze the behavior of the diode in forward bias, introducing the parameters that determine the recombination rates. Then, we observe the main features of current density–voltage curves. We study specific types of diodes like the LEDs, the molecular diodes represented by dye-sensitized solar cells (DSCs).

5.1 DARK EQUILIBRIUM OF ABSORPTION AND EMISSION OF RADIATION

In this chapter, we analyze the behavior of the fundamental solar cell model that was introduced in Chapter PSC.4. Here we investigate the features of the model in dark conditions, as shown in the upper row of Figure 5.1. In the subsequent chapters, the operation of the solar cell under illumination, shown in the bottom row of Figure 5.1, will be discussed step by step.

To analyze the interplay between light absorption, carrier densities, and photocurrents in a solar cell, it is important to first establish some basic properties that are settled in the dark equilibrium by detailed balance arguments that dictate recombination rates. Detailed balance of light absorption and emission is a rather fundamental physical restriction that must be realized by recombination processes in a semiconductor. Radiative recombination of an excitation is a necessary process that cannot be suppressed. As a matter of fact, at the molecular and atomic level, the light absorption event is subject to microscopic reversibility. Thus, if a molecule absorbs light it must also radiate. Therefore, the fundamental radiation rate of an atom or a molecule can be determined by the detailed balance of incoming blackbody radiation and outgoing radiation as was formulated by Einstein and applied to fluorescence by Kennard (1926). The ensuing relationships connect the linear optical spectra of luminescence and absorbance for homogeneous luminescent materials under thermal equilibrium (Strickler and Berg, 1962).

The traffic of electronic carriers in the operation of a solar cell is governed by three elementary processes: light absorption (leading to carrier generation), carrier recombination, and charge extraction, as shown in the bottom row of Figure 5.1. In the absence of any source of optical radiation, that is in the "dark" (the upper row of Figure 5.1), the semiconductor material is bathed in the Lambertian blackbody radiation coming from all directions that generates a flux of generated electrons termed j_0^{th}/q, where j_0^{th} is a current density. The voltage in the solar cell device can be controlled independently of the incident light, either by connecting the solar cell to a load resistor, or using a potentiostat. In Figure 5.1b, we show a situation in which the solar cell diode in the dark is forward biased so that carriers are injected to the semiconductor. In *steady state*, the net current extracted from the diode is

$$j_{inj} = j_0^{th} - j_{rec} \tag{5.1}$$

The recombination current density j_{rec} is defined as the integration of recombination rate across the film thickness

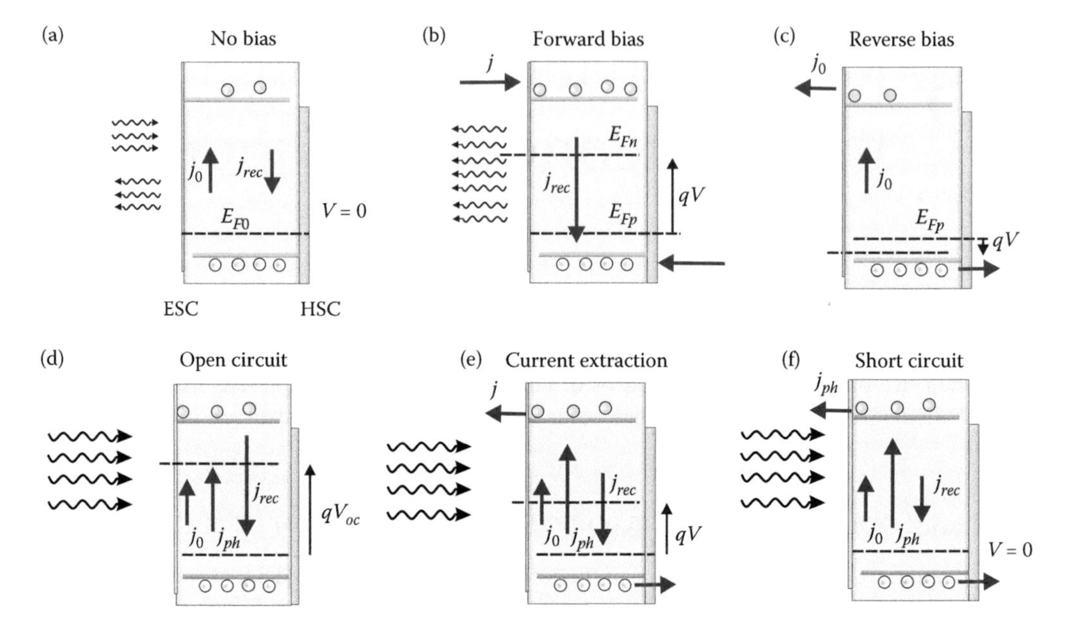

FIGURE 5.1 (a) Scheme of p-type semiconductor material with selective contacts to electrons and holes, ESC and HSC, a model of a solar cell (a–c) in dark. j_0 is the dark generation current (by incoming photons) that coincides with the diode reverse saturation current. j_{rec} is the recombination current that generates outgoing photons (neglecting photon recycling effects). (a) In dark equilibrium, showing the balance of incoming and outgoing photons absorbed or produced in the material, and the equilibrium Fermi level E_{F0}. (b) At forward bias, the minority carrier electrons Fermi level E_{Fn} increases and current is injected at the contacts. (c) At reverse bias, the carriers are extracted determining the reverse saturation current j_0. (d–f) Under illumination, absorbed photons promote excitation of electrons from the valence band to the conduction band. (d) Open circuit: The generation current achieves equilibrium with recombination. This raises the electron Fermi level and causes the production of the photovoltage. No current is extracted from the solar cell. (e) At voltage lower than V_{oc}, the recombination rate is less than the total generation, electrons and holes are extracted in the external circuit and this constitutes a photocurrent. (f) Short circuit: The voltage is zero, all the current created by photogeneration from the illumination in excess of thermal radiation is extracted and constitutes the short-circuit current j_{ph}.

$$j_{rec} = \int_0^d U_{rec}(n, p, x)\, dx \tag{5.2}$$

Note that the fluxes j_0^{th}/q and j_{rec}/q correspond to electron transitions between the semiconductor quantum states, the valence band, and the conduction band, and not to spatial flux of carriers. However, by virtue of transport in extended states, these local fluxes are able to produce real output currents at the contacts of the solar cell, quantified by j_{inj} as indicated in Equation 5.1 and Figure 5.1b. The effect of realistic finite mobilities on the transport features of solar cells will be discussed in Chapter PSC.10.

When a semiconductor attains thermal equilibrium in the dark with the surrounding blackbody radiation, electrons are continuously promoted to the conduction band by the absorption of the radiation. The detailed balance principle requires that the amount of absorbed external radiation is emitted at the same rate by photons resulting from radiative recombination of electrons and holes. The role of radiative recombination is therefore essential for expelling the energy without thermal losses. The fundamental solar cell model established in Section PSC.4.4 is complemented by the assumption that all recombination is purely radiative. The local rate of recombination is given by the expression of band-to-band radiative decay, indicated in Equation PSC.2.32:

$$U_{rad} = B_{rad}np \tag{5.3}$$

From the principle of detailed balance, in the situation of equilibrium at $V = 0$ shown in Figure PSC.4.7a and Figure 5.1a, the incoming and outgoing photon fluxes must be the same, otherwise a net flux of energy would be produced, contradicting the condition of thermodynamic equilibrium. Therefore,

$$j_0^{th} = j_{rec}(V = 0) \tag{5.4}$$

In consequence, if the semiconductor does absorb light, it also has to emit light at the rate fixed by the dark equilibrium balance. In Figure 5.1c, the diode is reverse biased so that recombination is suppressed and all generated carriers are extracted; hence, $j_{inj} = j_0^{th}$. It must be noted that j_0^{th} is the same parameter as the *reverse saturation current* mentioned in the diode equation PSC.4.1, as it is expected that the current in the diode will be saturated to this value in reverse voltage, see Figure ECK.5.13. Now we observe in Equation 5.4 that the thermal generation parameter is also a measure of the recombination rate in the solar cell. Hence, we drop the superscript that labels it as a generation quantity, and hereafter we simply write j_0. In addition, since the radiative recombination is unavoidable, we write $j_{0,rad}$ for the special case that establishes the minimum possible recombination rate in the solar cell.

As we will discuss later on, recombination is a loss process in the overall energy conversion scheme, hence radiative recombination parameter $j_{0,rad}$ forms the basis for the determination of the maximum optimized performance of a solar cell. This parameter is finally calculated in Equation PSC.7.3 and it gives the physical limit to the efficiency of performance of a semiconductor material as an energy converter from light to electricity.

However, a solar cell may have additional nonradiative recombination pathways as shown in Figure 5.2, which invariably produce the loss of additional carriers and a degradation of the performance. There may exist a number of additional mechanisms, as discussed in Section PSC.2.4, such as SRH and Auger recombination. Then, the rate of recombination can be increased with respect to Equation 5.3. These mechanisms can be studied from PL of the contactless films of absorbers as shown in Figure PSC.2.8, and additional recombination channels are usually introduced by the

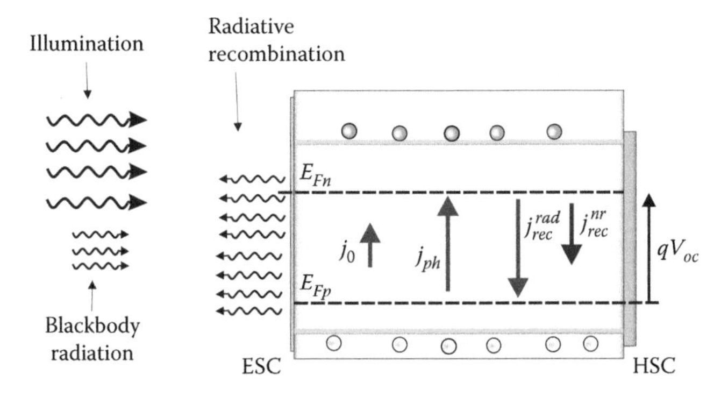

FIGURE 5.2 Scheme of a p-type semiconductor material, with selective contacts to electrons and holes, ESC and HSC. The voltage between the contacts corresponds to energy qV, associated with the splitting of the Fermi levels. The recombination rate is composed of both radiative j_{rec}^{rad} and nonradiative j_{rec}^{nr} mechanisms. The device can be operated as a solar cell or as an LED.

addition of contacts. Actual recombination current in the dark equilibrium, j_0, is larger than the radiative parameter due to the limitation of compensation of incoming photons by two effects: the nonradiative recombination and the inefficiency of photon extraction from the device. The additional components of j_0 have been described by Cuevas (2014).

Following Equation PSC.2.23, the relationship between ideal radiative current (the current $j_{0,rad}$ needed to cancel the absorbed photons) and the total recombination current j_0 is established by the external quantum efficiency of the diode operated as an LED:

$$j_0 = \frac{j_{0,rad}}{EQE_{LED}} \tag{5.5}$$

In summary, the central parameter determining the recombination properties in the solar cell, $j_{0,rad}$, can be obtained by optical and charge collection properties of the device. The specific methods to achieve this goal will be explained in Chapter PSC.7. The nonradiative recombination modes included in j_0 are accounted for by the quantity EQE_{LED}.

5.2 RECOMBINATION CURRENT

The properties of recombination current play a determinant role in the characteristics of a solar cell. To quantify the current, we use the fundamental model in which the forward bias creates the separation of flat Fermi levels and induces the recombination current by injection of carriers (Figure 5.1b). Solving Equation 5.2 gives the result

$$j_{rec} = qdU_{rec} \tag{5.6}$$

We consider the bimolecular recombination model of Equation 5.3. Thus,

$$j_{rec} = qdBnp \tag{5.7}$$

For the analysis of the performance of solar cells, it is necessary to determine $j_{rec}(V)$, so the dependence of carrier density on voltage is needed. This is normally an exponential dependence as discussed below. We can express the current–voltage behavior using m, the diode quality factor explained in Equation PSC.4.1. The recombination current is then written as

$$j_{rec} = j_0 e^{qV/mk_BT} \tag{5.8}$$

Usually, m takes values between 1 and 2. Note in Equation 5.8 that when the voltage is increased, the emission is simply enhanced by an exponential factor. According to the normal application of detailed balance described in Section ECK.6.1, the parameter j_0 is established in dark equilibrium, zero voltage conditions, as indicated in Equation 5.4, and it is assumed to hold when the system is biased in one direction, that of voltage-enhanced recombination.

In general, the dependence $n(V)$ required in Equation 5.7 is a device property that can be obtained solving a full transport-recombination model, as described in Chapter PSC.10. In our present simplified model with flat Fermi levels, we can establish the dependence of carrier density on voltage in two important types of situations.

1. If the generated carrier does not exceed majority carrier density, $p = p_0$, the minority carrier density n is connected to the voltage as

$$n = n_0 e^{qV/k_BT} \tag{5.9}$$

Hence, $k_{rec} = Bp_0$ and $m = 1$, with

$$j_0 = qd\, k_{rec} n_0 \tag{5.10}$$

2. When generated carrier number exceeds majority carrier density, we can express electron and hole densities as

$$n = n_0 e^{(E_{Fn} - E_{F0})/k_BT} = N_c e^{-(E_c - E_{Fn})/k_BT} \tag{5.11}$$

$$p = p_0 e^{-(E_{Fp} - E_{F0})/k_BT} = N_v e^{(E_v - E_{Fp})/k_BT} \tag{5.12}$$

Note the product

$$np = N_c N_v e^{-(E_g - qV)/k_BT} = n_0 p_0 e^{qV/k_BT} \tag{5.13}$$

The electroneutrality condition imposes equal concentration of oppositely charged carriers

$$n = p \tag{5.14}$$

Using Equation 5.7, we have

$$j_{rec} = qdBn^2 = j_0 e^{qV/k_BT} \tag{5.15}$$

The *saturation current* j_0 in this model has the value

$$\begin{aligned} j_0 &= qd\, Bn_i^2 \\ &= qdBn_0 p_0 \\ &= qdBN_c N_v e^{-E_g/k_BT} \end{aligned} \tag{5.16}$$

The carrier density depends on voltage as

$$n = n_i e^{qV/2k_BT} \tag{5.17}$$

Hence, $k_{rec} = B$ and $m = 1$.

Recombination may include a number of nonidealities or complex combination of mechanisms, thus one can use an effective general recombination order β as indicated in Equation PSC.2.42. Then, we have

$$j_{rec} = qdk_{rec} n^{\beta} \tag{5.18}$$

Thus, for model (a), $\beta = 1$ and for model (b), $\beta = 2$, as in Equation 5.15. The ideality factor may correspond to a variety of physical factors such as the voltage distribution at injection contacts, as discussed in Chapter PSC.4, and later on in Section PSC.9.3 in connection with the nonideal behavior of the photovoltage.

The expression in Equation 5.16 assumes that all photons generated by recombination, Equation 5.2, are expelled out of the cell and cancel the incoming blackbody radiation. However, as discussed in Section PSC.2.3, some of the photons may be reabsorbed and emitted again in the photon-recycling effect. Therefore, the derivation of the reverse saturation current requires a more detailed analysis that will be developed in Section PSC.7.4. Equation 5.16 can be considered as an expression valid under certain restrictions. We will see in Chapter PSC.7 that Equation 5.4 needs to be calculated from the incoming radiation flux.

5.3 DARK CHARACTERISTICS OF DIODE EQUATION

Let us consider the shape of the current density–voltage curve in the absence of external illumination. The recombination current has been described in Equation 5.8 as the parameter j_0 enhanced by the exponential term due to applied voltage. The current density–voltage characteristic, Equation 5.1, takes the form already given in Equation PSC.4.1:

$$j = j_0(e^{qV/mk_BT} - 1) \tag{5.19}$$

A number of examples of current–voltage characteristics of diodes in the dark are shown in Figure 5.3. We remark upon two different regimes of behavior:

1. Forward bias is the region of positive potential in Equation 5.19. At forward bias, the injection process produces an increase of the recombination current that depends exponentially with respect to voltage. In Figure 5.3a, we note very large differences in the onset voltage of forward current due to differences in j_0. A large j_0 favors recombination and produces strong forward current at relatively small bias. Figure 5.3b shows the current

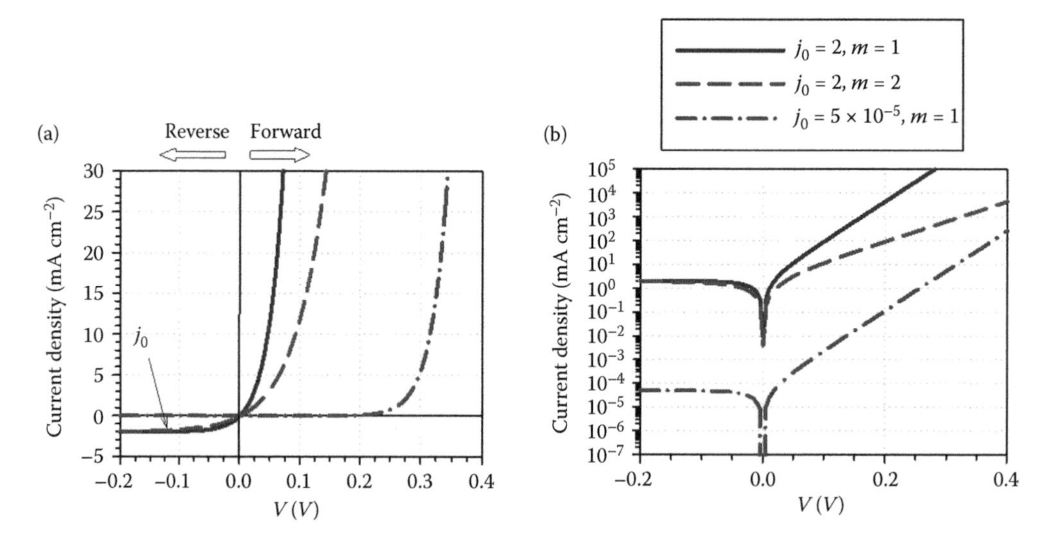

FIGURE 5.3 (a) Current–potential characteristics of a diode with dark saturation currents $j_0 = 2$ mA cm^{-2}, 5×10^{-5} mA cm^{-2}, and different diode quality factors. (b) Current–voltage characteristic in semilogarithmic plot.

density–potential characteristic in log vertical axis. The forward current branch can obtain different slopes according to the value of parameter m.

2. Reverse bias, shown in Figure 5.1c, inhibits the recombination current, so that total current tends to the constant value j_0, the reverse saturation current, which is a bulk property. In general, the reverse bias tends to extract carriers from the active layer. Due to the suppression of recombination, the carriers created by thermal generation are available for extraction. Therefore, the reverse saturation current is independent of the applied reverse voltage. However, in real diodes, in addition to the thermal generation of carriers, a leakage current occurs as well in reverse bias. Ascribing the reverse dark current to the same fundamental process associated with recombination at forward bias requires a careful investigation. For instance, thin organic diodes are very sensitive to shunt paths and edge effects that may dominate the reverse current.

5.4 LIGHT-EMITTING DIODES

The LED is a device that generates light by electroluminescence. Charge carriers of both signs are injected at forward bias in a semiconductor with appropriate selective contacts. As shown in Figure 5.1b, when a diode is forward biased, an enhanced recombination current is produced that results in photon generation. The radiative recombination process converts the carriers to photons. The basic structure of an LED based on organic materials (OLED) containing a light-emitting polymer is shown in Figure 5.4, and a more realistic structure including relevant optical pathways is shown in Figure PSC.2.1. The selectivity of contacts is realized using special transport layers at the cathode and anode as explained later in this section. These *injection* layers are very thin (10 nm) and optically inactive. They have the important role of confining the injected electrons and holes in the recombination layer where light is produced. Hence, efficient recombination occurs due to the overlapping distribution of electrons and holes (often facilitated through the formation of excitons). A very important feature of a practical LED is to avoid light trapping within the device by internal reflection, as indicated in Figure PSC.2.1.

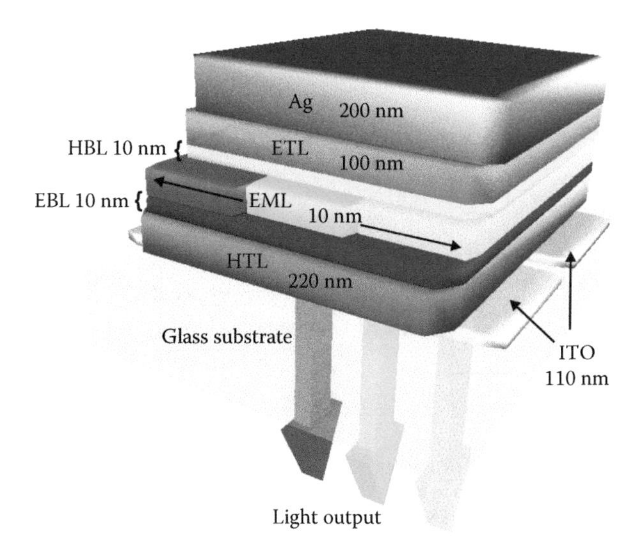

FIGURE 5.4 Basic structure of an OLED—a thin film stack consisting of indium-tin oxide (ITO) as transparent anode on a glass substrate, covered by a thermally evaporated stack of hole transport (HTL), electron blocking (EBL), emitting (EML), hole blocking (HBL), and electron transport layer (ETL), contacted by a silver (Ag) cathode. The EML consists of RGB layers prepared by co-evaporating matrix material and emitting phosphorescent dyes. (Adapted from Flämmich, M. et al. *Organic Electronics* 2011, 12, 1663–1668.)

The first practical LED was invented by Holonyak and Bevacqua (1962) based on the inorganic semiconductor GaAsP and became commercially available in the late 1960s. Electroluminescence from organic crystals was first observed in 1963 (Pope et al., 1968), applying a large voltage of several hundred volts to 10 μm thick crystals of anthracene. In the late 1980s, OLEDs requiring low-operation voltage were first developed using a double layer structure including organic fluorescent dyes based on the small organic molecule tris(8-hydroxy quinoline)Al (Alq_3) (Tang and VanSlyke, 1986). These results promoted extensive research on a wide variety of thin film OLED materials. Friend and coworkers were the first to develop a highly fluorescent conjugated polymer, poly(*p*-phenylene-vinylene) (PPV) as the active material in a single layer OLED (Burroughes et al., 1990), opening the way for the facile preparation of large area films by solution-processed deposition methods that has developed into a mature technology.

The materials that function as the hole-blocking layer in OLEDs should have both good electron-accepting and electron-transporting properties (Shirota, 2000). Wide bandgap electron-transporting materials with deep HOMO energies are used to block holes and wide bandgap hole-transporting materials with shallow LUMO energies block electrons (Sarasqueta et al., 2010). As an example, Figure 5.5 shows different configurations of contact layers in a diode formed with the emissive polymer F8T2 (poly(9,9-dioctylfluorene-alt-bithiophene)) (Jin et al., 2009). The devices are shown with and without TFB interlayers, using either Al or Ca as the cathode material to achieve weak or strong electron injection, respectively. Poly(9,9-dioctylfluorene-alt-*N*-(4-butylphenyl)-diphenylamine) (TFB) is a semiconductor of relatively high hole mobility 2×10^{-3} cm^2 V^{-1} s^{-1}, a bandgap of 3 eV and low-electron affinity, as shown in Figure 5.5a, so that it effectively serves as an electron-blocking

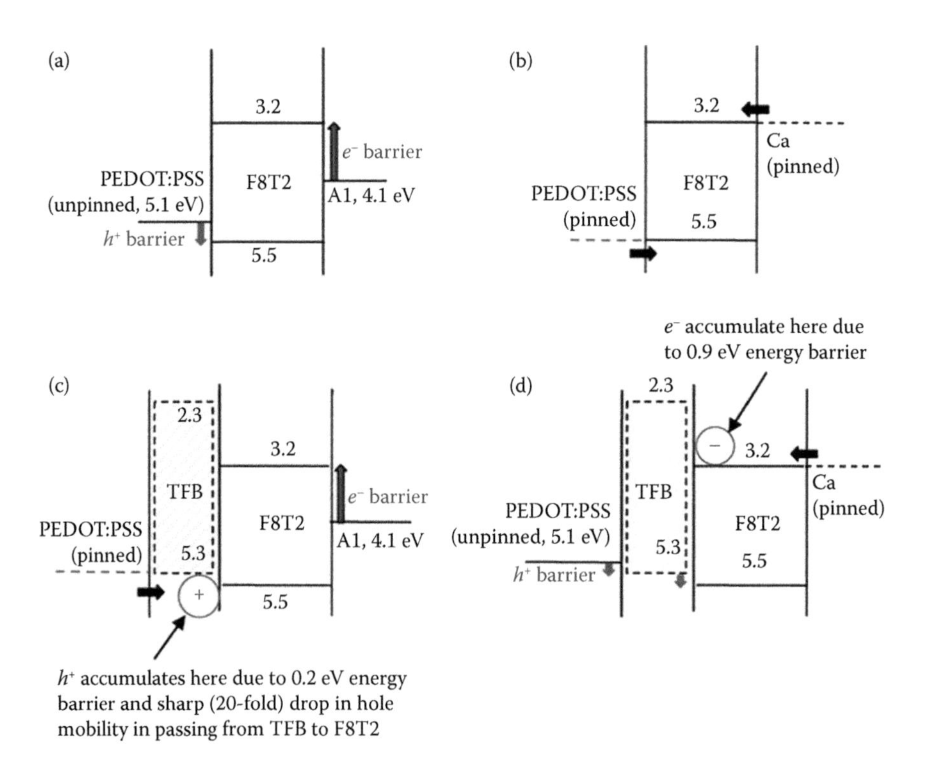

FIGURE 5.5 Energy level diagrams for four devices containing different types of contacts to F8T2. The Fermi level of the Ca cathode is pinned to the LUMO level of F8T2 in devices (b) and (d). The Fermi level of the PEDOT:PSS is pinned to the HOMO level of the adjacent organic layer in devices (b) and (c). (Reproduced with permission from Jin, R. et al. *Physical Chemistry Chemical Physics* 2009, 11, 3455–3462.)

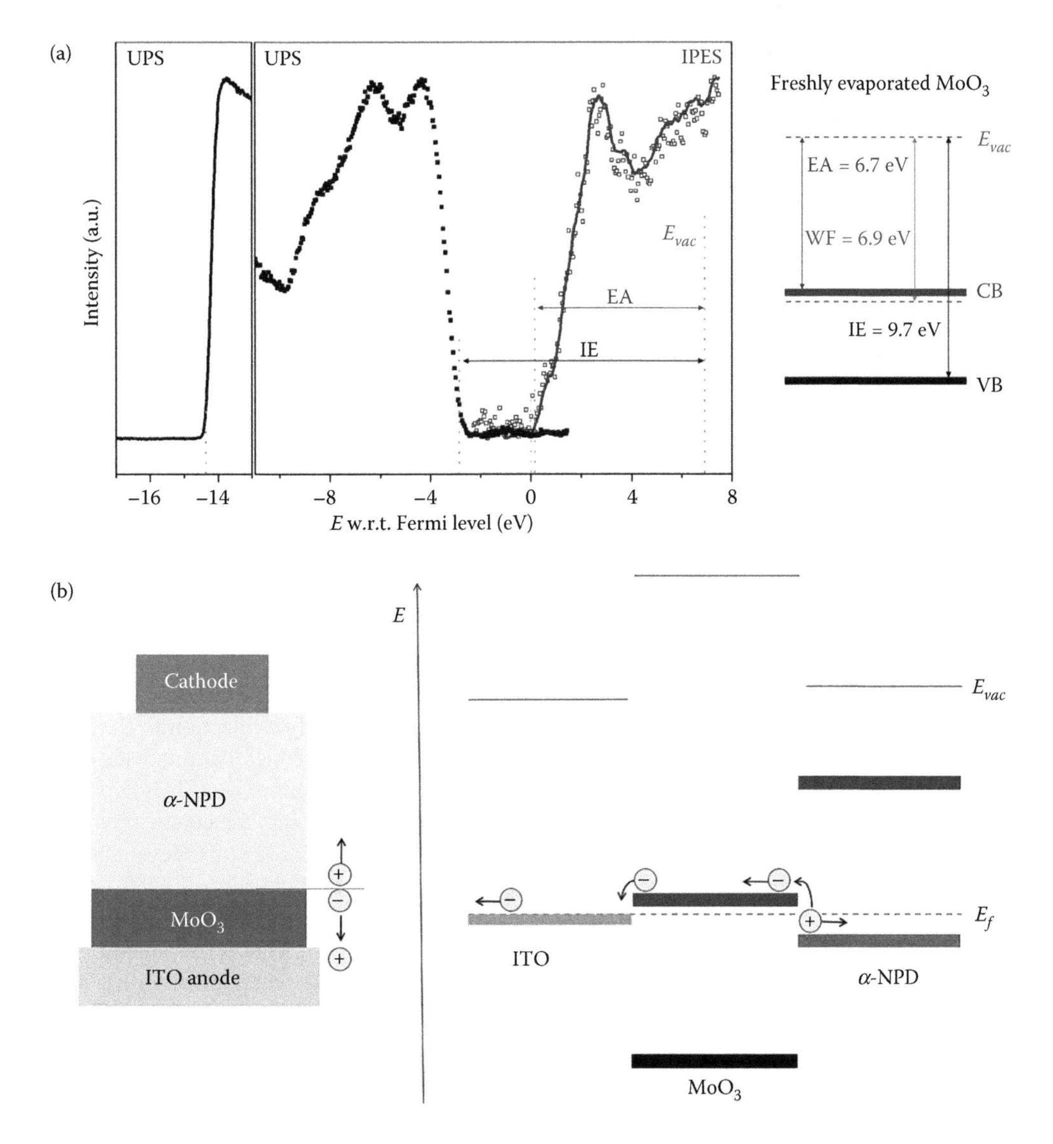

FIGURE 5.6 (a) (*Left*) Ultraviolet photoelectron spectroscopy (UPS) and inverse photoemission spectroscopy (IPES) spectra of clean MoO_3. (*Right*) Schematic energy diagram of the MoO_3 film. (b) Scheme of charge injection at a MoO_3/α-NPD interface. (Reproduced with permission from Meyer, J.; Kahn, A. *Journal of Photonics for Energy* 2011, 1, 011109.)

layer. Another approach is to use graded layers in which the p and n conductors are progressively mixed. This will have the effect of creating a gradient of donor–acceptor compositions that channels each carrier in the desired direction (Sullivan et al., 2004).

The use of high-work function transition metal oxides such as molybdenum (MoO_3), tungsten (WO_3), and vanadium (V_2O_5) oxide was first reported by Tokito et al. (1996) who demonstrated the use of evaporated MO_x (where M = vanadium, molybdenum, or ruthenium and x is the oxygen stoichiometry) to reduce the operating voltage of OLEDs. Many works have shown that transition metal oxide contacts such as MoO_3, WO_3, V_2O_3, and NiO form suitable hole selective contacts that can substitute PEDOT:PSS in organic solar cells and OLEDs. The photon emission spectroscopy measurements on interfaces between MoO_3 and organic hole transport materials indicate that MoO_3 has a very large work function and is strongly n-type, probably due to oxygen vacancies. As shown

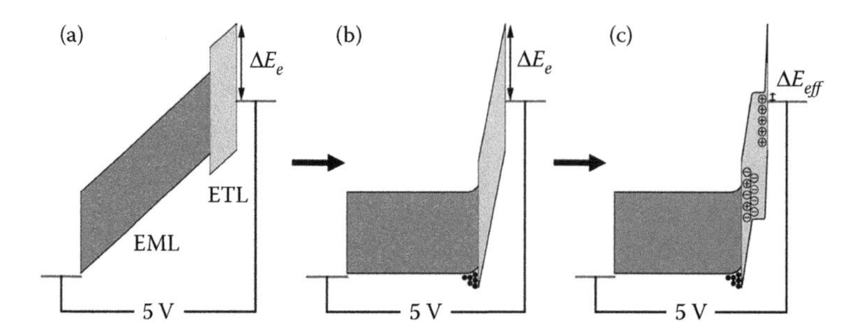

FIGURE 5.7 Schematic response of an anode/EML/ETL/cathode LED under 5 V applied bias. (a) The electric field (slope of the energy levels) is uniform across the device. (b) Injected holes (*black circles*) accumulate at the EML/ETL interface, screening the electric field to the ETL. (c) Ion charges within the conjugated polyelectrolyte ETL redistribute to screen the electric field to the two ETL interfaces. (Reproduced with permission from Hoven, C. V. et al. *Applied Physics Letters* 2009, 94, 033301–033303.)

in Figure 5.6, the electron affinity is about 6.7 eV, therefore the electron-transport states (LUMO) of organic materials in contact with MoO_3 are unlikely to overlap with the 3 eV bandgap of MoO_3 and hence cannot work as an electron-blocking layer. Despite the large difference between electron affinity of MoO_3 and ionization energy of the α-NPD, it is likely that the conduction band of MoO_3 aligns with the HOMO level of the organic layer, while the energy difference is taken by an interface dipole. The mechanism of hole extraction is indicated in Figure 5.6. Hole injection to the organic layer from the valence band of MoO_3 should not be facile, in view of the fact that the oxide valence band is separated several electron-volts (2.5–3.0 eV) below the indium-tin oxide (ITO) Fermi level. However, hole injection from ITO can occur via electron extraction through the conduction band of the oxide, which is situated only 0.6–0.7 eV above the HOMO of the organic film (Meyer and Kahn, 2011). This mechanism is observed in organic solar cells with MoO_3 hole contact layer (Yunlong and Russell, 2013).

Transport layers can readily modify not only the structure of the injection barrier but the full operation mode of the device under bias. An example of this effect is schematically presented in Figure 5.7, and it plays a double role on the device operation (Hoven et al., 2008, 2009). First, the injection of electrons is considerably improved. Second, the electrical field in the central emissive layer is nearly completely screened, as observed from the vanishing electroabsorption signal (Lane et al., 2003).

Electrons and holes injected electrically in a conjugated polymer get captured into excitons that form either the singlet or one of the three triplet states, with equal probability. Then, only 25% of the electron–hole pairs are useful for light production, so that $\chi_{op} = 0.25$ in Equation PSC.2.26.

5.5 DYE SENSITIZATION AND MOLECULAR DIODES

The fundamental structure of the recombination diode that has been discussed in Chapter PSC.4 requires an asymmetry of contacts to a medium that realizes the recombination of injected electrons and holes. It is conceivable to shrink this structure to the level of one molecule as indicated in Figure 5.8. This model represents a two-state molecular system contacted by n- and p-type semiconductors. In the forward mode, holes are injected to the ground state and electrons to the LUMO, so that the carriers recombine by downward electron transition to the hole-occupied ground state. The LUMO is closely related to the first excited state of the molecule; hence, the diode can emit radiation forming an electroluminescent device, a model for the OLED.

The key element for using a molecule as the active material of a recombination diode is the existence of a *kinetic* preference for the electrons to be injected to the LUMO and holes to the

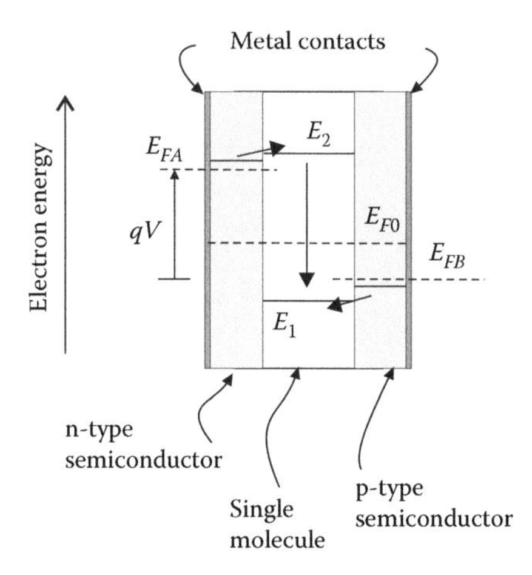

FIGURE 5.8 Basic energy structure of a single molecule diode formed by a molecule with a ground (E_1) and first excited state (E_2), n- and p-type semiconductors with the Fermi levels E_{FA} and E_{FB}, and metal contacts. The arrows indicate injection and recombination of electron and hole carriers.

ground state. This structure can be readily achieved using as one contact a metal or a semiconductor platform in which the organic molecule is grafted by appropriate linkers and an electrolytic contact at the other side. Electrochemical rectifiers formed by monolayers or thin molecular films grafted on electrodes have been widely studied (Abruña et al., 1981; Fujihira and Yamada, 1988). If the molecule is an efficient light-absorbing chromophore, then carriers can be separately extracted from the excited state to form a solar cell, as suggested by Gerischer et al. (1968). This structure uses an optically inactive wide bandgap metal oxide that electronically matches the excited state of a dye and also serves as electron conductor. On the other side, a very positive redox couple serves as selective contact for holes in the dye and realizes the regeneration of the oxidized chromophore, see Figure 5.9.

Photoinduced electron injection from a dye molecule anchored at the semiconductor surface constitutes a central model for heterogeneous electron transfer and has been extensively studied by experiment and simulation. Figure 5.10 shows the results of characterization of the excited states of the perylene molecule attached on the surface of TiO_2 by UPS and femtosecond time-resolved two-photon photoemission (2PPE). Due to limited extinction coefficient of organic dyes, a monolayer of dye on a flat electrode produces little effectiveness for sunlight energy harvesting. A major breakthrough toward an efficient dye-sensitized solar cell (DSC) was realized by O'Regan and Grätzel (1991) using a nanostructured TiO_2 electrode about 10 μm thick in combination with I_3^-/I^- redox mediator. The operation of the dye with respect to selective contacts to extraction of electrons and holes forming a diode structure is indicated in Figure ECK.8.12 and described in more detail in Chapter PSC.9. The nanostructured film greatly increases the internal area and facilitates charge shielding and transport. Figure 5.11 shows the kinetic timescales of the interfacial processes in a DSC using the paradigmatic dye termed N719, a polypyridyl-type ruthenium complex (*cis*-Ru(dcbpy)$_2$(NCS)$_2$) that yields overall solar-to-electric power conversion efficiencies close to 12%. This dye molecule has been designed to set an electron acceptor group close to the TiO_2 surface by two equivalent bipyridine ligands functionalized by carboxylic groups that ensure stable anchoring to the TiO_2 surface, allowing for the strong electronic coupling required for efficient excited-state charge injection. Under photoexcitation, there is a strong preference of the electron in the first excited state of the dye to be injected to TiO_2 due to an effective interaction of the dye with the titania levels, as shown by density functional theory calculations in Figure 5.12. In addition, the remaining hole stays away from the titania surface to facilitate regeneration by reduction from the

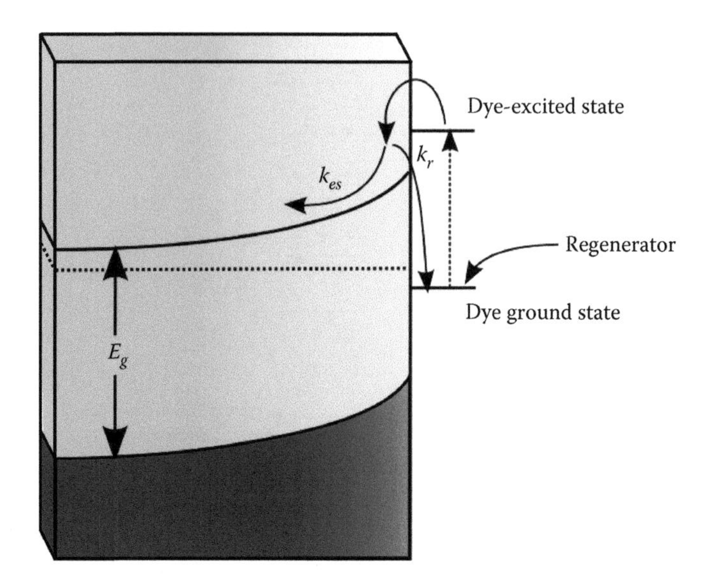

FIGURE 5.9 Schematic diagram of the competing rate processes when an electron is injected into the conduction band of an n-type semiconductor. The *dashed arrow* represents the photoexcitation process. k_{es} is the rate constant for escape from the dye into the semiconductor bulk to be collected as photocurrent and k_r is the rate constant for capture of a conduction band electron. (Reproduced with permission from Parkinson, B. A.; Spitler, M. T. *Electrochimica Acta* 1992, 37, 943–948.)

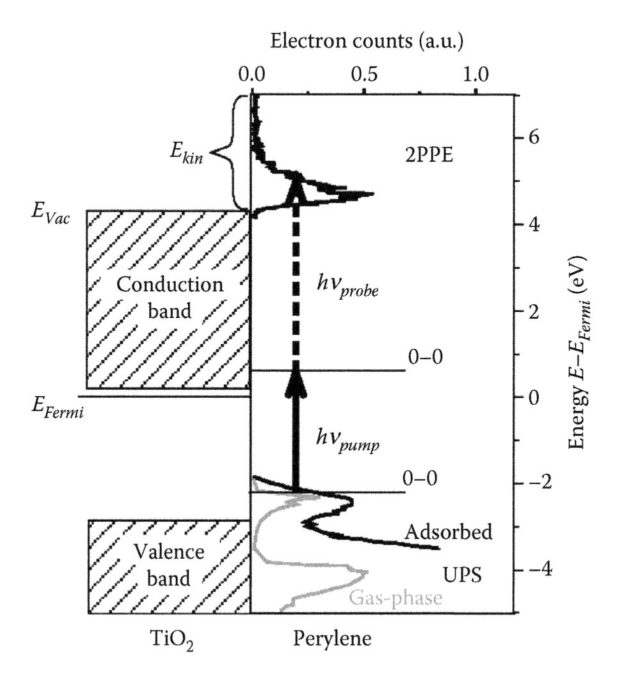

FIGURE 5.10 Alignment of the ground state and excited state of perylene with the tripod anchor/bridge group on the (110) rutile TiO_2 surface deduced from UPS and 2PPE measurements. The *gray curve* is a perylene gas-phase spectrum. (Reproduced with permission from Gundlach, L. et al. *Progress in Surface Science* 2007, 82, 355–377.)

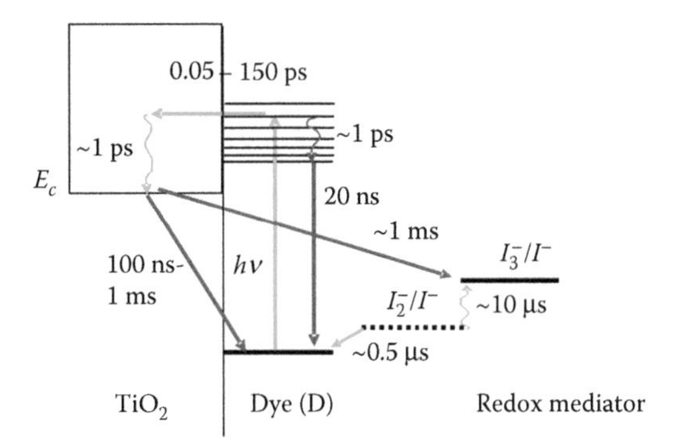

FIGURE 5.11 Kinetics of the N719 sensitized TiO_2 solar cell with I_3^-/I^- redox mediator. Typical time constants of the forward reactions (*green*) and recombination reactions (*red*) are indicated. (Reproduced with permission from Boschloo, G.; Hagfeldt, A. *Accounts of Chemical Research* 2009, 42, 1819–1826.)

redox electrolyte, and to avoid recombination of the photogenerated electron–hole pair. This simple view of the DSC neglects other significant recombination channels, but it effectively shows that the selectivity requirements of the diode structure of Figure 5.8 demand a specific control of intermolecular interactions at the interface. A combined design of the different elements that ensures rectification is required, by creating kinetic and energetic preference that directs carrier flows (Pastore and de Angelis, 2013).

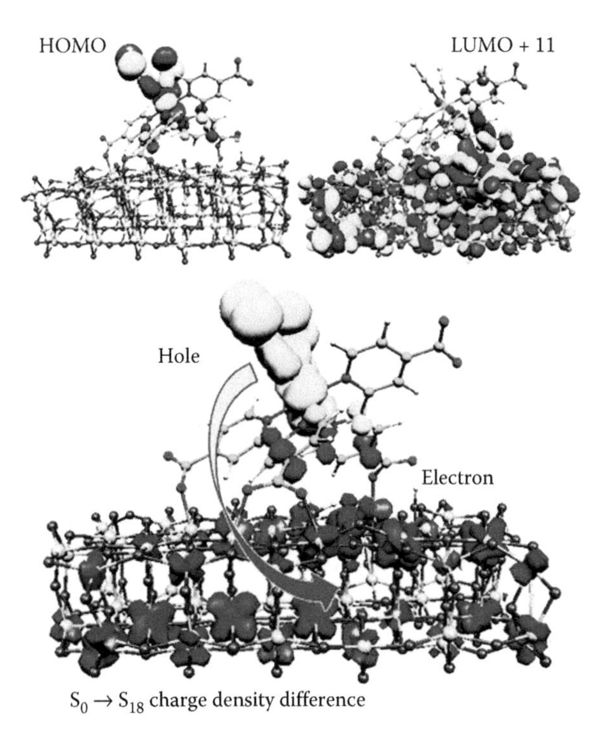

FIGURE 5.12 Isodensity plots of the HOMO and LUMO + 11 of the N719/TiO_2 system. *Bottom*: Charge density difference between the ground state (S_0) and the S_{18} excited state. A blue (*green*) color signifies an increase (decrease) of charge density upon electron excitation. (Reprinted with permission from De Angelis, F. et al. *The Journal of Physical Chemistry C* 2011, 115, 8825–8831.)

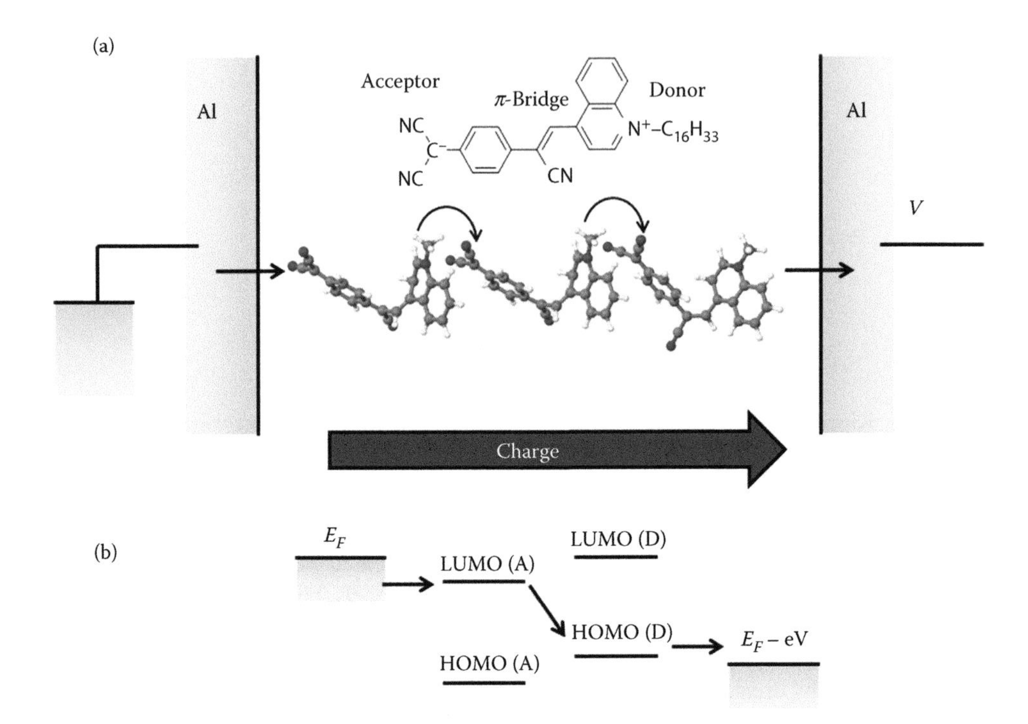

FIGURE 5.13 (a) Schematic representation of $C_{16}H_{33}Q$-3CNQ in contact with metal electrodes under applied bias voltage. (b) The Aviram–Ratner mechanism for molecular rectification. (Adapted from Krzeminski, C. et al. *Physical Review B* 2001, 64, 085405.)

In the absence of external selective contacts, it is much more difficult to build a diode using a single molecular unit. Aviram and Ratner (1974) proposed that a single organic molecule of the type D–σ–A could be a rectifier, see Figure 5.13. Here, D is a good organic one-electron donor (but poor acceptor), σ the covalent saturated bridge, and A is a good organic one-electron acceptor. In this model, system selectivity must be built internally in the molecular assembly in contact with two metal electrodes and the process involves electron transfer between molecular orbitals, whose probability amplitudes are asymmetrically placed within the molecule (Metzger, 2003). The realization of this idea has not been conclusively demonstrated. The bias potential between the electrodes not only injects carriers in the required molecular levels but also creates an electrical field that distorts those levels (Krzeminski et al., 2001). It was suggested that the main impediment to realize D–σ–A molecular rectifier is that the system loses unidirectionality of charge transfer under the applied bias (Mujica et al., 2002). Other work showed that the bias voltage actually aligns the D and A levels creating a resonant state delocalized over the entire molecule (Stokbro et al., 2003). This state provokes a large rise of current but no rectification. It is, therefore, likely that experimental observation of rectifying behavior in unimolecular systems is mainly a result of asymmetric coupling between the molecule and the electrodes, so that the rectification actually originates at the interface (Zhao et al., 2010).

GENERAL REFERENCES

Dye-sensitized solar cells: O'Regan and Grätzel (1991), Hagfeldt and Grätzel (1995), Hagfeldt and Grätzel (2000), Hagfeldt et al. (2010), Kalyanasundaram (2010), Barea and Bisquert (2013), and Bisquert and Marcus (2013).

Photoinduced electrons transfer from dye molecules: Rego and Batista (2003), Duncan et al. (2005), Duncan and Prezhdo (2007), and Gundlach et al. (2007).

REFERENCES

Abruña, H. D.; Denisevich, P.; Umana, M.; Meyer, T. J.; Murray, R. W. Rectifying interfaces using two-layer films of electrochemically polymerized vinylpyridine and vinylbipyridine complexes of ruthenium and iron on electrodes. *Journal of the American Chemical Society* 1981, 103, 1–5.

Aviram, A.; Ratner, M. A. Molecular rectifiers. *Chemical Physics Letters* 1974, 29, 277–283.

Barea, E. M.; Bisquert, J. Properties of chromophores determining recombination at TiO_2-dye-electrolyte interface. *Langmuir* 2013, 29, 8773–8781.

Bisquert, J.; Marcus, R. A. Device modeling of dye-sensitized solar cells. *Topics in Current Chemistry* 2013. doi: 10.1007/1128_2013_1471.

Boschloo, G.; Hagfeldt, A. Characteristics of the iodide/triiodide redox mediator in dye-sensitized solar cells. *Accounts of Chemical Research* 2009, 42, 1819–1826.

Burroughes, J. H.; Bradley, D. D. C.; Brown, A. R.; Marks, R. N.; MacKay, K.; Friend, R. H.; Burn, P. L.; Holmes, A. B. Light-emitting diodes based on conjugated polymers. *Nature* 1990, 347, 539–541.

Cuevas, A. The recombination parameter J_0. *Energy Procedia* 2014, 55, 53–62.

De Angelis, F.; Fantacci, S.; Mosconi, E.; Nazeeruddin, M. K.; Grätzel, M. Absorption spectra and excited state energy levels of the N719 Dye on TiO_2 in dye-sensitized solar cell models. *The Journal of Physical Chemistry C* 2011, 115, 8825–8831.

Duncan, W. R.; Prezhdo, O. V. Theoretical studies of photoinduced electron transfer in dye-sensitized TiO_2. *Annual Review of Physical Chemistry* 2007, 58, 143–184.

Duncan, W. R.; Stier, W. M.; Prezhdo, O. V. Ab initio nonadiabatic molecular dynamics of the ultrafast electron injection across the alizarin-TiO_2 interface. *Journal of the American Chemical Society* 2005, 127, 7941–7951.

Flämmich, M.; Frischeisen, J.; Setz, D. S.; Michaelis, D.; Krummacher, B. C.; Schmidt, T. D.; Brütting, W.; Danz, N. Oriented phosphorescent emitters boost OLED efficiency. *Organic Electronics* 2011, 12, 1663–1668.

Fujihira, M.; Yamada, H. Molecular photodiodes consisting of unidirectionally oriented amphipathic acceptor-sensitizer-donor triads. *Thin Solid Films* 1988, 160, 125–132.

Gerischer, H.; Michel-Beyerle, M. E.; Rebentrost, F.; Tributsch, H. Sensitization of charge injection into semiconductors with large band gap. *Electrochimica Acta* 1968, 13, 1509–1515.

Gundlach, L.; Ernstorfer, R.; Willig, F. Ultrafast interfacial electron transfer from the excited state of anchored molecules into a semiconductor. *Progress in Surface Science* 2007, 82, 355–377.

Hagfeldt, A.; Boschloo, G.; Sun, L.; Kloo, L.; Pettersson, H. Dye-sensitized solar cells. *Chemical Reviews* 2010, 110, 6595–6663.

Hagfeldt, A.; Grätzel, M. Light-induced redox reactions in nanocrystalline systems. *Chemical Reviews* 1995, 95, 49–68.

Hagfeldt, A.; Grätzel, M. Molecular photovoltaics. *Accounts in Chemical Research* 2000, 33, 269–277.

Holonyak, N.; Bevacqua, S. F. Coherent (visible) light emission from $Ga(As_{1-x}P_x)$ junctions. *Applied Physics Letters* 1962, 1, 82–83.

Hoven, C. V.; Peet, J.; Mikhailovsky, A.; Nguyen, T.-Q. Direct measurement of electric field screening in light emitting diodes with conjugated polyelectrolyte electron injecting/transport layers. *Applied Physics Letters* 2009, 94, 033301–033303.

Hoven, C. V.; Yang, R.; Garcia, A.; Crockett, V.; Heeger, A. J.; Bazan, G. C.; Nguyen, T.-Q. Electron injection into organic semiconductor devices from high work function cathodes. *Proceedings of the National Academy of Sciences* 2008, 105, 12730–12735.

Jin, R.; Levermore, P. A.; Huang, J.; Wang, X.; Bradley, D. D. C.; deMello, J. C. On the use and influence of electron-blocking interlayers in polymer light-emitting diodes. *Physical Chemistry Chemical Physics* 2009, 11, 3455–3462.

Kalyanasundaram, K. *Dye-Sensitized Solar Cells*; CRC Press: Boca Raton, 2010.

Kennard, E. H. On the interaction of radiation with matter and on fluorescent exciting power. *Physical Review* 1926, 28, 672–683.

Krzeminski, C.; Delerue, C.; Allan, G.; Vuillaume, D.; Metzger, R. M. Theory of electrical rectification in a molecular monolayer. *Physical Review B* 2001, 64, 085405.

Lane, P. A.; deMello, J. C.; Fletcher, R. B.; Bernius, M. Electric field screening in polymer light-emitting diodes. *Applied Physics Letters* 2003, 83, 3611–3613.

Metzger, R. M. Unimolecular electrical rectifiers. *Chemical Reviews* 2003, 103, 3803–3834.

Meyer, J.; Kahn, A. Electronic structure of molybdenum-oxide films and associated charge injection mechanisms in organic devices. *Journal of Photonics for Energy* 2011, 1, 011109.

Mujica, V.; Ratner, M. A.; Nitzan, A.: Molecular rectification: Why is it so rare? *Chemical Physics* 2002, 281, 147–150.

O'Regan, B.; Grätzel, M. A low-cost high-efficiency solar cell based on dye-sensitized colloidal TiO_2 films. *Nature* 1991, 353, 737–740.

Parkinson, B. A.; Spitler, M. T. Recent advances in high quantum yield dye sensitization of semiconductor electrodes. *Electrochimica Acta* 1992, 37, 943–948.

Pastore, M.; de Angelis, F. Intermolecular interactions in dye-sensitized solar cells: A computational modeling perspective. *Journal of Physical Chemistry Letters* 2013, 4, 956–974.

Pope, M.; Kallmann, H. P.; Magnante, P. Electroluminescence in organic crystals. *Journal of Chemical Physics* 1968, 38, 2042.

Rego, L. G. C.; Batista, V. S. Quantum dynamics simulations of interfacial electron transfer in sensitized TiO_2 semiconductors. *Journal of the American Chemical Society* 2003, 125, 7989–7997.

Sarasqueta, G.; Choudhury, K. R.; Subbiah, J.; So, F. Organic and inorganic blocking layers for solution-processed colloidal PbSe nanocrystal infrared photodetectors. *Advanced Functional Materials* 2010, 21, 167–171.

Shirota, Y.: Organic materials for electronic and optoelectronic devices. *Journal Materials Chemical* 2000, 10, 1–25.

Stokbro, K.; Taylor, J.; Brandbyge, M. Do Aviram and Ratner diodes rectify? *Journal of the American Chemical Society* 2003, 125, 3674–3675.

Strickler, S. J.; Berg, R. A. Relationship between absorption intensity and fluorescence lifetime of molecules. *Journal of Chemical Physics* 1962, 37, 814–820.

Sullivan, P.; Heutz, S.; Schultes, S. M.; Jones, T. S. Influence of codeposition on the performance of CuPc–C[sub 60] heterojunction photovoltaic devices. *Applied Physics Letters* 2004, 84, 1210–1212.

Tang, C. W.; VanSlyke, A. Organic electroluminescent diodes. *Applied Physics Letters* 1986, 51, 913–915.

Tokito, S.; Noda, K.; Taga, Y. Metal oxides as a hole-injecting layer for an organic electroluminescent device. *Journal of Physics D: Applied Physics* 1996, 29, 2750–2759.

Yunlong, Z.; Russell, J. H. Influence of a MoO_x interlayer on the open-circuit voltage in organic photovoltaic cells. *Applied Physics Letters* 2013, 103, 053302.

Zhao, J.; Yu, C.; Wang, N.; Liu, H. Molecular rectification based on asymmetrical molecular electrode contact. *The Journal of Physical Chemistry C* 2010, 114, 4135–4141.

6 Radiative Equilibrium in a Semiconductor

The materials for the conversion of solar photons to electrical flux require to be spectrally matched for optimal energy harvesting. Using a model absorber material with a step-like bandgap, we discuss the central features of maximal photocurrent and electrical power that can be obtained from the incident sunlight. We then move to fundamental considerations that have a large impact on the properties of solar cells. Semiconductors that absorb light also emit radiation by radiative recombination. Detailed balance is established in thermal equilibrium (no applied voltage in the dark) between the absorption of incoming blackbody radiation and the emission of photons by recombination of excited carriers, which imposes fundamental constraints to the rate of recombination. Furthermore, when the semiconductor receives extra illumination, the quasi-Fermi levels of electrons and holes are split. A reciprocity relationship between the absorbed and emitted radiation allows us to accurately determine the internal photovoltage and the light absorption characteristics of the material from PL measurements.

6.1 UTILIZATION OF SOLAR PHOTONS

The light absorption characteristics of the materials that convert radiant energy to useful electrical energy must be adapted optimally to the spectral properties of solar radiation. The principal feature of the available radiation is the number of photons per energy (or frequency) interval dE denoted as $\phi_{ph}(E)dE$ in Section PSC.1.7. As discussed in Section PSC.1.8, sunlight shows a rather broad spectrum including photons in the energy range from ultraviolet to infrared (280–2500 nm, 0.5–4.4 eV). Figure 6.1a shows the spectral shape of terrestrial solar irradiation represented by blackbody radiation at $T_S = 5800$ K, and equivalent projected solid angle $\varepsilon_S^{1.5AM} = 4.8 \times 10^{-5}$ with a total power $\Phi_{E,tot} = 1000$ W m^{-2}.

The photocurrent in a solar cell consists of the extraction of the photogenerated electrons and holes and is measured by the short-circuit photocurrent density, j_{ph}. Solar energy converters utilize the photons *one by one,* and each absorbed photon creates one electron–hole pair exactly. Therefore, the maximal value of the electrical current that can be extracted in photovoltaic cells is given by the number of photons that are absorbed, times the elementary charge. The primary limitations to the photocurrent are quantified by the spectral absorptivity or absorptance of the device, $a(E)$. In addition, in order to convert the incoming photons into an electron current in the external circuit, several steps that determine the internal quantum efficiency (IQE_{PV}) must function efficiently, as commented in Section PSC.2.3; note the relationship in Equation PSC.2.28, $EQE_{PV}(E) = a(E)IQE_{PV}(E)$. If the photovoltaic external quantum efficiency EQE_{PV} of the solar cell has been determined, the short-circuit photocurrent under arbitrary illumination corresponds to

$$j_{ph} = q\int_0^\infty EQE_{PV}(E)\phi_{ph}(E)\,dE \tag{6.1}$$

Some matcrials have been proposed in which high-energy photons create more than one electron–hole pair, showing quantum efficiency larger than 1, although these have so far produced very low-power conversion efficiencies (Sambur et al., 2010; Semonin et al., 2011).

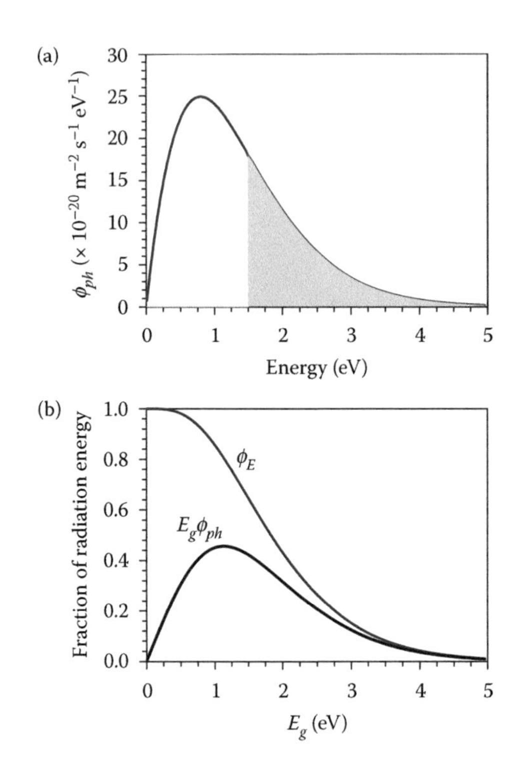

FIGURE 6.1 (a) Photon flux emitted by a blackbody at temperature 5800 K, with an *equivalent* projected solid angle $\varepsilon_S^{1.5AM} = D\pi = 4.8\times10^{-5}$ that gives a total power of 1000 W m^{-2}. The photons absorbed for a bandgap $E_g = 1.5$ eV are indicated. (b) Fraction of the incoming power absorbed as a function of the absorber gap. The lower line considers the thermalization to E_g of photon energy of each absorbed photon.

For an approximate evaluation of photovoltaic properties, it is useful to assume an ideal thick semiconductor that absorbs all photons of energy larger than the semiconductor bandgap. The step absorptance takes the values

$$\begin{aligned} a(E) &= 0, \quad \text{for} \quad E < E_g \\ a(E) &= 1, \quad \text{for} \quad E \geq E_g \end{aligned} \tag{6.2}$$

This is a useful simplification for assessing the fundamental efficiency limit of solar cells described by just two properties: the semiconductor bandgap and the temperature T_A of the absorber. Note that it is assumed that the optical pathway for above bandgap radiation is shorter than the film thickness, so that $a = 1$. Furthermore, we assume optimal charge collection at electrodes giving $IQE_{PV} = 1$. These two conditions are somewhat contradictory regarding the required film thickness, as total collection requires a short film, but nonetheless, we adopt this model as a hypothetical reference for optimal photovoltaic performance that will be qualified later on in Chapter PSC.10. Obviously, the model works best for semiconductors that have a nearly step-like spectral absorption like GaAs and $CH_3NH_3PbI_3$ perovskite rather than for indirect semiconductors, see Figure PSC.3.4.

Based on Equation 6.2, the maximal current that can be obtained as a function of the bandgap of the absorber is

$$j_{ph} = q\Phi_{ph}(E_g,\infty) = q\int_{E_g}^{\infty} \phi_{ph}(E)\,dE \tag{6.3}$$

The spectral flux for blackbody radiation with the étendue ε is indicated in Equation PSC.1.32. The integral then takes the form

$$j_{ph} = q\frac{\varepsilon}{\pi}b_{\pi}\int_{E_g}^{\infty}\frac{E^2}{e^{E/k_BT}-1}dE \tag{6.4}$$

where the prefactor for a hemisphere $\varepsilon = \pi$ is the constant $b_{\pi} = 2\pi/h^3c^2$. If $E_g \gg k_BT$, the integral is evaluated as follows:

$$\int_{E_g}^{\infty} E^2 e^{-E/k_BT}\,dE = e^{-E_g/k_BT}k_BT(E_g^2 + 2E_gk_BT + 2k_B^2\,T^2) \tag{6.5}$$

We define the function

$$\Gamma_{ph}(E,T) = E^2 + 2Ek_BT + 2k_B^2T^2 \tag{6.6}$$

Hence, the photocurrent is approximately given by

$$j_{ph} = q\frac{\varepsilon}{\pi}b_{\pi}k_BT\Gamma_{ph}(E_g,T_S)e^{-E_g/k_BT_S} \tag{6.7}$$

For sunlight modeled in terms of the étendue $\varepsilon_S^{AM1.5}$ in Equation PSC.1.45, the photocurrent is

$$j_{ph} = q\frac{\varepsilon_S^{AM1.5}}{\pi}b_{\pi}k_BT\Gamma_{ph}(E_g,T_S)e^{-E_g/k_BT_S} \tag{6.8}$$

The maximum current density that can be obtained from the blackbody radiation at $T_S = 5800$ K ($k_BT_S = 0.5$ eV) and $\varepsilon_S^{1.5AM}$, that is taking the integral from $E_g = 0$, is $j_{ph}^{max} = 73$ mA cm^{-2}. The fraction of the incoming photon flux absorbed by a semiconductor of bandgap E_g can be observed in Figure 6.1a. The current as a function of the absorption edge, that is the semiconductor bandgap, is shown in Figure 6.2a both for the case of a 5800 K blackbody spectrum and for the *AM*1.5*G* spectrum. The actual current obtained in record inorganic cells of GaAs and Si and a lead-halide perovskite solar cell are also presented. We observe that the photocurrent in top quality devices is quite close to the value theoretically achievable from *AM*1.5*G*. The maximal efficiencies for more realistic absorption conditions are discussed in Figure PSC.8.10.

The benchmark feature for a solar energy converter is the *power conversion efficiency* (PCE), that is the ratio of the electrical power P_{el} produced by the conversion device, with respect to the total incident radiant power $\Phi_{E,tot}$

$$\eta_{PCE} = \frac{P_{el}}{\Phi_{E,tot}} \tag{6.9}$$

In principle, it is desirable to use a material that absorbs a large fraction of the spectrum, implying a small bandgap, in order to generate a large photocurrent. But a second main aspect of spectral solar energy conversion relates to the utilization of the *energy* contained in each energy interval of the

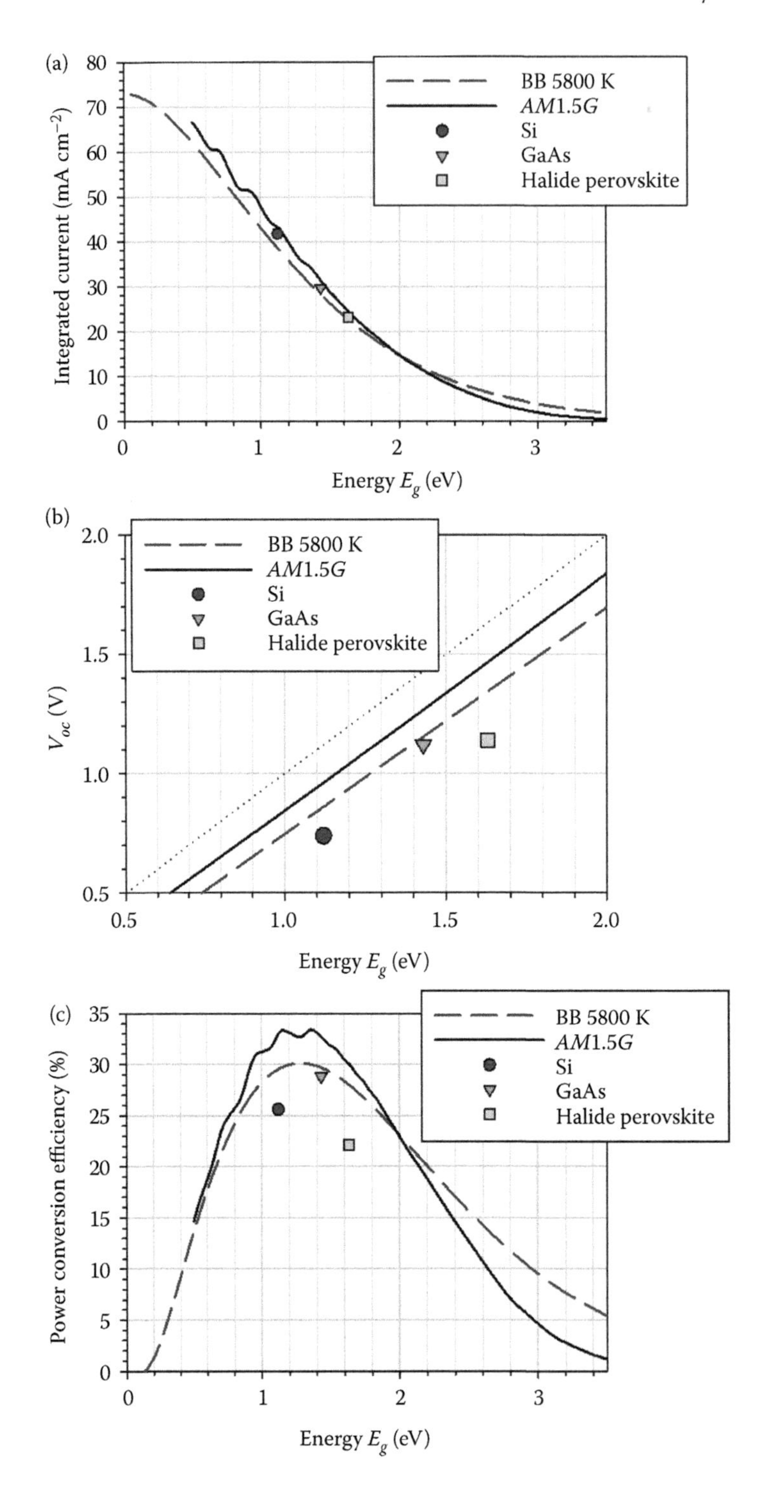

FIGURE 6.2 Photovoltaic parameters as a function of the bandgap with the assumption of sharp absorption edge and total charge collection. The quantities are shown for incident blackbody radiation at temperature $T_S = 5800$ K, with total energy flux $\Phi_{E,tot} = 1000$ W m^{-2}, and for actual *AM*1.5*G* solar irradiance. (a) Integrated current density. (b) The radiative voltage. (c) The maximal photovoltaic Shockley–Queisser efficiencies. The points indicate the values for record cells.

radiation field, $\phi_E(E)dE = E\phi_{ph}(E)dE$. The carriers initially obtain a separation of energy corresponding to that of the absorbed photon, see Figure PSC.3.1, but they are rapidly thermalized to the edges of the bands corresponding to the quasiequilibrium energy configuration, E_g, that precedes recombination. Energy relaxation of the generated electrons and holes to the bandgap has been previously commented on in Section ECK.5.6. After the fast thermalization in the absorber, the energy available is E_g for each absorbed photon. Since the quantity of energy $h\nu - E_g$ per photon is lost, the PCE is severely reduced. Meanwhile, as taken into account in Equation 6.2, the photons with energy less than E_g are not absorbed.

Let us find the flux of energy contained in the incoming blackbody radiation that is absorbed by the semiconductor:

$$\Phi_E(E_g,\infty) = \int_{E_g}^{\infty} E\phi_{ph}(E)dE \tag{6.10}$$

$$\Phi_E(E_g,\infty) = \frac{\varepsilon}{\pi} b_\pi \int_{E_g}^{\infty} \frac{E^3}{e^{E/k_BT} - 1} dE \tag{6.11}$$

If $E_g \gg k_BT$, the integral is evaluated as follows:

$$\Phi_E(E_g,\infty) = \frac{\varepsilon}{\pi} b_\pi e^{-E_g/k_BT} k_BT \chi_E(E_g,T) \tag{6.12}$$

using the function

$$\chi_E(E,T) = E^3 + 3k_BTE^2 + 6k_B^2T^2E + 6k_B^3T^3 \tag{6.13}$$

The lost power due to the transparent range is

$$\Phi_E(0,E_g) = \Phi_E(0,\infty) - \Phi_E(E_g,\infty) \tag{6.14}$$

The total flux is given in Equation PSC.1.39

$$\Phi_{E,tot} = \frac{\varepsilon}{\pi} b_\pi \frac{\pi^4}{15} (k_BT)^4 \tag{6.15}$$

If the total flux is calculated from Equation 6.12, a prefactor 6 (instead of $\pi^4/15$) is obtained.

We can estimate the so-called Trivich–Flinn efficiency, which assumes that every electron–hole pair is extracted with a voltage equal to E_g/q (Green, 2012). It is

$$\eta = \frac{E_g \Phi_{ph}(E_g,\infty)}{\Phi_{E,tot}} \tag{6.16}$$

Taking the approximate expressions for blackbody radiation at the temperature T_S of the sun's surface, Equations 6.6 and PSC.1.39, and the normalized energy $x_g = E_g/k_BT_S$, we obtain

$$\eta = \frac{15}{\pi^4} e^{-x_g} x_g \left(x_g^2 + 2x_g + 2 \right) \tag{6.17}$$

This preliminary efficiency is shown in Figure 6.1b as a function of bandgap. It is *not* a well-defined conversion efficiency in the sense of Equation 6.9, since each electron–hole pair in a solar cell is extracted at a much smaller voltage than E_g, as discussed in Section PSC.8.3. Nonetheless, Equation 6.17 interestingly reveals that maximal efficiency occurs at intermediate wavelength within the solar spectrum due to the two competing effects that determine solar energy conversion into electrical energy using a single semiconductor: (1) The transparency of the semiconductor to long wavelength photons and (2) the loss of a quantity of energy $E-E_g$ per each absorbed photon. These energy loss effects are severe. Even in the most favorable condition, which is $E_g \cong 1.1$ eV, more than 50% of the energy content of blackbody radiation at T_S is lost.

Additional considerations required for the physically correct theoretical efficiency of the ideal converter will be described in Section PSC.8.4. The complete approach is based on reciprocity principles that are derived in the following sections. We anticipate that the maximal efficiency is developed from a conservation argument. The number of photons per unit time that are absorbed from the incident spectrum minus the photons that are emitted by radiative recombination determines the electrical current, and the power is calculated at the voltage where the electrical output is optimal. The result of the calculation is the (Shockley and Queisser, 1961) (SQ) efficiency shown in Figure 6.2c for model blackbody radiation and $AM1.5G$ solar irradiance. The first step toward a determination of the maximal efficiency of a solar cell is to obtain the fundamental radiative lifetime of a semiconductor as we explain in the next section.

6.2 FUNDAMENTAL RADIATIVE CARRIER LIFETIME

The detailed balance of absorbed and emitted radiation in a semiconductor that has attained thermal equilibrium in the dark with the surrounding blackbody radiation enables one to establish a fundamental equation for the radiative lifetime, as first derived by van Roosbroeck and Shockley (1954), see also Pankove (1971). The lifetime is determined only by the absorption coefficient $\alpha(E)$ and the refraction index of the semiconductor n_r.

We consider a small fragment of a semiconductor, as shown in Figure 6.3. In equilibrium, the volume element exchanges radiation with the surroundings. The density of photons in the volume element is determined by Planck's law established in Equation PSC.1.19, as follows:

$$\frac{dn_{ph}(E)}{dE} = \frac{4b_\pi n_r^3}{c} \frac{E^2}{e^{E/k_BT}-1} \tag{6.18}$$

The photon absorption rate in the volume $U_{abs}(E)$ is the product of the photon density and the absorption probability, which is given by the lifetime of the photons of energy E, $\tau_{ph}(E)$

$$U_{abs}(E)dE = \frac{1}{\tau_{ph}(E)} dn_{ph}(E) \tag{6.19}$$

The lifetime of the photons can be established analyzing their mean free path in the medium according to the velocity of the photons:

$$\tau_{ph}(E) = \frac{1}{\alpha(E)c_\gamma} = \frac{n_r}{\alpha(E)c} \tag{6.20}$$

The recombination rate $U_{rec}(E)$ in the volume causes the emission of photons of energy E, as indicated in Equation PSC.2.32. Using the intrinsic density, $n_i^2 = n_0p_0$, we have the expression

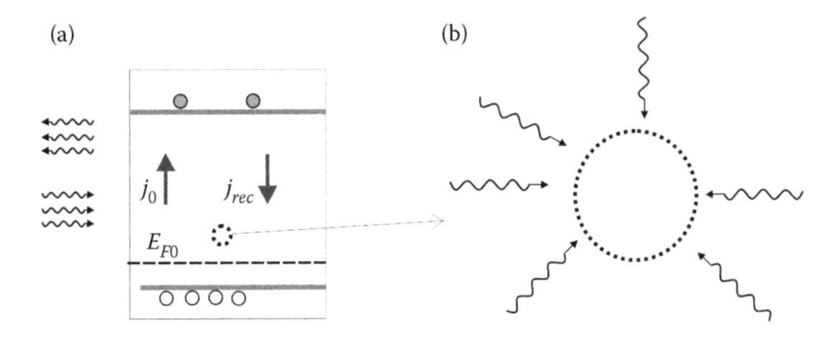

FIGURE 6.3 (a) Scheme of p-type semiconductor material in dark equilibrium, showing the balance of incoming and outgoing photons absorbed or produced in the material, and the equilibrium Fermi level E_{F0}. The generation by incoming thermal radiation produces a generation flux j_0/q and radiative recombination internal flux is j_{rec}/q. (b) In equilibrium, the volume element corresponding to a fragment of the semiconductor sees a black body radiation coming from its surroundings. The density of photons in the volume element results from two sources, the external radiation and the radiation from the other parts of the semiconductor.

$$U_{rec}(E)\,dE = B_{rad}(E)n_i^2\,dE \tag{6.21}$$

Equating the rates, we obtain

$$B_{rad}(E)n_i^2 = 4b_\pi n_r^2 \alpha(E)\frac{E^2}{e^{E/k_BT}-1} \tag{6.22}$$

Integrating over the entire energy spectrum, we arrive at the radiative recombination constant

$$B_{rad}n_i^2 = 4b_\pi n_r^2 \int_0^\infty \alpha(E)E^2 e^{-E/k_BT}\,dE \tag{6.23}$$

Here, we have assumed that $E \gg k_BT$ for the parts of the spectrum where the absorption coefficient has nonnegligible values. The determination of the recombination rate in Equation 6.23 is shown in Figure 6.4b for a perovskite $CH_3NH_3PbI_3$ film, as explained later in more detail.

We have concluded in Equation PSC.1.32 that the flux of photons emitted into a hemisphere by a blackbody is

$$\phi_{ph}^{bb,hemi}(E) = b_\pi \frac{E^2}{e^{E/k_BT}-1} \tag{6.24}$$

We may express Equation 6.22 more compactly

$$B_{rad}(E)n_i^2 = 4n_r^2\alpha(E)\phi_{ph}^{bb,hemi}(E) \tag{6.25}$$

6.3 RADIATIVE EMISSION OF A SEMICONDUCTOR LAYER

The result of Equation 6.22 has been achieved for a small fragment of semiconductor in equilibrium with its surrounding. However, in actual measurements, we deal with the absorption and emission of an entire semiconductor film, as indicated in Figure PSC.4.7a. We then need to extend

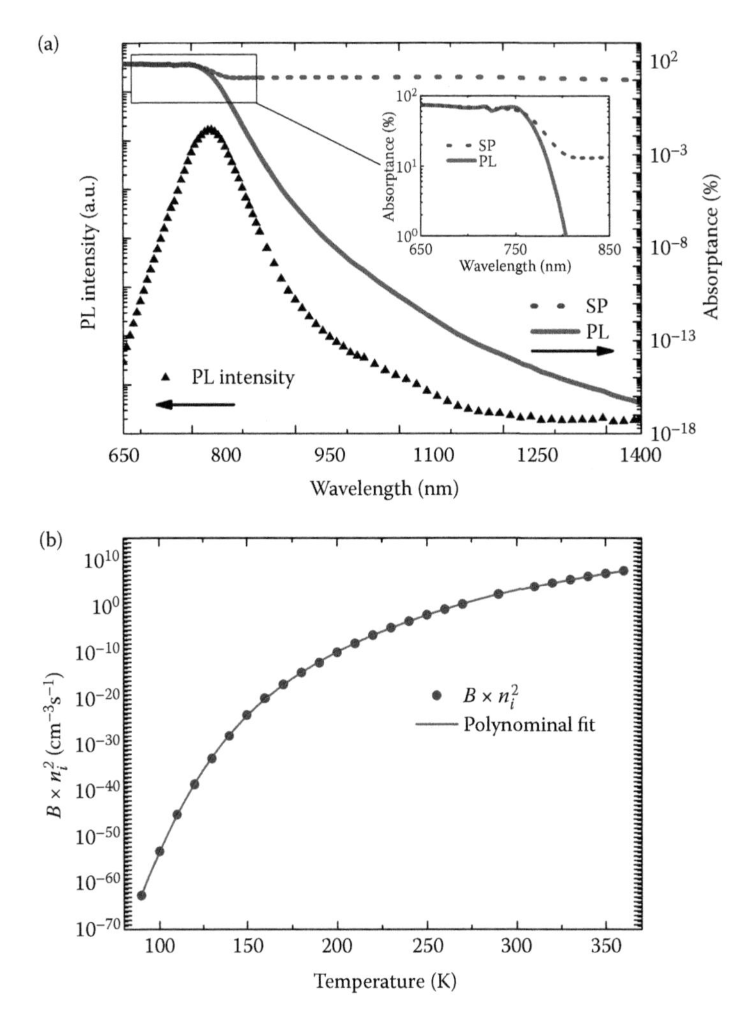

FIGURE 6.4 (a) PL spectra of 300 nm perovskite $CH_3NH_3PbI_3$ film (*left* y-axis). Spectrophotometer-measured absorptance and PL-extracted absorptance of $CH_3NH_3PbI_3$ as a function of wavelength (*right* y-axis). (b) Product of the radiative recombination coefficient and the square of the intrinsic carrier density $B_{rad} \times n_i^2$ plotted as a function of temperature. (Reproduced with permission from Barugkin, C. et al. *The Journal of Physical Chemistry Letters* 2015, 6, 767–772.)

the reciprocity reasoning to the light absorbed and emitted by the complete film or solar cell device. We recall that the quantity describing the amount of absorbed radiation is the absorptance $a(E)$, which depends not only on the absorption coefficient of the medium, but also on the macroscopic characteristics of the film such as the thickness and reflectivity of the surfaces, see Equation PSC.2.13.

In Section PSC.1.5, we have defined a blackbody as one that absorbs all incoming radiation such that the absorptivity is $a = 1$ at all photon energies. The radiation emitted by a blackbody is thermal radiation (at zero chemical potential) at the given temperature, Equation 6.24. In bodies that are not perfect absorbers, the rate of emission of radiant energy is reduced with respect to the radiation of a blackbody. The radiant power of a surface is described by the *emissivity* $\varepsilon_{em}(E)$. It is the spectral ratio of the energy or photon flux emitted by the body to that of a blackbody of a similar temperature:

$$\varepsilon_{em}(E) = \frac{\phi_E^{em}(E)}{\phi_E^{bb}(E)} = \frac{\phi_{ph}^{em}(E)}{\phi_{ph}^{bb}(E)} \tag{6.26}$$

Consider the solar cell to be in thermal equilibrium with its surroundings. Then, the fraction of the incident energy $\phi_E^{bb}(E)dE$ absorbed per unit area per unit time is given by

$$a(E)\phi_E^{bb}(E)dE \tag{6.27}$$

while the energy radiated per unit surface is

$$\phi_E^{em}(E)dE = \varepsilon_{em}(E)\phi_E^{bb}(E)dE \tag{6.28}$$

Therefore,

$$\varepsilon_{em}(E) = a(E) \tag{6.29}$$

This is Kirchoff's law, which states that the emissivity of a body equals its absorptivity. Thus, for a blackbody, $\varepsilon_{em} = 1$, implying that it is a perfect radiator as well as a perfect absorber. A more general proof of Kirchhoff's law can be established including directionality in a projected solid angle (or étendue ε) for any type of surface (Greffet and Nieto-Vesperinas, 1998). Thus, we have the relationship

$$\phi_{ph}^{em}(E,\varepsilon)d\varepsilon\, dE = a(E)\phi_{ph}^{bb}(E,\varepsilon)d\varepsilon\, dE \tag{6.30}$$

In order to discuss a particular and relevant form of a light-absorbing film in the photovoltaic device, we consider a planar geometry with lateral dimensions that are much larger than the thickness. At the front surface, we assume a perfect antireflecting coating that causes the zero reflectivity and at the rear surface a perfect reflecting coating that gives unity reflectivity. Under these conditions, absorptive and emissive fluxes stem only from the front surface of the solar cell and the étendue for both absorption of the blackbody radiation and emission from recombination is $\varepsilon = \pi$. From Equations PSC.1.35, 6.28, and 6.29, we obtain the expression of the spectral emission of the film

$$\phi_{ph}^{em}(E) = b_\pi a(E)\frac{E^2}{e^{E/k_BT} - 1} \tag{6.31}$$

This central result provides the basis for determining a number of photovoltaic properties as will be explained in the Chapters PSC.7 and PSC.8. Shockley and Queisser (1961) established a radiative balance that enables to predict the maximal efficiency of a solar cell based on a semiconductor with a sharp bandgap, as commented before. In a seminal work, Ross (1967) showed that the equilibrium between the distribution of photons and the recombination allows one to predict a photovoltage in terms of the quantum yield of any photochemical converter. Smestad and Ries (1992) developed this idea for the photovoltage of a solar cell. Lasher and Stern (1964), Ruppel and Wurfel (1980), and Würfel (1982) established the equilibrium for the external flux in a semiconductor layer. Rau (2007) and Kirchartz and Rau (2007) determined the influence of the internal properties of the solar cell such as charge collection efficiency on the reciprocity between LED and photovoltaic properties.

6.4 PHOTONS AT NONZERO CHEMICAL POTENTIAL

We have previously discussed in Section PSC.1.5 the thermalization of radiation in a *cavity*. This process leads to an equilibrium distribution determined by the temperature of the walls and the Planck spectrum, which yields the spectral photon flux in Equation PSC.1.35. When a semiconductor film characterized by the absorptance $a(E)$ interacts with the radiation in a cavity, a stationary situation is reached by reciprocal absorption and emission. The radiation field corresponds to an equilibrium spectral distribution according to the generalized Planck's law based on the occupation function of Equation PSC.1.13, consisting of photons at nonzero chemical and electrochemical potential $\eta_{ph} = \mu_{ph}$. The spectral photon flux, shown in Figure 6.5, is

$$\phi_{ph}^{bb}(E, \eta_{ph}) = b_\pi \frac{E^2}{e^{(E-\mu_{ph})/k_BT} - 1} \tag{6.32}$$

and the emitted photon flux has the form

$$\phi_{ph}^{em}(E) = b_\pi a(E) \frac{E^2}{e^{(E-\mu_{ph})/k_BT} - 1} \tag{6.33}$$

In the illuminated semiconductor, the electrochemical potential difference of electrons and holes is the net chemical potential μ_{np} as stated in Equation PSC.4.6 and shown in Figure PSC.4.7b. The interaction of the electron–hole gas and the photons can be viewed as a chemical reaction with a chemical equilibrium in which the chemical potential of the photons coincides with that of the electrons and holes, implying $\mu_{ph} = \mu_{np}$.

Under moderate illumination conditions, $\mu_{ph} = \mu_{np} < E_g$ is satisfied; therefore, the Wien approximation can be used in Equation 6.33 and the result is

$$\phi_{ph}^{em}(E, V) = b_\pi a(E) E^2 e^{-E/k_BT} e^{\mu_{np}/k_BT} \tag{6.34}$$

As an illustration to show the connection between light absorption and emission of a semiconductor film, we use the model of a sharp bandgap (Equation 6.2, Figure 6.6a). We observe that

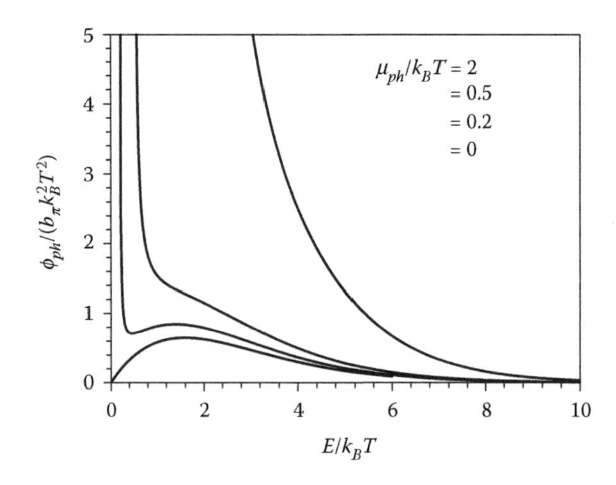

FIGURE 6.5 Dimensionless photon flux $\phi_{ph}^{bb}(E, \eta_{ph})/(b_\pi k_B^2 T^2)$ as a function of dimensionless energy E/k_BT, for the indicated values of the dimensionless photon chemical potential.

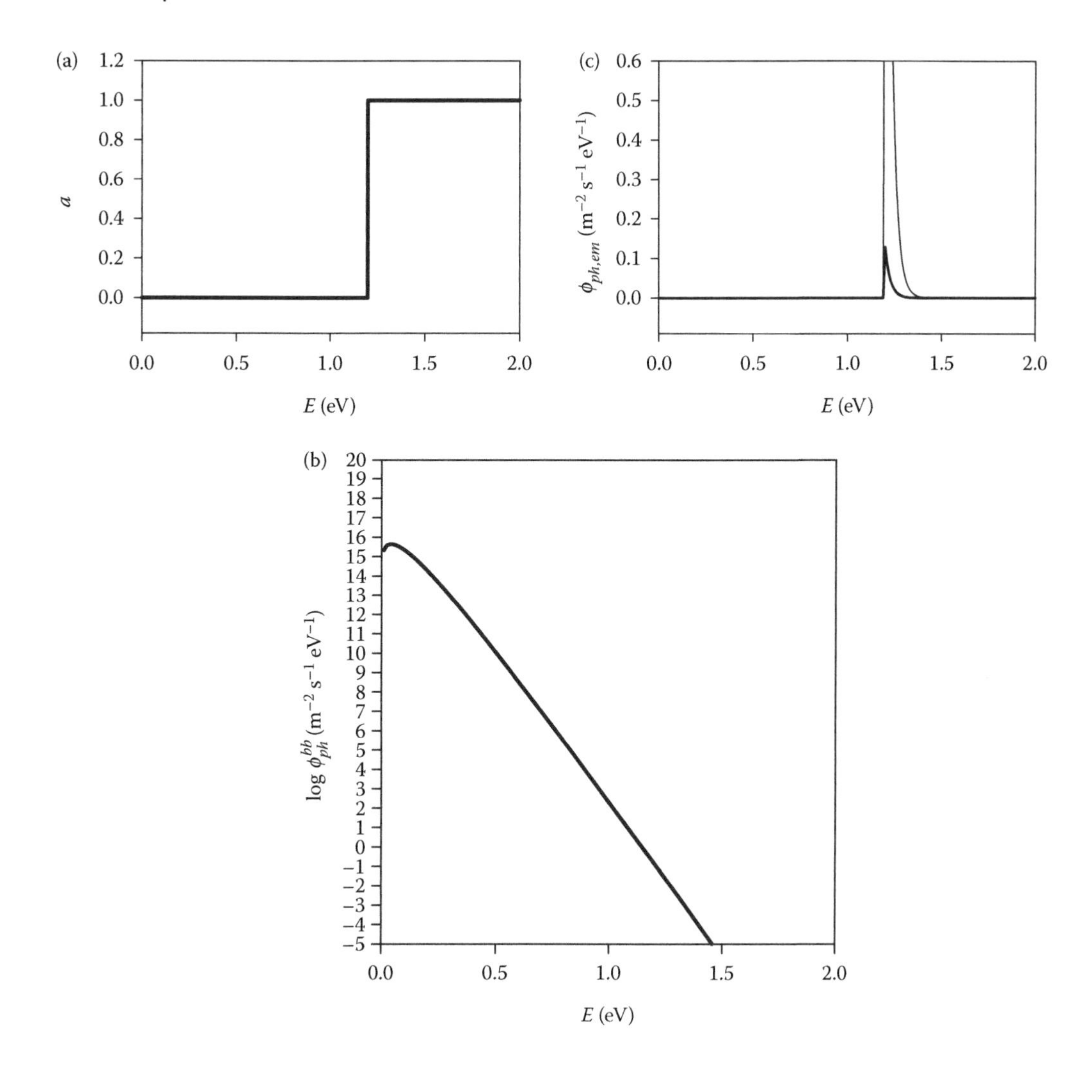

FIGURE 6.6 (a) Model semiconductor with sharp absorptance function at the bandgap value $E_g = 1.2$ eV. (b) The thermal radiation at $k_BT = 0.026$ eV received by the semiconductor. (c) The spectral light emission of the diode in thermal equilibrium (*thick line*) and at small applied voltage $qV = 3k_BT$ (*thin line*).

there is no subbandgap emission while for energies above the bandgap, the emission is given by the blackbody radiation shown in Figure 6.6b. The combination of both factors in Equation 6.34 implies the spectral shape of the emitted radiation field, as shown in Figure 6.6c. If the splitting of Fermi levels produces an internal voltage, then the photon density is increased, as shown in Figure PSC.1.8, and spectral emission is exponentially enhanced, as indicated in Figure 6.6c (Herrmann and Würfel, 2005).

The reciprocity between light absorption and emission in a semiconductor established in Equation 6.34 enables the determination of the absorption coefficient from PL. The method was originally established for silicon (Daub and Würfel, 1995; Trupke et al., 1998). This approach exclusively provides the band-to-band absorption coefficient and achieves enormous sensitivity in the spectral region where α is rather small as was mentioned in Section PSC.3.1. A model such as that indicated in Equation PSC.2.13 is used to determine the absorption coefficient taking into account the reflection at interfaces. As it is often difficult to calibrate the exact amount of photon emission, the method only gives relative values of the absorption coefficient as a function of photon energy. The procedure to overcome this limitation is to match the results from PL with the absolute

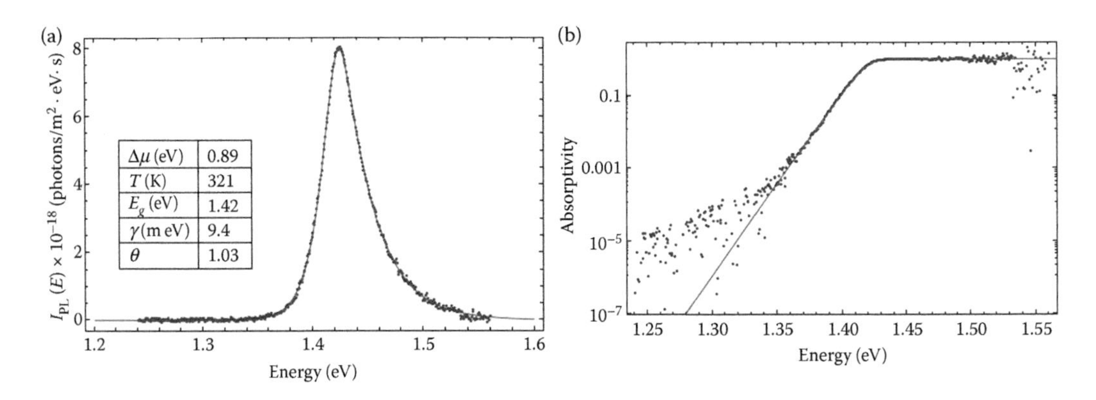

FIGURE 6.7 (a) PL (*points*) and full spectrum fit (*curve*) of p-GaAs. (b) Absorptivity extracted from the PL data (*points*). (Reproduced with permission from Katahara, J. K.; Hillhouse, H. W. *Journal of Applied Physics* 2014, 116, 173504.)

measurements obtained directly by light absorption at shorter wavelengths. An example for the lead halide perovskite is shown in Figure 6.4a, where the absorption obtained from PL is compared with the absolute values determined by De Wolf et al. (2014) and shown in Figure PSC.3.4. From the absorption coefficient, the total radiative recombination rate of Equation 6.23 is calculated, as shown in Figure 6.4b.

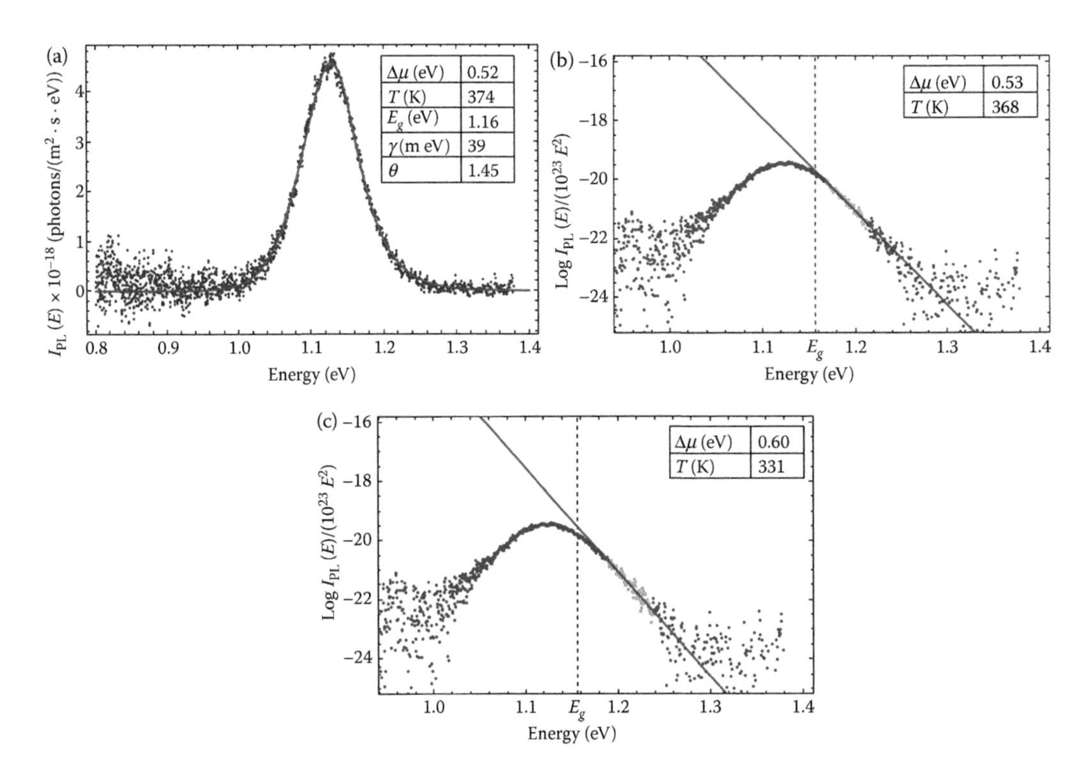

FIGURE 6.8 (a) Full-spectrum fitting of PL from CIGSe device. (b) High-energy tail-only fitting (*red line*) over the region 1.17–1.21 eV (*green points*) assuming that $a = 1$ above bandgap. (c) High-energy tail-only fitting (*red line*) over the region 1.19–1.24 eV (*green points*) and assuming that $a = 1$ above bandgap. (Reproduced with permission from Katahara, J. K.; Hillhouse, H. W. *Journal of Applied Physics* 2014, 116, 173504.)

The reciprocity relation (6.34) can be exploited in another direction. Based on the measurement of the absorption coefficient, it is possible to accurately predict the internal splitting of the quasi-Fermi levels (photovoltage) in the semiconductor (Bauer and Gütay, 2007). The transition region of the absorptance function plays a dominant role for the accurate application of Equation 6.34, due to the fact that the term e^{-E/k_BT} grows very rapidly as the energy decreases. In particular, subbandgap states with significant absorption cross sections cause a large contribution to the emission spectrum (Katahara and Hillhouse, 2014). One example for the radiative emission of a silicon diode with a textured surface is shown in Figure PSC.2.2b. The light emission matches well with the blackbody spectrum for wavelengths shorter than the bandgap (at 1102 nm), but rather than abruptly terminating it, the luminescence continues at longer wavelengths. Additional examples are indicated in Figures 6.4 and 6.7.

A simple procedure to determine the photovoltage from Equation 6.34 is to fit only the high-frequency part of the spectrum, where it can be assumed that $a = 1$ so that

$$\phi_{ph}^{em}(E,V) = b_\pi E^2 e^{-E/k_BT} e^{\mu_{np}/k_BT} \tag{6.35}$$

Here, the PL is dominated by the exponential function commented previously in Figure PSC.1.8. However, this method can be applied only if a full exponential dependence is experimentally observed, as shown in Figure 6.8.

GENERAL REFERENCES

Kirchoff's law: Landsberg (1978) and Howell et al. (2010).

Photons at nonzero chemical potential: Landsberg (1981), Würfel (1982), Ries and McEvoy (1991), and Herrmann and Würfel (2005).

Determination of Fermi level splitting from PL: Bauer and Gütay (2007), Unold and Gütay (2011), Katahara and Hillhouse (2014), and El-Hajje et al. (2016).

REFERENCES

Barugkin, C.; Cong, J.; Duong, T.; Rahman, S.; Nguyen, H. T.; Macdonald, D.; White, T. P.; Catchpole, K. R. Ultralow absorption coefficient and temperature dependence of radiative recombination of $CH_3NH_3PbI_3$ perovskite from photoluminescence. *The Journal of Physical Chemistry Letters* 2015, 6, 767–772.

Bauer, G. H.; Gütay, L. Analyses of local open circuit voltages in polycrystalline Cu(In,Ga)Se_2 thin film solar cell absorbers on the micrometer scale by confocal luminescence. *CHIMIA International Journal for Chemistry* 2007, 61, 801–805.

Daub, E.; Würfel, P. Ultralow values of the absorption coefficient of Si obtained from luminescence. *Physical Review Letters* 1995, 74, 1020–1023.

De Wolf, S.; Holovsky, J.; Moon, S.-J.; Löper, P.; Niesen, B.; Ledinsky, M.; Haug, F.-J.; Yum, J.-H.; Ballif, C. Organometallic halide perovskites: Sharp optical absorption edge and its relation to photovoltaic performance. *The Journal of Physical Chemistry Letters* 2014, 5, 1035–1039.

El-Hajje, G.; Momblona, C.; Gil-Escrig, L.; Avila, J.; Guillemot, T.; Guillemoles, J.-F.; Sessolo, M.; Bolink, H. J.; Lombez, L. Quantification of spatial inhomogeneity in perovskite solar cells by hyperspectral luminescence imaging. *Energy & Environmental Science* 2016, 9, 2286–2294.

Green, M. A. Analytical treatment of Trivich–Flinn and Shockley–Queisser photovoltaic efficiency limits using polylogarithms. *Progress in Photovoltaics: Research and Applications* 2012, 20, 127–134.

Greffet, J.; Nieto-Vesperinas, M. Field theory for generalized bidirectional reflectivity: Derivation of Helmholtz's reciprocity principle and Kirchhoff's law. *Journal of the Optical Society of America* 1998, 15, 2735–2744.

Herrmann, F.; Würfel, P. Light with nonzero chemical potential. *American Journal of Physics* 2005, 73, 717–720.

Howell, J. R.; Siegel, R.; Menguc, M. P. *Thermal Radiation Heat Transfer*; CRC Press: New York, 2010.

Katahara, J. K.; Hillhouse, H. W. Quasi-Fermi level splitting and sub-bandgap absorptivity from semiconductor photoluminescence. *Journal of Applied Physics* 2014, 116, 173504.

Kirchartz, T.; Rau, U. Electroluminescence analysis of high efficiency Cu(In,Ga)Se_2 solar cells. *Journal of Applied Physics* 2007, 102, 104510.

Landsberg, P. T. *Thermodynamics and Statistical Mechanics*; Dover: New York, 1978.

Landsberg, P. T. Photons at non-zero chemical potential. *Journal of Physics C* 1981, 14, L1025.

Lasher, G.; Stern, F. Spontaneous and stimulated recombination radiation in semiconductors. *Physical Review* 1964, 133, A553–A563.

Pankove, J. I. *Optical Processes in Semiconductors*; Prentice-Hall: Englewood Cliffs, NJ, 1971.

Rau, U. Reciprocity relation between photovoltaic quantum efficiency and electroluminescent emission of solar cells. *Physical Review B* 2007, 76, 085303.

Ries, H.; McEvoy, A. J. Chemical potential and temperature of light. *Journal of Photochemistry and Photobiology A: Chemistry* 1991, 59, 11–18.

Ross, R. T. Some thermodynamics of photochemical systems. *The Journal of Chemical Physics* 1967, 46, 4590–4593.

Ruppel, W.; Wurfel, P. Upper limit for the conversion of solar energy. *IEEE Transactions on Electron Devices* 1980, 27, 877–882.

Sambur, J. B.; Novet, T.; Parkinson, B. A. Multiple exciton collection in a sensitized photovoltaic system. *Science* 2010, 330, 63–66.

Semonin, O. E.; Luther, J. M.; Choi, S.; Chen, H.-Y.; Gao, J.; Nozik, A. J.; Beard, M. C. Peak external photocurrent quantum efficiency exceeding 100% via MEG in a quantum dot solar cell. *Science* 2011, 334, 1530–1533.

Shockley, W.; Queisser, H. J. Detailed balance limit of efficiency of p-n junction solar cells. *Journal of Applied Physics* 1961, 32, 510–520.

Smestad, G.; Ries, H. Luminescence and current-voltage characteristics of solar cells and optoelectronic devices. *Solar Energy Materials and Solar Cells* 1992, 25, 51–71.

Trupke, T.; Daub, E.; Würfel, P. Absorptivity of silicon solar cells obtained from luminescence. *Solar Energy Materials and Solar Cells* 1998, 53, 103–114.

Unold, T.; Gütay, L. Photoluminescence analysis of thin-film solar cells. In *Advanced Characterization Techniques for Thin Film Solar Cells*; Abou-Ras, D., Kirchartz, T., Rau, U. (Eds.); Wiley-VCH Verlag: Berlin, 2011; pp 151–175.

van Roosbroeck, W.; Shockley, W. Photon-radiative recombination of electrons and holes in germanium. *Physical Review* 1954, 94, 1558–1560.

Würfel, P. The chemical potential of radiation. *Journal of Physics C* 1982, 15, 3967–3985.

7 Reciprocity Relations and the Photovoltage

The connection between incoming and outgoing radiation, complemented by quantum yields for carrier collection and light emission, provides important relationships between photovoltaic and LED operation modes of a semiconductor diode that can be checked experimentally. This approach introduces fundamental aspects of solar cell operation, which serve as a benchmark for the evaluation of the quality of a class of devices. We develop a detailed study of the photovoltage that can be obtained in a solar cell, with particular attention to the maximal photovoltage in the limit of pure radiative recombination, and we investigate the causes for a reduced photovoltage. For cells that operate close to the radiative recombination limit, the management of photons plays an important role in the performance, demanding control over the photon reabsorption and emission processes in the set of phenomena termed as photon recycling. We also describe the physical constraints of the phenomenon of luminescent refrigeration based on the fact that the radiative emission of a biased semiconductor can be applied to remove heat in the form of photons. Finally, we discuss the application of the reciprocity relation on an organic solar cell where light absorption is associated with the CT between different materials.

7.1 THE RECIPROCITY BETWEEN LED AND PHOTOVOLTAIC PERFORMANCE PARAMETERS

In Chapters PSC.5 and PSC.6, it has been explained that the radiative saturation current density $j_{0,rad}$ describes the minimum recombination situation, and hence contains the main information about the optimal performance of a class of solar cells. We have arrived at important results using arguments of reciprocity of light absorption and emission, and we now progress toward the determination of $j_{0,rad}$ by the same type of reasoning, which will allow us to calculate central performance features, such as the maximal open-circuit voltage in the radiative limit.

In Chapter PSC.6, we derived the rate of absorption and emission of radiation in a semiconductor film based on the measurement of the spectral absorptance $a(E)$, Equation PSC.6.31. This result, which was obtained from the analysis of a semiconductor film as represented in Figure PSC.4.7, does not yet take into account all material properties of a solar cell concerning the charge collection limitations. For the production of electricity, the solar cell contains selective contacts that make the conversion between a separation of Fermi levels caused by the radiation field, and the external voltage and electrical current. This property goes beyond the equilibrium of incoming and outgoing photons. There exists a connection between charge injection by the absorption of a photon and charge collection properties at the outer contacts, as shown in Figure 7.1. A photon, absorbed at a point x inside the semiconductor absorber, generates an electron–hole pair that requires suitable processes of transport and (sufficiently slow) recombination to produce a certain voltage at the contacts. We recall from Equation PSC.2.28 that the photovoltaic external quantum efficiency (i.e., that of the diode operated as a solar cell) is $EQE_{PV}(E) = a(E)IQE_{PV}(E)$, where the internal quantum efficiency incorporates the limitations of separation and collection of photogenerated charge, which will be studied in detail in Chapter PSC.10.

In the other direction, when the diode is operated as an LED, application of voltage to the selective contacts must split the Fermi levels so that the carrier concentration is increased at the internal point x. Physically, there is an equivalency between the two processes. It is thus necessary to extend Equation PSC.6.31 to include the effect of IQE_{PV}, related to the formation of gradients of the Fermi

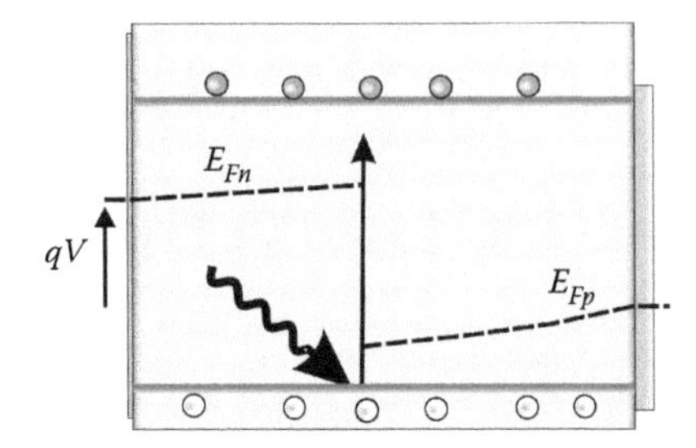

FIGURE 7.1 Scheme of p-type semiconductor material with selective contacts to electrons and holes, a model of a solar cell. The scheme shows the effect of absorption of a photon at a point x in creating a voltage in the external circuit. The Fermi levels are tilted due to limitations of charge collection that require a driving force for the extraction of carriers. Note that the diagram is illustrative of the reciprocity effect; however, the Fermi levels are ensemble properties not associated with a single electron–hole pair.

level in Figure 7.1. The extension is given by the following expression that relates the spectral radiative emission of a solar cell to the blackbody radiation field via the photovoltaic EQE_{PV}:

$$\phi_{ph}^{em}(E) = EQE_{PV}(E)\phi_{ph}^{bb}(E)e^{qV/k_BT} \tag{7.1}$$

This result has been shown by Rau (2007) under certain restrictive conditions. The exponent in Equation 7.1 takes into account the enhanced recombination by the applied voltage V, as indicated in Equation PSC.6.34.

The application of the reciprocity relation (7.1) in a solar cell, in order to obtain the light emission of the device, is not immediately obvious. The conditions in which the reciprocal relation between internal concentration and voltage at the contacts is strictly satisfied have been discussed by Kirchartz and Rau (2008) and Kirchartz et al. (2016). The reciprocity is directly justified by the Donolato theorem that relates carrier collection and generation in a solar cell, see Section PSC.10.5. However, the latter result only applies when recombination is linear in minority carrier density and when drift-currents can be neglected. This fails under high-injection conditions (because recombination becomes nonlinear in charge carrier density and minority carriers are no longer well defined), or when a substantial part of the absorber volume is depleted implying that the diffusion-only approximation for transport is violated. Therefore, the exact application of Rau's relationship needs to be carefully investigated for the measurement of external radiative efficiency (Wang and Lundstrom, 2013).

Since the light emission in a solar cell occurs in a restricted energy interval at $E \approx E_g \gg k_BT$, we can use Equation PSC.1.34 to describe the thermal radiation field and Equation 7.1 can be simplified as

$$\phi_{ph}^{em}(E,V) = b_\pi EQE_{PV}(E)E^2e^{-E/k_BT}e^{qV/k_BT} \tag{7.2}$$

with $b_\pi = 2\pi/h^3c^2$ as defined in Chapter PSC.1. This formula is the generalization of Equation PSC.6.34 for operating photovoltaic devices. A calculation of the radiative properties of different classes of solar cells using the EQE_{PV} based on Equation 7.2 is shown in Figure 7.2.

In order to derive the reverse saturation current of the solar cell, we integrate Equation 7.2 over all photon energies. The radiative saturation current for emission of the planar device to a hemisphere is

$$j_{0,rad} = qb_\pi \int_0^\infty EQE_{PV}(E)E^2e^{-E/k_BT}dE \tag{7.3}$$

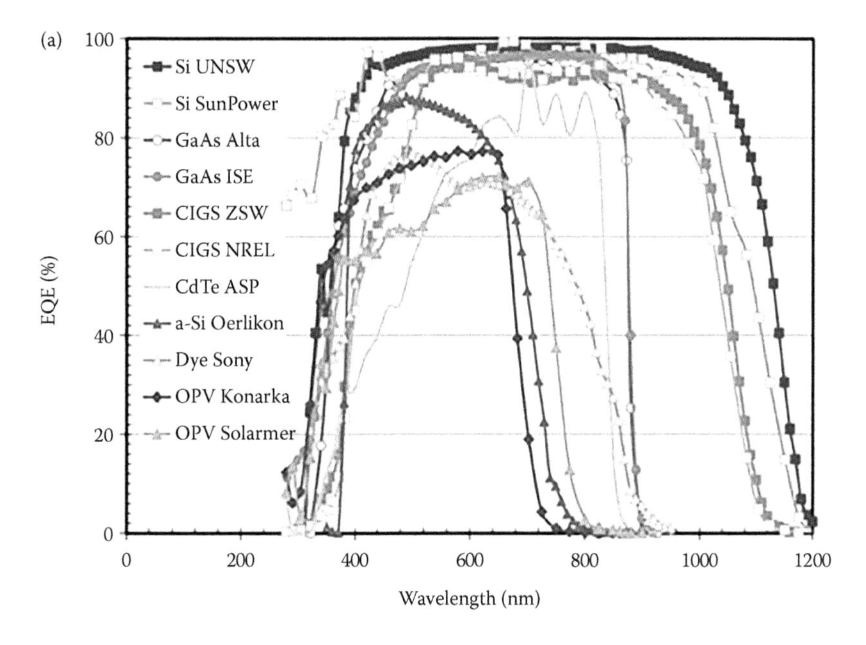

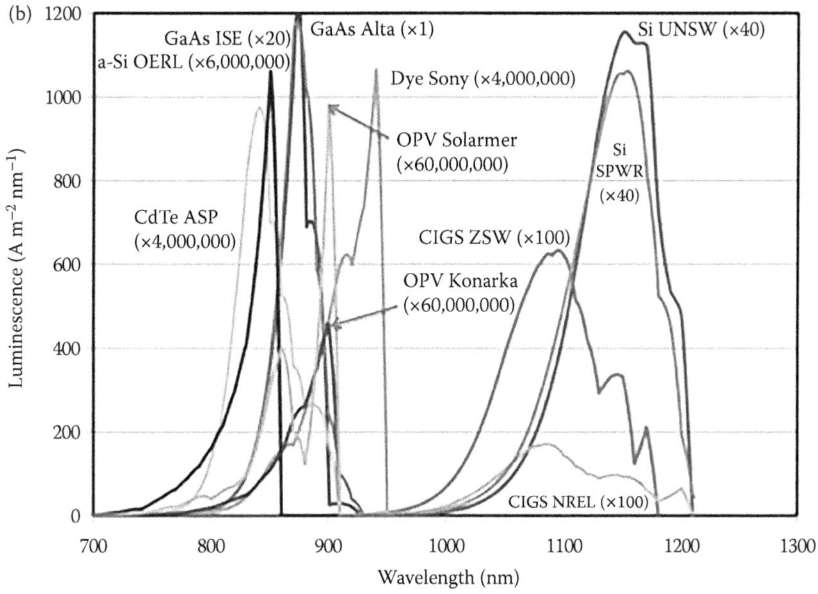

FIGURE 7.2 (a) Measured EQE_{PV} for different types of solar cells. (b) The calculated spectral luminescence by application of Equation 7.1, $\phi_{ph}^{em}(E) = EQE_{PV}(E)\phi_{ph}^{bb}(E)e^{qV/k_BT}$. (Reproduced with permission from Green, M. A. *Progress in Photovoltaics: Research and Applications* 2012, 20, 472–476.)

This is the central parameter for evaluation of the materials and device performance, as stated in Equation PSC.5.4.

We consider some simplified versions of Equation 7.3. For the ideal model of a semiconductor with a sharp absorption edge and $IQE_{PV} = 1$, we obtain from Equation PSC.6.7 and Equation 7.3

$$j_{0,rad} = q\frac{\varepsilon_{out}}{\pi} b_\pi k_B T \Gamma_{ph}(E_g, T) e^{-E_g/k_BT} \tag{7.4}$$

For $E_g \gg k_BT$, we can simplify $\Gamma_{ph}(E, T) \approx E^2$, see Equation PSC.6.6, and in the case $\varepsilon_{out} = \pi$, we have the value

$$j_{0,rad} = qb_{\pi}k_BTE_g{}^2e^{-E_g/k_BT} \tag{7.5}$$

Application of Equation 7.3 to calculate $j_{0,rad}$ in experimental conditions requires an accurate measurement of the EQE_{PV} over a wide spectral band and especially at subbandgap values, whose contribution becomes dominant due to the rapid increase of the exponential in Equation 7.3 toward low energies, as remarked earlier. High-sensitivity EQE_{PV} can be obtained with Fourier-transform based photocurrent spectroscopy (FTPS) technique (Vandewal et al., 2009; Tress et al., 2015; Müller and Kirchartz, 2016).

The reciprocal pathway, that is to obtain the EQE_{PV} starting from a measurement of the LED emission of the solar cell is also possible. One can determine the low values of $EQE_{PV}(E)$ far from the major absorption onset using a calibration method. From EL data at voltage V, the quantum efficiency denoted EQE_{PV-EL} can be obtained as follows:

$$EQE_{PV-EL}(E) = f_g\, \phi_{ph}^{em}(E)E^{-2}e^{E/k_BT}e^{-qV/k_BT} \tag{7.6}$$

This last result provides the spectral shape but not the absolute values of $EQE_{PV}(E)$. However, the normalizing factor f_g may be adjusted from one point of the directly measured EQE_{PV}, in order to establish that the shapes of EQE_{PV-EL} and EQE_{PV} are equal.

The match between EQE_{PV-EL} and EQE_{PV} is illustrated for a number of classes of solar cells: Figure 7.3 for dye-sensitized solar cell, Figure 7.4 for a Cu(In,Ga)Se_2 solar cell, and Figure 7.5 for a perovskite solar cell.

In the experimental analysis of photovoltaic devices, the recombination parameter must take into account all recombination pathways as outlined in Figure PSC.5.2. By Equation PSC.5.5, the actual diode saturation current takes the form

$$j_0 = \frac{1}{EQE_{LED}} qb_{\pi} \int_0^{\infty} EQE_{PV}(E)E^2e^{-E/k_BT}dE \tag{7.7}$$

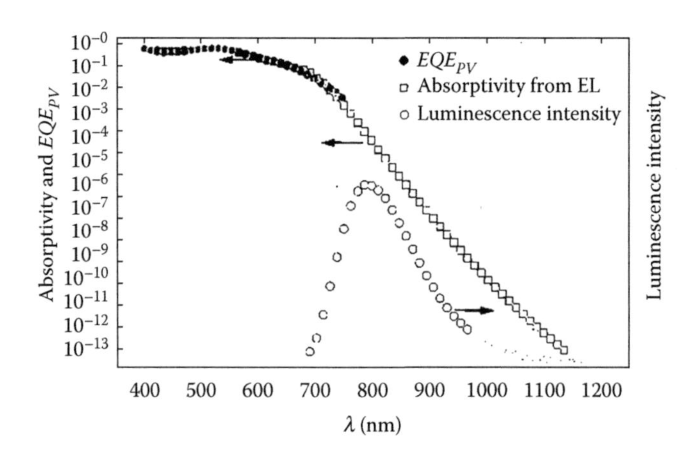

FIGURE 7.3 EL measured on a dye-sensitized solar cell and the absorptance that is calculated by its division by E^2e^{-E/k_BT} as in Equation 7.6. The obtained relative values for the absorptance have been fitted to the measured EQE_{PV} (*full circles*). (Reproduced with permission from Trupke, T. et al. *The Journal of Physical Chemistry B* 1999, 103, 1905–1910.)

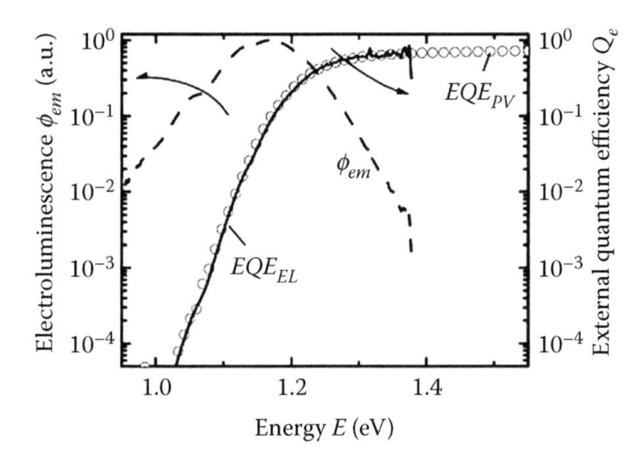

FIGURE 7.4 Measurement of an EL spectrum $\phi_{ph}^{em}(E)$ of a $Cu(In,Ga)Se_2$ solar cell, which allows us to compute the solar cell quantum efficiency EQE_{PV-EL}. Thus, EQE_{PV-EL} is derived from the EL measurement in relative units. A direct measurement of the quantum efficiency EQE_{PV} allows one to scale EQE_{PV-EL} and to verify that the relative shape of EQE_{PV-EL} and EQE_{PV} is equal. (Reproduced with permission from Kirchartz, T.; Rau, U. *Physica Status Solidi (A)* 2008, 205, 2737–2751.)

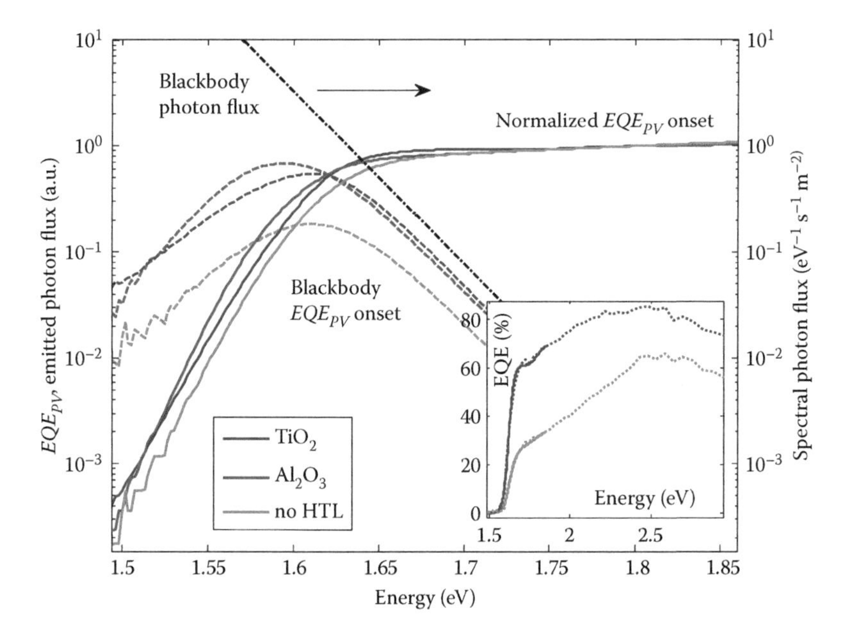

FIGURE 7.5 Analysis of the open-circuit voltage of $CH_3NH_3PbI_3$ perovskite solar cells. Onset of the incident-photon-to-electron-conversion efficiency (EQE_{PV}) measured with Fourier-transform photocurrent spectroscopy and normalized at 1.8 eV (*lines*). Data shown for devices with TiO_2 or Al_2O_3 scaffold and for one device without hole-transport layer. Also shown is the calculated emitted spectral photon flux when the device is in equilibrium with the blackbody radiation of the surroundings (*dash-dotted*). The inset compares the EQE_{PV} onset from FTPS with the overall EQE_{PV} measured with monochromatic light and a lock-in amplifier. The EQE_{PV} of the device without HTL is lower due to optical and electrical losses, whereas the EQE_{PV} for Al_2O_3 and TiO_2 devices with HTL is comparable. (Reproduced with permission from Tress, W.; Marinova, N. et al. *Advanced Energy Materials* 2015, 5, 10.1002/aenm.201400812.)

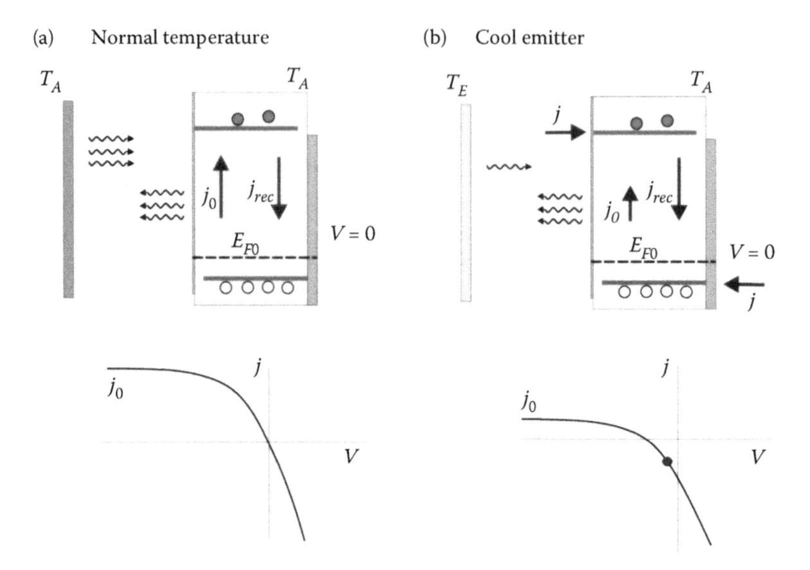

FIGURE 7.6 "Negative illumination" of a semiconductor diode. (a) The solar cell is at the same temperature T_A as the surrounding thermal radiation. We show the equilibrium in the dark of carrier generation by blackbody thermal radiation and recombination, which balances to net zero current at short circuit. At reverse voltage, recombination is suppressed and the diode saturation current j_0 corresponds to the extraction of the carriers generated by the thermal radiation. (b) The solar cell is at the ambient temperature T_A but now it receives thermal radiation from a colder source at T_E. The generation rate that determines j_0 decreases, while the recombination rate remains the same. Hence, a negative current occurs at short circuit, and power can be produced by operating the cell at negative voltage as shown by a point in the current density–voltage characteristic.

where EQE_{LED} is the external quantum efficiency of an LED, defined in Equation PSC.2.23. Note that EQE_{LED} is a scalar quantity unlike the solar cell EQE_{PV}, which is a function of wavelength or photon energy.

So far, it was naturally assumed that the environment provides a steady background of the Planckian radiation at the same temperature of the solar cell, T_A. We now consider a different type of environment in which the solar cell device operates in an unconventional way.

From the property of equilibrium of outgoing and incoming radiation in dark equilibrium, we derived important relationships such as PSC.5.4 and PSC.6.31. However, by setting the thermal radiation coming from an emitter at a *lower* temperature than that of the solar cell device, $T_E < T_A$, an imbalance of incoming and outgoing radiation can be forced, as shown in Figure 7.6. Thus, the parameter of thermal generation is reduced and Equation PSC.5.4 is displaced from equilibrium as $j_0^{th} < j_{rec}(V = 0)$. Since the solar cell is still emitting radiation at T_A, in this situation of "negative radiation," we obtain at short circuit an observable negative current in the dark (Santhanam and Fan, 2016). The diode can be operated at reverse bias resulting in thermal-to-electrical energy conversion.

7.2 FACTORS DETERMINING THE PHOTOVOLTAGE

We now focus our attention on the open-circuit voltage V_{oc}, which is the voltage produced autonomously by the solar cell when it is irradiated with light. The open-circuit voltage is rather easy to measure and it is a main performance parameter of a solar cell. In Section PSC.6.1, we have discussed the properties of the photocurrent. This is a straightforward quantity, in that the absorbed photons are converted to electron–hole pairs, which are measured as an electrical current j_{ph} at short circuit, as stated in Equation PSC.6.1. In contrast to this, the analysis of the photovoltage involves subtle aspects of the operation of the solar cell, which we treat in the following sections.

In the present analysis, we continue to use the reference model of Figure PSC.5.1 in which the solar cell is composed of a single homogeneous light-absorbing material with the property of very high-carrier mobilities so that the application of either external illumination or bias voltage produces uniformly the flat Fermi levels throughout the absorber. The solar cell possesses ideal selective contacts that transform the splitting of the Fermi levels to the voltage as Equation PSC.4.7. The physical and materials properties of the contacts can impose further limitations to the photovoltage in the solar cell, which are ignored in this model. The band bending, surface recombination, and other usual features that occur during operation will be discussed in Chapter PSC.10.

We consider the solar cell to be illuminated by a light source in open-circuit conditions, as shown in Figure PSC.5.2. The optical radiation over the background thermal radiation creates excess carriers in the transport bands of the semiconductor. Since no carriers are extracted at the contacts, the balance equation reads

$$j_0 + j_{ph} = j_{rec} \tag{7.8}$$

At the left-hand side of Equation 7.8, we count the generation fluxes from the thermal and external sources, and at the right-hand side, we have the internal recombination current, defined in Section PSC.5.2, which removes the photogenerated electrons and holes.

Under standard conditions, the solar cell receives solar $AM1.5G$ radiation as explained in Section PSC.1.8, denoted $\phi_{ph}^{AM1.5}(E)$, with an integrated photon flux $\Phi_{ph}^{AM1.5}$ given by Equation PSC.1.48. In the following analysis, we will be mainly interested in the case in which a substantial illumination with a power flux of the order of 1 sun impacts the solar cell. Therefore, we will label the generation flux in terms of a photocurrent that clearly offsets the thermal generation, $j_{ph}^{sun} \gg j_0$, and the latter will be neglected. The recombination current at open circuit can be obtained from Equation PSC.5.8 and the balance equation becomes

$$j_{ph}^{sun} = j_0 e^{qV_{oc}/mk_BT} \tag{7.9}$$

Therefore, we get the result

$$V_{oc} = \frac{mk_BT}{q} \ln\left(\frac{j_{ph}^{sun}}{j_0}\right) \tag{7.10}$$

Clearly, to obtain a high photovoltage, it is necessary to achieve a large separation of electron and hole Fermi levels, which means that the carrier densities have to be as large as possible, as suggested in Figure PSC.5.2. Therefore, it is obvious that recombination is the main process limiting the photovoltage in a solar cell. According to Equation 7.10, the photovoltage is controlled only by the recombination parameter j_0, given the photogeneration rate determined by j_{ph}^{sun}.

For the evaluation of solar cell materials, it is very interesting to establish the largest V_{oc} that can be possibly obtained in a class of materials. The maximal open-circuit voltage is reached when the recombination parameter j_0 takes the minimal value, with all nonradiative pathways being suppressed. This situation is reached exactly if the saturation current density j_0 is equal to the radiative saturation current density $j_{0,rad}$ defined in Equation 7.3. Assuming an ideality factor $m = 1$, the *radiative limit* to the photovoltage is

$$V_{oc}^{rad} = \frac{k_BT}{q} \ln\left(\frac{j_{ph}^{sun}}{j_{0,rad}}\right) \tag{7.11}$$

As mentioned earlier, Equation 7.3 is the critical parameter to establish the radiative voltage.

In the preliminary arguments about the solar cell efficiency presented in Section PSC.6.1, we regarded the output electrical power as the product of the photocurrent and the voltage corresponding to bandgap energy. However, a better estimation of the extracted electrical power is given by the voltage that carriers can produce when they are accumulated in the solar cell, which is the V_{oc}. Therefore, it is central to the discussion of power conversion efficiency to obtain the difference between qV_{oc} and E_g, whose magnitude causes a significant loss of extracted power. We explore now the main physical effects governing the radiative photovoltage of a solar cell.

We first analyze the value of the radiative voltage under the assumption of a planar cell with an opaque backside, sharp bandgap, and excellent charge collection ($IQE = 1$). The sun is modeled as a blackbody at high temperature T_S. The current density has been previously described in Equation PSC.6.7, and the saturation current at the cell temperature T_A is given in Equation 7.4. Substituting in Equation 7.11, we obtain the following equation, where we have expressed the radiative voltage in successive terms that reduce the carrier extraction energy with respect to the semiconductor bandgap:

$$qV_{oc}^{rad} = E_g - \frac{T_A}{T_S}E_g - k_BT_A\ln\left(\frac{\varepsilon_{out}}{\varepsilon_{in}}\right) + k_BT_A\ln\left(\frac{T_S\Gamma(E_g,T_S)}{T_A\Gamma(E_g,T_A)}\right) \tag{7.12}$$

The first term of voltage reduction is the decrease of energy due to the Carnot factor $(1 - T_A/T_S)$, accounting for the equilibration of entropy necessary for extraction of work in a thermal machine, as discussed in Section ECK.5.9. The second loss term in Equation 7.12 represents entropy generation by the occupation of available states associated with expansion of the étendue of the incoming and outgoing photon fluxes. Note that the entropy loss can be eliminated by a concentrator that makes the solid angle of escape of the light to be the same as that of the light arriving to the device. The third term describes an *increase* of free energy per carrier as a result of the temperature of the absorbed and emitted photon distributions, see Markvart (2007) and Hirst and Ekins-Daukes (2011) for further details. These components of the voltage loss are shown in Figure 7.7a. The following approximation assuming $E_g \gg k_BT_S(\Gamma \approx E_g^2)$ is widely used (Ruppel and Wurfel, 1980).

$$qV_{oc}^{rad} = E_g\left(1 - \frac{T_A}{T_S}\right) + k_BT_A\ln\left(\frac{\varepsilon_S}{\pi}\right) + k_BT_A\ln\left(\frac{T_S}{T_A}\right) \tag{7.13}$$

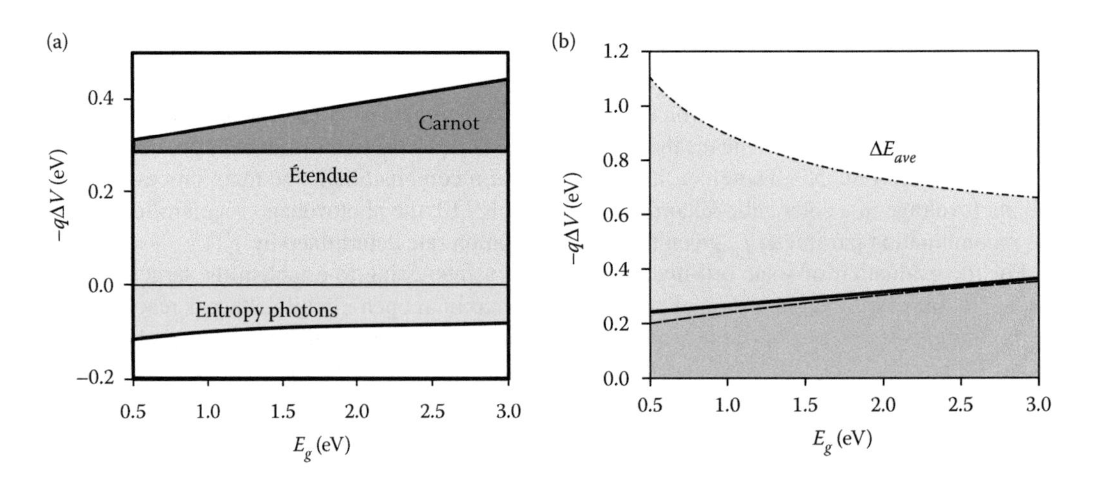

FIGURE 7.7 (a) Components of the thermodynamic loss in the radiative photovoltage for a solar cell that absorbs blackbody radiation at $T_S = 5800$ K, $T_A = 300$ K, according to Equation 7.13, using $\varepsilon_{in} = \varepsilon_S^{1.5AM} = 4.8\times10^{-5}$, $\varepsilon_{out} = \pi$. (b) The total voltage loss (*thick line*). The dashed line is the formula $qV_{oc}^{rad} = E_g(1 - T_A/T_S) + k_BT_A\ln(\varepsilon_S/\pi) - k_BT_A\ln(T_A/T_S)$. The *dashed-dot* line is the average energy of absorbed photons.

This approximation to the radiative losses is shown in Figure 7.7b, and this is the expression used for the radiative photovoltage in Figure PSC.6.2b. The average energy of absorbed photons is (Würfel, 2009)

$$\bar{E}(E_g,T_S)=\frac{\Phi_E(E_g,T_S)}{\Phi_{ph}(E_g,T_S)}=\frac{\chi_E(E_g,T_S)}{\Gamma_{ph}(E_g,T_S)} \tag{7.14}$$

In order to obtain more explicit expressions related to measured quantities, we write Equation 7.11 as

$$qV_{oc}^{rad}=k_BT\ln\left(j_{ph}^{sun}\right)-k_BT\ln\left(qb_\pi\int_0^\infty a(E)E^2e^{-E/k_BT}dE\right) \tag{7.15}$$

For a solar cell with a sharp bandgap and constant absorptivity a, we have from Equation PSC.6.5

$$qV_{oc}^{rad}=E_g+k_BT\ln(j_{ph}^{sun})-k_BT\ln\left(qb_\pi ak_BT\,E_g^{\,2}\right) \tag{7.16}$$

We remark that the V_{oc}^{rad} shows the following general properties:

1. It directly correlates with the value of the semiconductor bandgap.
2. It increases logarithmically with the photogeneration rate.
3. It decreases logarithmically with the recombination rate, as indicated by the negative dependence on absorptance parameter a in Equation 7.16. The reason for this is that higher absorption implies a larger rate of radiative emission that causes the loss of carriers. However, a higher absorption will produce a larger j_{ph} that offsets by far the loss of V_{oc} in the calculation of the efficiency.

In general, if the $EQE_{PV}(E)$ is known over a broad spectral range for a solar cell device, the radiative voltage of a device in standard calibration conditions can be obtained as follows:

$$V_{oc}^{rad}=\frac{k_BT}{q}\ln\left(\frac{\int_0^\infty EQE_{PV}(E)\phi_{ph}^{AM1.5}(E)dE}{\int_0^\infty EQE_{PV}(E)\phi_{ph}^{bb}(E)dE}\right) \tag{7.17}$$

A similar procedure is used based on the absorptance $a(E)$ for deriving the intrinsic limitations to the photovoltage of a contactless material.

So far, we have described the photovoltage corresponding to a high-performance solar cell for a material specified by its charge collection properties, in which charge generation is balanced exclusively by radiative recombination. However, additional loss processes will cause a reduction of the photovoltage in practice. In order to quantify the departure of the voltage from the radiative limit, we can use Equation 7.7 to express the actual recombination parameter j_0 in terms of the efficiency of the solar cell operated as an LED, EQE_{LED}. Thus, we have

$$V_{oc}=\frac{k_BT}{q}\ln\left(\frac{j_{ph}}{j_{0,rad}}EQE_{LED}\right) \tag{7.18}$$

Therefore, the actual open-circuit voltage relates to the ideal radiative limit as

$$V_{oc} - V_{oc}^{rad} = \frac{k_B T}{q} \ln(EQE_{LED}) \tag{7.19}$$

This is the result derived by Ross (1967). The conclusion is that when comparing cells of similar characteristics such as the bandgap of the semiconductor, the solar cell with the highest radiative efficiency (large EQE_{LED}) will also show the highest photovoltage because it contains less nonradiative losses, and consequently, less recombination. Contrary to intuition, a solar cell must be a good radiator in order to approach the theoretical limits of energy conversion efficiency from sunlight to electricity (Miller et al., 2012).

Taking the typical form for the IQE_{LED} in Equation PSC.2.24 and neglecting outcoupling effects, we can write Equation 7.19 as

$$V_{oc}^{rad} - V_{oc} = -\frac{k_B T}{q} \ln\left(\frac{j_{rad}}{j_{rad} + j_{nrad}}\right) \tag{7.20}$$

This equation indicates the fraction of recombination events that are radiative. Normally, for solar cells that are not yet close to top quality performance, nonradiative recombination is the major problem to increase the obtained photovoltage and the factor j_{rad}/j_{nrad} can be a very small number. Nevertheless, every increase of a factor 10 in EQE_{LED} improves the photovoltage by 60 mV. The correlation of the power conversion efficiency to the EQE_{LED} is shown in Figure 7.8 for a number of inorganic, perovskite, and organic solar cells (Yao et al., 2015).

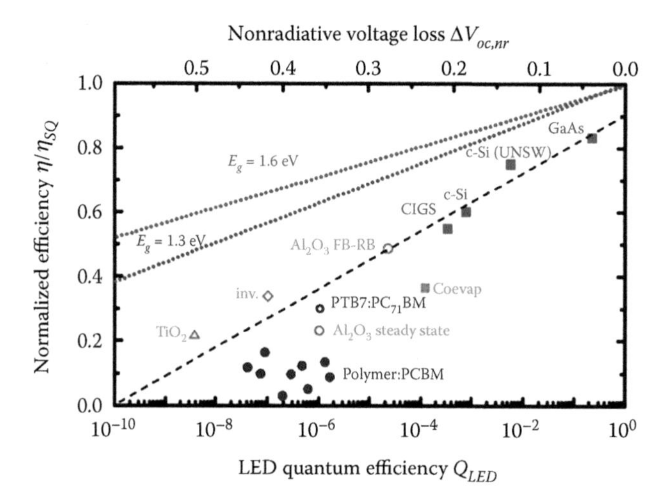

FIGURE 7.8 Power conversion efficiency for different solar cell technologies normalized to the Shockley–Queisser limit as a function of LED quantum efficiency EQE_{LED}. The *dotted lines* define the theoretical limits of various EQE_{LED} at two different optical bandgaps: 1.6 and 1.3 eV. The top x-axis is the nonradiative voltage loss over the range of EQE_{LED}, referring to Equation 7.19. The data points for inorganic solar cells are shown in *red squares*. Different perovskite fabrication technologies are shown in *green*. *Open green* and *circle points* represent the solution-processed perovskite devices made on mesoporous TiO_2 and Al_2O_3 films, respectively. The open green diamond point is the solution-processed inverted device using PEDOT:PSS and PCBM interlayers. The solid square point is a coevaporated film. Different organic solar cells are shown in *blue*. PTB7:PC_{71}BM, the organic system with the highest PCE, is shown as the *open blue circle*. The *dashed line* is a guide to the eye representing the approximate experimental trend. (Reprinted with permission from Yao, J. et al. *Physical Review Applied* 2015, 4, 014020.)

However, when the recombination is mostly radiative, the efficient extraction of photons by the optical design of the cell becomes a dominant issue as will be analyzed in the following sections.

7.3 EXTERNAL RADIATIVE EFFICIENCY

So far, we have used information about the light absorption and recombination features to predict the performance of the photovoltage. As mentioned in Chapter PSC.5, it is also interesting to analyze the features of the radiative emission of the cell. Normally, the radiative saturation current density $j_{0,rad}$ is rather small, but one can operate the solar cell as an LED and obtain readily visible EL when a sufficient voltage is applied. We continue to assume that the applied voltage is translated into homogeneous quasi-Fermi levels, which requires high mobilities and other conditions that will be studied in the subsequent chapters. The total radiative flux at V_{oc} is

$$\Phi_{ph}^{em} = \frac{j_{0,rad}}{q} e^{qV_{oc}/k_BT} \tag{7.21}$$

The radiative emission in the radiative limit of the photovoltage would be

$$\Phi_{ph}^{em,rad} = \frac{j_{0,rad}}{q} e^{qV_{oc}^{rad}/k_BT} \tag{7.22}$$

Therefore, the LED quantum efficiency is

$$EQE_{LED} = \frac{\Phi_{ph}^{em}}{\Phi_{ph}^{em,rad}} = \frac{e^{qV_{oc}/k_BT}}{e^{qV_{oc}^{rad}/k_BT}} \tag{7.23}$$

Note that this result follows directly from Equation 7.19. Using Equation 7.11, we can write Equation 7.23 as

$$EQE_{LED} = \frac{j_{0,rad}e^{qV_{oc}/k_BT}}{j_{ph}^{sun}} \tag{7.24}$$

Equation 7.24 compares the theoretical radiative current at open circuit to the actual photocurrent. It can be written also as

$$EQE_{LED} = \frac{e^{qV_{oc}/k_BT}\displaystyle\int_0^\infty EQE_{PV}(E)\phi_{ph}^{bb}(E)dE}{\displaystyle\int_0^\infty EQE_{PV}(E)\phi_{ph}^{AM1.5}(E)dE} \tag{7.25}$$

As mentioned earlier, for high-efficiency solar cells, the radiative efficiency makes a major difference in the performance of otherwise similar devices. For example, two reported GaAs cells (Green, 2012) show a substantial difference of radiative efficiency by a factor of 20, with $EQE_{LED} = 22.5\%$ for the best cell. This improvement produces a change of the respective cell efficiencies from 26.4% to 27.6%. Since these cells are rather close to the radiative limit, the gain of photonic quality is mainly responsible for the increase of efficiency.

The EQE_{LED} of different types of solar cells shown in Figure 7.8 has been tabulated by Yao et al. (2015). This figure shows that a good solar cell is a good LED too. Note that the solar cells based on organic materials have the lowest EQE_{LED}. This behavior is primarily due to the existence of several parallel recombination pathways in addition to recombination through the absorber, which, by itself, represents only 10^{-6} of the total recombination.

7.4 PHOTON RECYCLING

At this point, we wish to compute the actual rate of radiative emission from a solar cell, and we adopt the usual planar geometry for a recombination diode of thickness d with flat Fermi levels in which all recombination is radiative, that is $IQE_{LED} = 1$ and $EQE_{PV}(E) = a(E)$. Based on the previously obtained detailed balance considerations, the emitted photon flux at the semiconductor surface is

$$\Phi_{ph}^{em,ext} = \frac{j_0^{ext}}{q} = b_\pi \int_0^\infty a(E) E^2 e^{-E/k_BT}\, dE \tag{7.26}$$

While Equation 7.26 is obtained viewing the cell from outside and is, therefore, labeled external emission, we may as well obtain the generated photons from within, starting from the diode model of Figure PSC.5.1. The total photon production is determined by integration of the recombination rate U_{rec} over the whole diode thickness as previously shown in Equation PSC.5.16. We obtained

$$j_0^{\text{int}} = qdB_{rad}n_i^{\,2} \tag{7.27}$$

Here, n_i is the intrinsic carrier density.

The radiative lifetime has been calculated earlier in Equation PSC.6.23 in terms of the absorption coefficient α and the refraction index n_i. The internally emitted flux has the expression

$$\Phi_{ph}^{em,\text{int}} = \frac{j_0^{\text{int}}}{q} = 4dn_r^{\,2} b_\pi \int_0^\infty \alpha(E) E^2 e^{-E/k_BT} dE \tag{7.28}$$

Antonio Martí identified the Shockley paradox in that both formulae 7.26 and 7.28 do not match each other (Martí et al., 1997). It is remarked that using the van Roosbroeck–Shockley radiative lifetime in the fundamental diode equation, one cannot obtain the actual radiative emission of the solar cell, Equation 7.26. The latter equation is derived from the fundamental restriction of detailed balance of incoming and outgoing fluxes in equilibrium, as in the Shockley–Queisser approach. The failure of Equation 7.28 to produce the output emission is due to the fact that it assumes that all photons resulting from radiative recombination are automatically expelled out of the cell. But this is far from true, since luminescent photons can be reabsorbed producing a fresh electron–hole pair, in a process termed photon recycling (Asbeck, 1977), previously commented upon in Section PSC.2.3. The significance of photon reabsorption and reemission in the operation of a solar cell is shown schematically in Figure 7.9. Equation 7.28 predicts that the emitted flux grows in proportion to the solar cell thickness but in reality, the radiative emission saturates when the absorptance of the cell becomes $a(E) = 1$ for a thick layer, as indicated in Equation 7.26.

Since the photons created by radiative recombination must be either expelled out of the cell or reabsorbed, an emission probability p_e can be defined as the fraction between the rate of production of radiative photons and the rate of the external radiative emission, thus (Kirchartz et al., 2016a,b)

$$p_e = \frac{\Phi_{ph}^{em,ext}}{\Phi_{ph}^{em,int}} = \frac{\int_0^\infty a(E) E^2 e^{-E/k_BT}\, dE}{4dn_r^{\,2} \int_0^\infty \alpha(E) E^2 e^{-E/k_BT}\, dE} \tag{7.29}$$

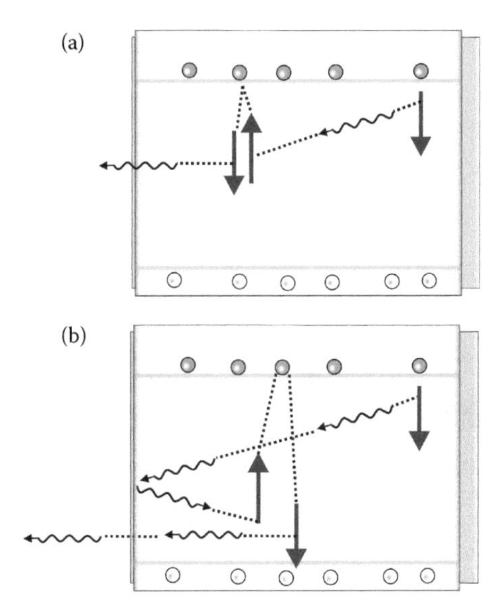

FIGURE 7.9 Illustration of photon recycling processes. (a) A photon is generated inside the absorber away from the transparent contact. It is reabsorbed and reemitted and then expelled out of the layer, thereby contributing to the luminescent flux. (b) The photon is reflected at the transparent contact, then reabsorbed and reemitted in a direction inside the escape cone of the semiconductor.

The contribution of photon recycling to the power output of the solar cell is only significant when recombination is mainly radiative, which so far has been achieved by GaAs and lead halide perovskites. Thomas Kirchartz has compared in Figure 7.10 two cases of lead halide perovskite solar cells with different scattering properties of the front surface, where both cells have similar antireflective coating and total reflecting back surface, as in the usual model adopted in previous sections. In Figure 7.10a, the Lambert–Beer model allows only a double pass of the light ray, while in Figure 7.10b, the Lambertian diffuser scatters the rays randomly and increases the light trapping as commented in Section PSC.2.1. Let us discuss the ratio

$$P(E) = \frac{a(E)}{4dn_r^2\alpha(E)} \tag{7.30}$$

which determines the emission probability according to Equation 7.29. The spectral shape of absorption coefficient $\alpha(E)$ of the lead iodide perovskite is shown in Figure PSC.3.5.

For weak absorption of subbandgap wavelengths ($\alpha(E)d \ll 1$), the photon will hit the surface many times before it has a chance to be internally reabsorbed. From Equation PSC.2.12, the Lambert–Beer model gives the absorptance $a(E) = 2\alpha(E)d$, hence $P = 1/2n_r^2$. This is because the percentage of photons that have an angle that allows them to escape is just $(2n_r^2)^{-1}$, as discussed in Section PSC.2.1. In the case of the light trapping model, Equation PSC.2.18 gives the absorptance $a(E) = 4n_r^2 d\alpha(E)$ in the weak absorption range. Since the escape cone from a semiconductor is quite narrow, the external emission requires repeated attempts. The photon impacting the surface is randomized many times until it hits the escape cone and hence the probability of emission is $P(E) = 1$ in the limit of weak absorption.

However, for the domain of suprabandgap photons with strong optical absorption, the behavior is rather different. In this range of photon energies, $\alpha(E)d > 1$ is satisfied. It is, therefore, not feasible to directly extract these photons (except those generated close to the transparent contact) as they have a very large chance of reabsorption, so that Equation 7.28 grossly overestimates the flux that can be extracted. The absorption coefficient grows steadily, as shown in Figure PSC.3.5 while the absorptance stabilizes to $a(E) = 1$, so that $P(E) < 1$ makes a contribution to the integrals in

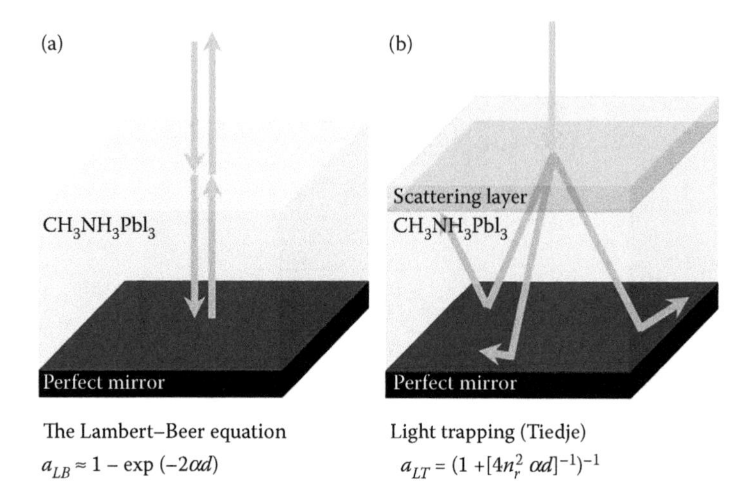

FIGURE 7.10 Schematic depiction of two optical models. Panel (a) describes perfect light incoupling and a perfect back reflector but no scattering. The absorptance is described for direct incidence by the Lambert–Beer equation. The integration over all angles of absorption and emission has very little effect on the total result. Panel (b) still considers perfect incoupling and a perfect back reflector but in addition includes a perfect Lambertian scattering layer that leads to a Lambertian distribution of angles. (Reproduced with permission from Kirchartz, T. et al. *ACS Energy Letters* 2016, 731–739.)

Equation 7.26 that decreases p_e. The relevant terms in the integral and their ratios are depicted in Figure 7.11, see Kirchartz et al. (2016a,b) for a more detailed discussion. In conclusion, the smaller the p_e, the larger the need for efficient photon recycling. If $p_e = 1$, then photon recycling is nonexistent, but when it is smaller than 1, then multiple reabsorption events are needed to maximize the external luminescence efficiency. This analysis shows that light trapping and light emission are correlated processes (Miller et al., 2012).

Consider the outcoupling efficiency defined in Section PSC.2.3, which is the probability that a photon is emitted and not reabsorbed. In the absence of parasitic absorption, it is simply

$$\eta_{outco} = p_e \tag{7.31}$$

where p_e is given in Equation 7.29. Every cycle of reabsorbed photons increases the external quantum efficiency in the geometric series

$$\begin{aligned} EQE_{LED} &= IQE_{LED}\eta_{outco} + IQE_{LED}(1-\eta_{outco})IQE_{LED}\eta_{outco} + \cdots \\ &= \frac{IQE_{LED}\eta_{outco}}{1 - IQE_{LED}(1-\eta_{outco})} \end{aligned} \tag{7.32}$$

An extension of the voltage losses including the photon recycling can be formulated using Equation 7.32 in 7.19 (Rau and Kirchartz, 2014; Rau et al., 2014).

In summary, for solar cells that are close to the radiative recombination limit, the optical features play an important role in increasing the photovoltage and the power conversion efficiency. Most of the photons may undergo total internal reflection upon reaching the semiconductor–air interface producing a much larger photon flux inside the semiconductor film than that emitted at the surface. In these conditions, with plenty of long-lived photons inside the cell, the output power of the solar cell becomes rather sensitive to a small nonradiative recombination current. Efficient external

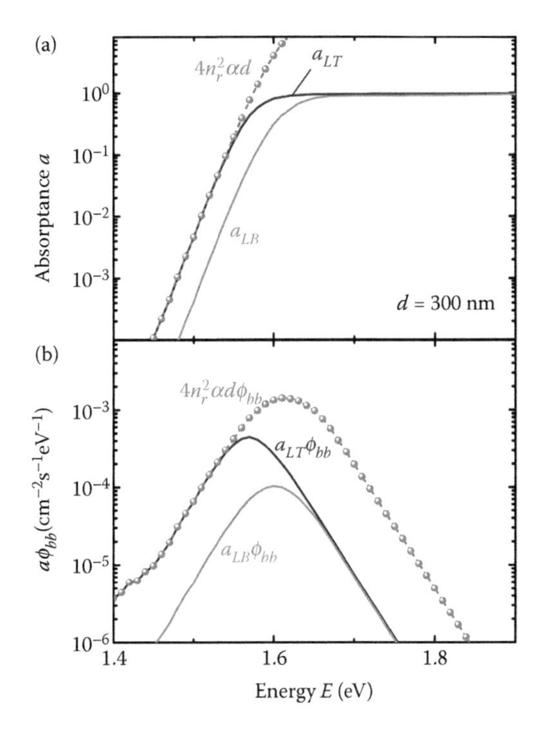

FIGURE 7.11 (a) Absorptance versus photon energy for different configurations (the Lambert–Beer or Lambertian light trapping) compared to the $4n_r^2\alpha d$ that defines the upper limit for the absorptance for $\alpha d \ll 1$. (b) Same as panel (a) but multiplied by the blackbody spectrum. The integral over the curves in panel (b) controls the ratio j_0^{int}/j_0^{ext} of the saturation current densities and thereby the probability of emission p_e. (Reproduced with permission from Kirchartz, T. et al. *ACS Energy Letters* 2016, 731–739.)

fluorescence is an indicator of low-internal optical losses, which will avoid the recycling of photons and successive loss by any remnant nonradiative pathways.

7.5 RADIATIVE COOLING IN EL AND PHOTOLUMINESCENCE

In Section ECK.5.12, we remarked that a semiconductor diode can be operated as a Carnot heat engine in which thermal radiation and radiative emission play the role of heat baths (Rose, 1960; Berdahl, 1985). Emission of photons in EL takes away a quantity of heat (and entropy), which allows to operate an LED as a radiative refrigerator. Let us discuss the physical constraints to realize cooling by photon emission.

The energy that must be supplied to produce an electron–hole pair in a biased semiconductor diode is qV. As a result of the recombination, a photon of energy $h_\nu \approx E_g$ is emitted. The photon removes a quantity of heat from the semiconductor lattice

$$Q_{rh} = h\nu - qV \tag{7.33}$$

Therefore, cooling a semiconductor by EL photon emission requires that most of the photons be emitted at energy $h_\nu > qV$. The removal of one electron–hole pair by recombination decreases the entropy in the semiconductor by the quantity

$$\Delta s = -\frac{Q_{rh}}{T} \tag{7.34}$$

The production of the photon incurs a creation of the amount of entropy

$$\Delta s' = \frac{h\nu}{T*} \tag{7.35}$$

where $T*$ is the temperature of the radiation, which may be different from T. The process removes the heat from a source at temperature T to a sink at temperature $T*$ (Dousmanis et al., 1964). Since the total entropy production must be positive, we have

$$h\nu - qV < \frac{T}{T*} h\nu \tag{7.36}$$

Consider the case of the reversible process in which the entropy production is zero. It corresponds to the equality

$$\frac{h\nu - qV}{T} = \frac{h\nu}{T*} \tag{7.37}$$

Comparing Equations PSC.6.32 and 7.37, we note that the latter corresponds to the equilibrium of emission and absorption when the radiation field obtains the temperature corresponding to a nonzero chemical potential $\mu_{ph} = qV$. This equilibrium is further discussed in Section PSC.11.1.

We derive the rate of heat emission by radiative decay in the LED. If τ_{rad} is the radiative lifetime, after an initial excitation, the population of electrons and holes decreases by recombination as

$$\frac{dn}{dt} = \frac{dp}{dt} = -\frac{n}{\tau_{rad}} \tag{7.38}$$

By the definition of the chemical potential ζ in terms of the configurational entropy S of electrons and holes, Equations ECK.5.4 and ECK.5.50, we have the rate of the total entropy production by the removal of electrons and holes as

$$\frac{d(S_n + S_p)}{dt} = -\frac{\zeta_n + \zeta_p}{T} \frac{dn}{dt} \tag{7.39}$$

Applying the radiative decay process, we have (Weinstein, 1960)

$$\frac{d(S_n + S_p)}{dt} = \frac{\zeta_n + \zeta_p}{T} \frac{n}{\tau_{rad}} \tag{7.40}$$

From Equations ECK.2.35 and ECK.2.42, the sum of chemical potentials can be written as

$$\begin{aligned} \zeta_n + \zeta_p &= E_{Fn} - E_{Fp} - E_c + E_v \\ &= qV - E_g \end{aligned} \tag{7.41}$$

Finally,

$$\begin{aligned} \frac{dS}{dt} &= -\frac{E_g - qV}{T} \frac{n}{\tau_{rad}} \\ &\approx -\frac{Q_{rh}}{T} \frac{n}{\tau_{rad}} \end{aligned} \tag{7.42}$$

The entropy decreases when $qV < E_g$.

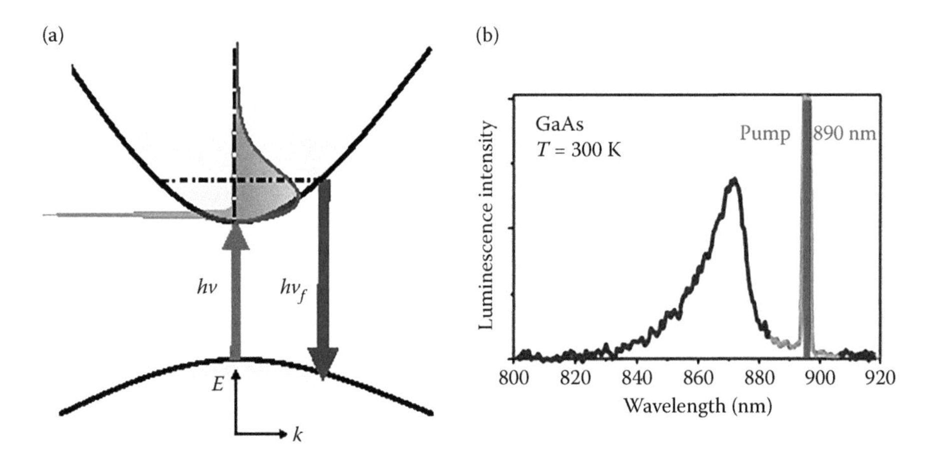

FIGURE 7.12 (a) Cooling cycle in laser refrigeration of a semiconductor in which absorption of laser photons with energy $h\nu$ creates a cold distribution of electron–hole carriers (only electron distribution is shown). The carriers then heat up by absorbing phonons followed by an up-converted luminescence at $h\nu_F$ (b). Typical anti-Stokes' luminescence observed in GaAs/GaInP double heterostructure. (Reproduced with permission from Sheik-Bahae, M.; Epstein, R. I. Laser cooling of solids. *Laser & Photonics Reviews* 2009, 3, 67–84.)

For the radiative cooling to be efficient, the EQE_{LED} must be large as otherwise, the cooling energy gain Q_{rh} per photon is lost by the fraction of photons that are reabsorbed or reflected at the edges and not extracted from the device. Photon recycling can enhance considerably the efficiency of the semiconductor EL cooling. Refrigeration can also be achieved with thermionic emission.

Refrigeration by photon emission in PL is called *laser cooling*. It requires anti-Stokes' emission so that the frequency of the pump source is lower than the mean luminescence frequency. A monochromatic source pumps photons in the low energy tail of the absorption spectrum of a material. The photogenerated electrons are thermalized in the conduction band so that they obtain an increase of energy of order k_BT from the semiconductor lattice. Subsequently, more energetic blue-shifted photons are emitted, see Figure 7.12. This emission process is very difficult to achieve in practice, since the cooling effects associated with nonradiative decay and multiphonon emission processes dominate over internal heating in the conduction band.

7.6 RECIPROCITY OF ABSORPTION AND EMISSION IN A CT BAND

The photoinduced transitions between organic donor (D) and acceptor (A) materials have been described in Figure ECK.6.5. In Figure PSC.3.26, we introduce the CT transition across the interface between the D and the A. The combination of D–A and the resulting CT and charge separation processes form the basis for the organic solar cells (Bisquert and Garcia-Belmonte, 2011). The energy diagram is shown in Figure 7.13. The purpose of using the interface to produce a photovoltage is to overcome the poor charge separation of the main absorber material, which would lead to recombination of most photogenerated carriers.

As indicated in Figure 7.13, there are two main pathways to create a photogenerated electron in A leaving the corresponding hole in D. First is a photon absorption across the intrinsic gap E_D, followed by fast injection down the energy gradient, the same process as that shown in Figure ECK.6.5. The second possibility is a CT transition across the optical gap that in this case is given by $E_0 = E_{LUMO(A)} - E_{HOMO(D)}$. The optical gap may actually be smaller than this difference due to excitonic effects, so that E_0 is defined as the energy difference between the CT complex ground state and the CT-excited state. However, the specific energy value is of little relevance for the following considerations.

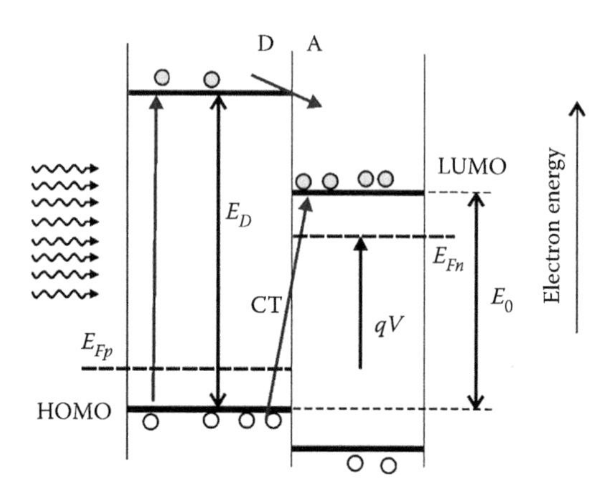

FIGURE 7.13 Schematic representation of a solar cell based on donor (D)–acceptor (A) system. Incoming light is absorbed both in the donor absorber and in the CT process across the interface. Excited electrons from the first type of absorber are transferred by charge separation to A. However, the CT process directly promotes an electron to the LUMO of the A, red shifted with respect to the intrinsic absorption. If the charge separation is efficient, most of the recombination occurs across the optical gap E_0.

Here, we focus our attention on the characteristics of CT process between molecular materials. In these materials, strong reorganization effects occur that have been described in Figure ECK.6.5. We will analyze the impact of reorganization in the reciprocity relation, dominated by CT rather than the intrinsic transition, by the following reasoning. In Figure 7.14, we show the $EQE_{PV}(E)$ of bulk heterojunction polymer-fullerene devices. It is observed that the relevant values of $EQE_{PV}(E)$ for photovoltaic operation, that is between 0.1 and 1, occur at the larger spectral energies due to the fact that most of the photons are absorbed at energies larger than E_D in the interior of the polymer domains. But additional features are observed at lower photon energies, when the EQE_{PV} is reduced by several orders of magnitude, and these must correspond to the E_0 transition. While the effect of the latter is negligible in terms of photocurrent generation, it can play a large role in the radiative equilibrium, which is often dominated by subbandgap features as noticed previously. Therefore, the properties of this transition are relevant for the solar cell operation.

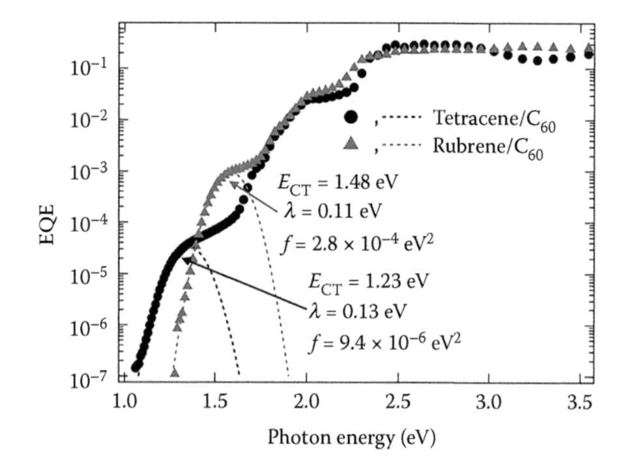

FIGURE 7.14 EQE_{PV} spectra for rubrene/C_{60} and tetracene/C_{60} bilayer organic solar cells. (Reproduced with permission from Graham, K. R. et al. *Advanced Materials* 2013, 25, 6076–6082.)

The thermal effects of reorganization are closely related to the maximum of absorbance and fluorescence, as previously suggested in Figure ECK.6.9. According to Marcus theory, the spectral lineshape of the CT absorption coefficient $\alpha(E)$ at photon energy E is given by the following expression (Marcus, 1989; Gould et al., 1993; Vandewal et al., 2010):

$$\alpha(E) = \frac{f_{abs} N_{CTC}}{E\sqrt{4\pi\lambda k_B T}} \exp\left(\frac{-(E_0 + \lambda - E)^2}{4\lambda k_B T}\right) \tag{7.43}$$

Here λ is the reorganization energy, N_{CTC} is the number of CTCs per unit volume, and f_{abs} is a constant proportional to the square of the electronic coupling element. Equation 7.43 can be derived from the picture of Figure ECK.6.9 using the following assumptions: (a) the Boltzmann population of vibronic states within the ground and excited states, (b) standard relation between Einstein A and B coefficients, and (c) the reorganization energies for ground and excited states are the same (Koen Vandewal, private communication).

The external quantum efficiency for a photovoltaic cell based on the CT absorption in the weak absorption limit can be written as

$$EQE_{PV}(E) = \frac{f_{CT}}{E\sqrt{4\pi\lambda k_B T}} \exp\left(\frac{-(E_0 + \lambda - E)^2}{4\lambda k_B T}\right) \tag{7.44}$$

where the prefactor is

$$f_{CT} = IQE_{PV}(E) N_{CTC} 2d f_{abs} \tag{7.45}$$

in terms of the internal quantum efficiency for charge separation and collection, and the thickness of the layer d, assuming a back reflector so that the optical pathway is $2d$ (Vandewal et al., 2010). Equation 7.44 is represented in Figure 7.15a. The experimental data in Figure 7.14 suggest the observation of the CT band in measurements of EQE_{PV} of molecular organic solar cells.

In order to derive the relationship between light emission efficiency and absorbance, we now apply the reciprocity relationship (7.2). The spectral emission flux in dark equilibrium can be stated as

$$\phi_{ph}^{em}(E) = b_\pi \frac{f_{CT}}{\sqrt{4\pi\lambda k_B T}} E e^{-E/k_B T} \exp\left(\frac{-(E_0 + \lambda - E)^2}{4\lambda k_B T}\right) \tag{7.46}$$

This last equation can be rewritten as follows:

$$\phi_{ph}^{em}(E) = b_\pi \frac{f_{CT}}{\sqrt{4\pi\lambda k_B T}} E e^{-E_0/k_B T} \exp\left(\frac{-(E_0 - \lambda - E)^2}{4\lambda k_B T}\right) \tag{7.47}$$

Note that the sign of λ is inverted in Equation 7.47 with respect to Equation 7.46. Therefore, representation of a reduced normalized EQE, $\overline{\eta} \propto E \times EQE_{PV}$, and a reduced normalized photon flux that can be measured by EL, $\overline{\phi} \propto \phi_{ph}^{em}/E$, produces mirror images, as illustrated in Figure 7.15b. The intersection of the spectra is separated by a distance λ from the maxima of absorption and emission. An observation of the mirror images for the reduced quantities for CT absorption and emission in molecular complexes is shown in Figure 7.16 (Gould et al., 1993). The measurements of organic solar cell devices for donor–acceptor blend materials have been presented in Figure 7.14. Further examples of the CTC in organic solar cells are shown in Figure 7.17 (Guan et al., 2016).

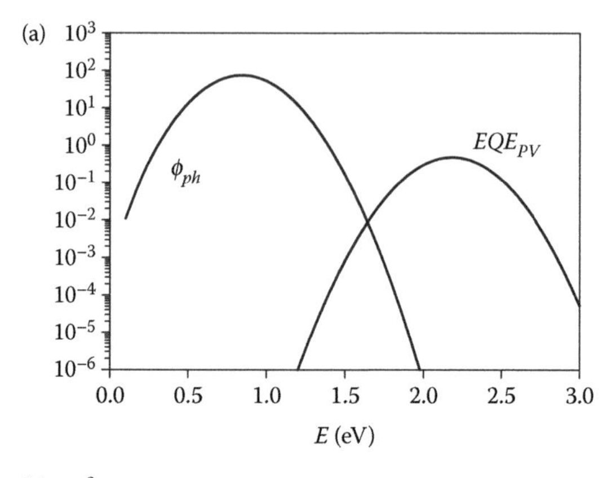

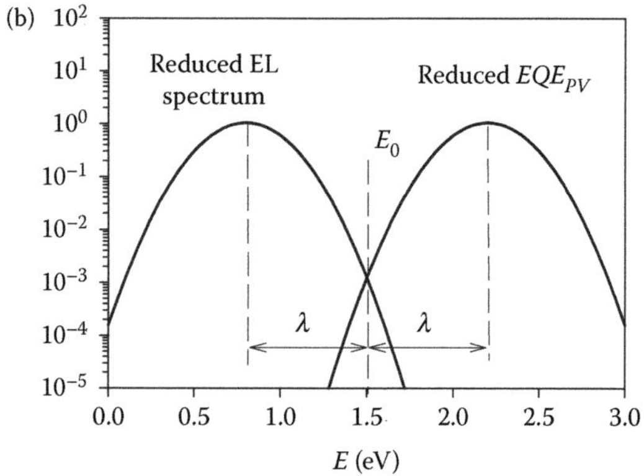

FIGURE 7.15 (a) EQE_{PV} and photon emission (m^{-2} s^{-1} eV^{-1}) at equilibrium with blackbody radiation of a CT complex that obeys the Marcus model with CT gap energy E_0 and reorganization energy λ. (b) Reduced *EQE*, $\bar{\eta} \propto E \times EQE_{PV}$, and photon emission spectra $\bar{\phi} \propto \phi_{ph}^{em}/E$.

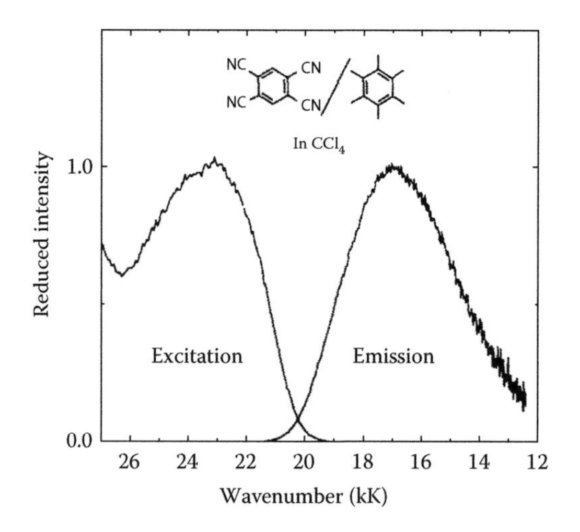

FIGURE 7.16 Reduced excitation and emission spectra for the excited CT complex of 1,2,4,5-tetracyano-benzene/hexamethylbenzene in carbon tetrachloride at room temperature. (Reproduced with permission from Gould, I. R. et al. *Chemical Physics* 1993, 176, 439–456.)

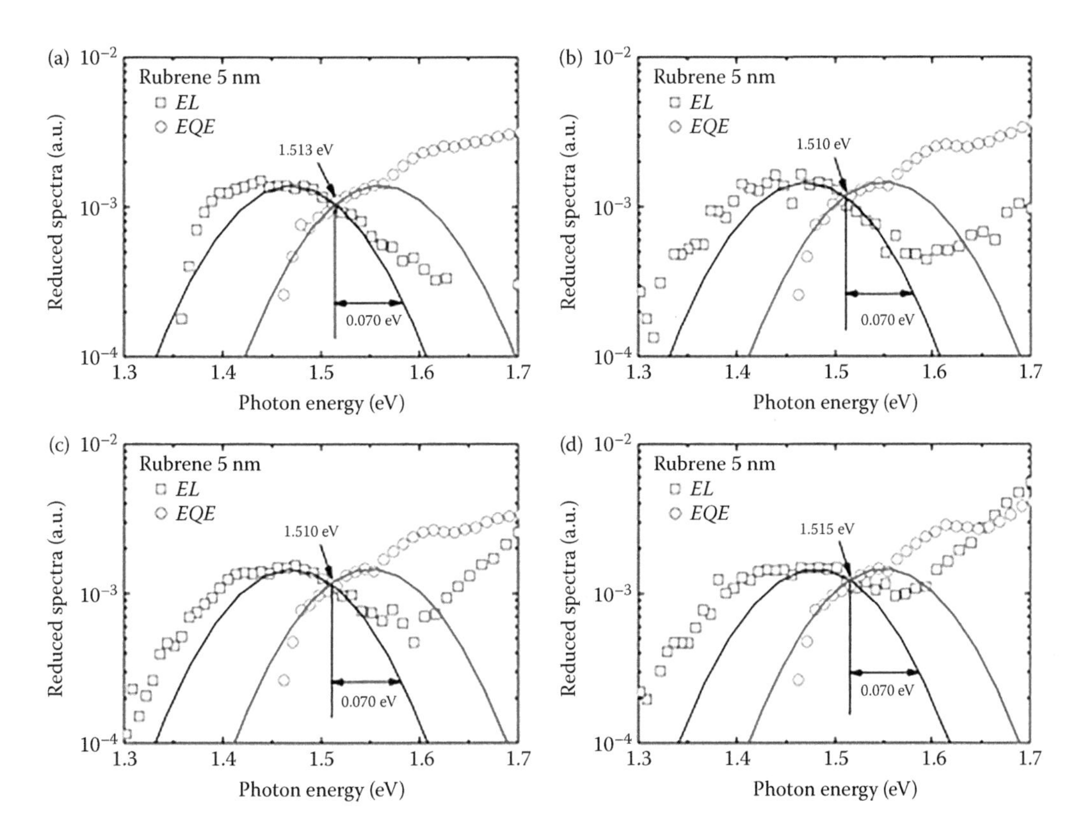

FIGURE 7.17 OPV devices of structure of ITO/MoO_3/rubrene (5, 10, 15, 20 nm)/C_{60} (30 nm)/bathocuproine (7 nm)/Ag (120 nm). Reduced spectra of OPV devices with rubrene thickness of (a) 5, (b) 10, (c) 15, and (d) 20 nm at the rubrene/C_{60} interface. The single-headed arrows along with text indicate CT. The double-headed arrows along with text indicate λ. (Reproduced with permission from Su, W.-C. et al. *ACS Applied Materials & Interfaces* 2016.)

These results show that the reciprocity relationship between absorption and emission is well obeyed, as assumed in the original derivation (Marcus, 1989). From Equation 7.47, the saturation current of the organic diode can be determined as follows:

$$j_{0,rad} = qb_\pi \frac{f_{\text{CT}}}{\sqrt{4\pi\lambda k_B T}} e^{-E_0/k_B T} \int_0^\infty E \exp\left(\frac{-(E_0-\lambda-E)^2}{4\lambda k_B T}\right) dE \tag{7.48}$$

It can be shown that $j_{0,rad}$ can be calculated by the approximated expression (Vandewal et al., 2010)

$$j_{0,rad} = qb_\pi (E_0 - \lambda) f_{\text{CT}} e^{-E_0/k_B T} \tag{7.49}$$

Finally, the radiative limit to the photovoltage is given as

$$qV_{oc}^{rad} = E_0 + k_B T \ln(j_{ph}^{sun}) - k_B T \ln(qb_\pi (E_0 - \lambda) f_{\text{CT}}) \tag{7.50}$$

However, as mentioned in Figure 7.8, the photovoltage is usually quite far from the radiative limit in this type of cell due to additional nonradiative interfacial recombination.

GENERAL REFERENCES

Reciprocity relationships: Ross (1967), Ruppel and Wurfel (1980), Weber and Dignam (1984), Baruch (1985), Smestad and Ries (1992), Markvart and Landsberg (2002), Markvart (2007), Rau (2007), and Kirchartz and Rau (2008).

Radiative voltage in perovskite solar cells: Tress (2017).

Experimental relation of EQE to electroluminescence: Kirchartz and Rau (2008), Vandewal et al. (2009), Tvingstedt et al. (2014), Tress et al. (2015), and Yao et al. (2015).

Donolato theorem: Donolato (1985), Donolato (1989), and Green (1997).

Photon recycling in solar cell devices: Asbeck (1977), Martí et al. (1997), Miller et al. (2012), Kirchartz et al. (2016a,b), and Pazos-Outón et al. (2016).

Radiative cooling: Mahan (1994), Sheik-Bahae and Epstein (2004), Rakovich et al. (2009), Yen and Lee (2010), and Lee and Yen (2012).

REFERENCES

Asbeck, P. Self-absorption effects on the radiative lifetime in GaAs-GaAlAs double heterostructures. *Journal of Applied Physics* 1977, 48, 820–822.

Baruch, P. A two level system as a model for a photovoltaic solar cell. *Journal of Applied Physics* 1985, 57, 1347–1355.

Berdahl, P. Radiant refrigeration by semiconductor diodes. *Journal of Applied Physics* 1985, 58, 1369–1374.

Bisquert, J.; Garcia-Belmonte, G. On voltage, photovoltage, and photocurrent in bulk heterojunction organic solar cells. *Journal of Physical Chemistry Letters* 2011, 2, 1950–1964.

Donolato, C. A reciprocity theorem for charge collection. *Applied Physics Letters* 1985, 46, 270–272.

Donolato, C. An alternative proof of the generalized reciprocity theorem for charge collection. *Journal of Applied Physics* 1989, 66, 4524–4525.

Dousmanis, G. C.; Mueller, C. W.; Nelson, H.; Petzinger, K. G. Evidence of refrigerating action by means of photon emission in semiconductor diodes. *Physical Review* 1964, 133, A316.

Gould, I. R.; Noukakis, D.; Gomez-Jahn, L.; Young, R. H.; Goodman, J. L.; Farid, S. Radiative and nonradiative electron transfer in contact radical-ion pairs. *Chemical Physics* 1993, 176, 439–456.

Graham, K. R.; Erwin, P.; Nordlund, D.; Vandewal, K.; Li, R.; Ngongang Ndjawa, G. O.; Hoke, E. T. et al. Re-evaluating the role of sterics and electronic coupling in determining the open-circuit voltage of organic solar cells. *Advanced Materials* 2013, 25, 6076–6082.

Green, M. A. Generalized relationship between dark carrier distribution and photocarrier collection in solar cells. *Journal of Applied Physics* 1997, 81, 268–271.

Green, M. A. Radiative efficiency of state-of-the-art photovoltaic cells. *Progress in Photovoltaics: Research and Applications* 2012, 20, 472–476.

Guan, Z.; Li, H.-W.; Cheng, Y.; Yang, Q.; Lo, M.-F.; Ng, T.-W.; Tsang, S.-W.; Lee, C.-S. Charge-transfer state energy and its relationship with open-circuit voltage in an organic photovoltaic device. *The Journal of Physical Chemistry C* 2016, 120, 14059–14068.

Hirst, L. C.; Ekins-Daukes, N. J. Fundamental losses in solar cells. *Progress in Photovoltaics: Research and Applications* 2011, 19, 286–293.

Kirchartz, T.; Nelson, J.; Rau, U. Reciprocity between charge injection and extraction and its influence on the interpretation of electroluminescence spectra in organic solar cells. *Physical Review Applied* 2016, 5, 054003.

Kirchartz, T.; Rau, U. Detailed balance and reciprocity in solar cells. *Physica Status Solidi A* 2008, 205, 2737–2751.

Kirchartz, T.; Staub, F.; Rau, U. Impact of photon recycling on the open-circuit voltage of metal halide perovskite solar cells. *ACS Energy Letters* 2016, 731–739.

Lee, K.-C.; Yen, S.-T. Photon recycling effect on electroluminescent refrigeration. *Journal of Applied Physics* 2012, 111, 014511.

Mahan, G. D. Thermionic refrigeration. *Journal of Applied Physics* 1994, 76, 4362–4366.

Marcus, R. A. Relation between charge transfer absorption and fluorescence spectra and the inverted region. *The Journal of Physical Chemistry* 1989, 93, 3078–3086.

Markvart, T. Thermodynamics of losses in photovoltaic conversion. *Applied Physics Letters* 2007, 91, 064102.

Markvart, T.; Landsberg, P. T. Thermodynamics and reciprocity of solar energy conversion. *Physica E* 2002, 14, 71–77.

Martí, A.; Balenzategui, J. L.; Reyna, R. F. Photon recycling and Shockley's diode equation. *Journal of Applied Physics* 1997, 82, 4067.
Miller, O. D.; Yablonovitch, E.; Kurtz, S. R. Strong internal and external fluorescence as solar cells approach the Shockley-Queisser limit. *IEEE Journal of Photovoltaics* 2012, 2, 303–311.
Müller, T. C. M.; Kirchartz, T. Absorption and photocurrent spectroscopy with high dynamic range. In *Advanced Characterization Techniques for Thin Film Solar Cells*; Abou-Ras, D., Kirchartz, T., and Rau, U. (Eds); Wiley-VCH Verlag GmbH & Co. KGaA: Berlin, 2016; pp. 189–214.
Pazos-Outón, L. M.; Szumilo, M.; Lamboll, R.; Richter, J. M.; Crespo-Quesada, M.; Abdi-Jalebi, M.; Beeson, H. J. et al. Photon recycling in lead iodide perovskite solar cells. *Science* 2016, 351, 1430–1433.
Rakovich, Y. P.; Donegan, J. F.; Vasilevskiy, M. I.; Rogach, A. L. Anti-Stokes cooling in semiconductor nanocrystal quantum dots: A feasibility study. *Physica Status Aolidi A* 2009, 206, 2497–2509.
Rau, U. Reciprocity relation between photovoltaic quantum efficiency and electroluminescent emission of solar cells. *Physical Review B* 2007, 76, 085303.
Rau, U.; Kirchartz, T. On the thermodynamics of light trapping in solar cells. *Nature Materials* 2014, 13, 103–104.
Rau, U.; Paetzold, U. W.; Kirchartz, T. Thermodynamics of light management in photovoltaic devices. *Physical Review B* 2014, 90, 035211.
Rose, A. L. Photovoltaic effect derived from the Carnot cycle. *Journal ofApplied Physics* 1960, 31, 1640–1641.
Ross, R. T. Some thermodynamics of photochemical systems. *The Journal of Chemical Physics* 1967, 46, 4590–4593.
Ruppel, W.; Wurfel, P. Upper limit for the conversion of solar energy. *IEEE Transactions on Electron Devices* 1980, 27, 877–882.
Santhanam, P.; Fan, S. Thermal-to-electrical energy conversion by diodes under negative illumination. *Physical Review B* 2016, 93, 161410.
Sheik-Bahae, M.; Epstein, R. I. Can laser light cool semiconductors? *Physical Review Letters* 2004, 92, 247403.
Sheik-Bahae, M.; Epstein, R. I. Laser cooling of solids. *Laser & Photonics Reviews* 2009, 3, 67–84.
Smestad, G.; Ries, H. Luminescence and current-voltage characteristics of solar cells and optoelectronic devices. *Solar Energy Materials and Solar Cells* 1992, 25, 51–71.
Su, W.-C.; Lee, C.-C.; Li, Y.-Z.; Liu, S.-W. Influence of singlet and charge-transfer excitons on the open-circuit voltage of rubrene/fullerene organic photovoltaic device. *ACS Applied Materials & Interfaces* 2016, 8, pp. 28757–28762.
Tress, W. Perovskite solar cells on the way to their radiative efficiency limit—Insights into a success story of high open-circuit voltage and low recombination. *Advanced Energy Materials* 2017, 1602358-n/a.
Tress, W.; Marinova, N.; Inganäs, O.; Nazeeruddin, M. K.; Zakeeruddin, S. M.; Graetzel, M. Predicting the open-circuit voltage of $CH_3NH_3PbI_3$ perovskite solar cells using electroluminescence and photovoltaic quantum efficiency spectra: The role of radiative and non-radiative recombination. *Advanced Energy Materials* 2015, 5, 10.1002/aenm.201400812.
Trupke, T.; Würfel, P.; Uhlendorf, I.; Lauermann, I. Electroluminescence of the dye-sensitized solar cell. *The Journal of Physical Chemistry B* 1999, 103, 1905–1910.
Tvingstedt, K.; Malinkiewicz, O.; Baumann, A.; Deibel, C.; Snaith, H. J.; Dyakonov, V.; Bolink, H. J. Radiative efficiency of lead iodide based perovskite solar cells. *Scientific Reports* 2014, 4, 6071.
Vandewal, K.; Tvingstedt, K.; Gadisa, A.; Inganas, O.; Manca, J. V. Relating the open-circuit voltage to interface molecular properties of donor:acceptor bulk heterojunction solar cells. *Physical Review B* 2010, 81, 125204.
Vandewal, K.; Tvingstedt, K.; Gadisa, A.; Inganäs, O.; Manca, J. V. On the origin of the open-circuit voltage of polymer–fullerene solar cells. *Nature Materials* 2009, 8, 904–909.
Wang, X.; Lundstrom, M. S. On the use of Rau reciprocity to deduce external radiative efficiency in solar cells. *IEEE Journal of Photovoltaics* 2013, 3, 1348–1353.
Weber, M. F.; Dignam, M. J. Efficiency of splitting water with semiconducting electrodes. *Journal of the Electrochemical Society* 1984, 131, 1258–1265.
Weinstein, M. A. Thermodynamics of radiative emission processes. *Physical Review* 1960, 119, 499.
Würfel, P. *Physics of Solar Cells. From Principles to New Concepts*, 2nd edition; Wiley: Weinheim, 2009.
Yao, J.; Kirchartz, T.; Vezie, M. S.; Faist, M. A.; Gong, W.; He, Z.; Wu, H. et al. Quantifying losses in open-circuit voltage in solution-processable solar cells. *Physical Review Applied* 2015, 4, 014020.
Yen, S.-T.; Lee, K.-C. Analysis of heterostructures for electroluminescent refrigeration and light emitting without heat generation. *Journal of Applied Physics* 2010, 107, 054513.

8 Basic Operation of Solar Cells

We describe the main characteristics of the operation of a solar cell with respect to voltage, electrical current, and illumination. We aim to explain the central feature of the solar cell performance in terms of the current density–voltage curve, considering the photovoltage, photocurrent, and the fill factor (FF), and how these are combined to produce the power conversion efficiency (PCE). In Section 8.4, we derive the limits to efficiency conversion based on the Shockley–Queisser approach and we present a detailed discussion of the different fundamental intrinsic losses of light to electrical power conversion that lead to the maximal theoretical efficiency as a function of the bandgap energy of the absorber.

8.1 CURRENT–VOLTAGE CHARACTERISTICS

We analyze the operational properties of a solar cell under a substantial incident illumination using the basic model shown in Figure PSC.5.1. We start with Figure PSC.5.1d that shows the situation *of open circuit* in which no current is extracted. The excitation by light produces excess electrons and holes in the carrier bands, which causes the separation of the corresponding Fermi levels, as commented previously in Section PSC.4.4. By controlling the voltage V that exists between the two contacts of the solar cell, the amount of recombination can be changed. In the case of Figure PSC.5.1e, the voltage is $V < V_{oc}$. Here, the solar cell generates a current and acts as a battery. Another important operational variable is the current extracted from the solar cell, already indicated in Section PSC.6.1. The nonrecombined carriers, by conservation, produce a photocurrent density, as indicated in Figure PSC.5.1f.

In Equation PSC.5.1, we established that the current injected to a diode in the dark is the difference between recombination current and the dark generation current j_0. The photocurrent, j_{ph}, arising from the excess radiation that impacts the solar cell, is added to the recombination and dark generation processes as represented in Figure PSC.5.1. In summary, we have

$$j = j_{ph} - j_{rec} + j_0 \tag{8.1}$$

The balance between the three terms on the right-hand side of Equation 8.1 gives rise to the current that flows in the external circuit in terms of electrons leaving the excited state through their selective contact and entering the ground state through the hole selective contact. This process can be, therefore, viewed as a recombination through the external circuit and it is the current in which we are directly interested, as it is the one capable of doing useful work on external elements.

From Equation PSC.5.8, the current density–voltage (j–V curve) characteristic of the solar cell is

$$j = j_{ph} - j_0(e^{qV/mk_BT} - 1) \quad (V_{forward} > 0) \tag{8.2}$$

Figure 8.1a shows the shape of the j–V curve corresponding to the ideal diode performance. We observe that under illumination, the dark characteristic is shifted up in the current density axis. At open circuit ($j = 0$), the recombination current balances the photogeneration rate from the external sources. The open-circuit voltage V_{oc} is given by the expression

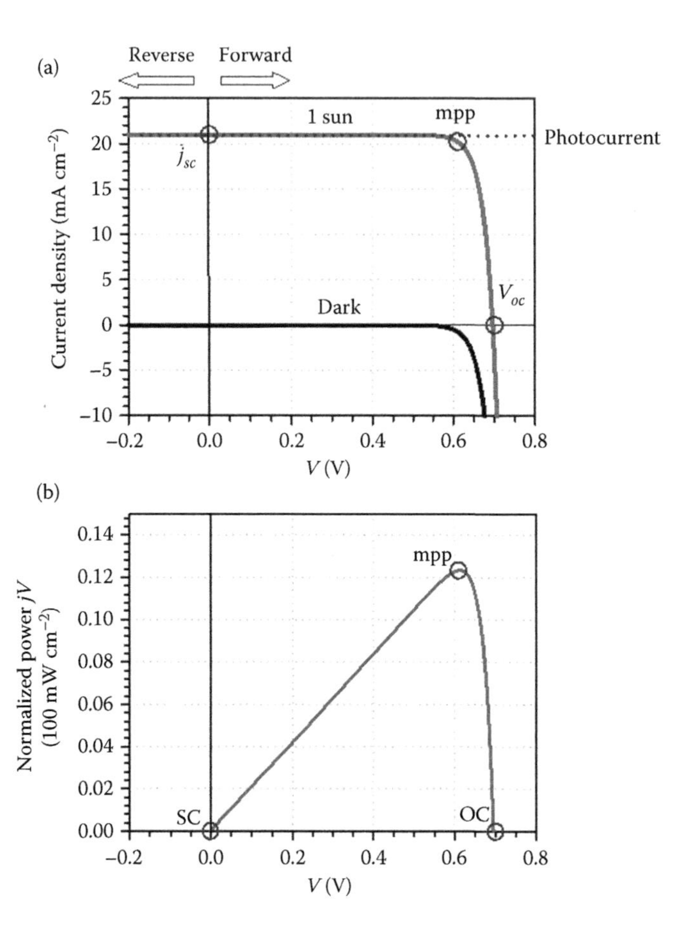

FIGURE 8.1 (a) Current density–potential characteristics of a solar cell that consists of a fundamental diode with dark saturation current $j_0 = 4.34 \times 10^{-11}$ mA cm^{-2} and diode quality factor $m = 1$. Shown are the short-circuit current and open-circuit voltage points. (b) The electrical power supplied by the solar cell as a function of voltage. The power is normalized to 100 mW cm^{-2} (corresponding to 1 sun light intensity). Shown is the maximum power point.

$$V_{oc} = \frac{mk_BT}{q} \ln\left(\frac{j_{ph}}{j_0} + 1\right) \tag{8.3}$$

The second point of interest is the *short-circuit* condition, shown in Figure PSC.5.1e. In this case, we have $j_{rec} = j_0$ as indicated in Equation PSC.5.4. Therefore, all the excess generated carriers contribute to the external current, the *short-circuit photocurrent*,

$$j_{sc} = j_{ph} \tag{8.4}$$

Thus, we obtain an operational definition of the quantity j_{ph}: it is the current density at short circuit.

Between open circuit and short circuit, the voltage can have any value $0 \le V \le V_{oc}$, and this situation is indicated in Figure PSC.5.1e. Finally, at strong forward bias, a large recombination current is induced.

Usually, the current in the solar cell grows very rapidly when V departs from V_{oc} toward lower voltages, see Figure. 8.1a, but when the voltage further decreases, the current becomes stabilized to $j \approx j_{ph}$. This is because at $V \approx V_{oc}$, we induce a large internal recombination current that compensates the photogeneration, but when V decreases considerably, the recombination current is negligible with respect to j_{ph}. At strong reverse bias, the current in the ideal model is

$$j_{sat} = j(V \ll 0) = j_{ph} + j_0 \tag{8.5}$$

Figure 8.2 shows the j–V and EQE_{PV} characteristics of two lead halide perovskite solar cells, including one high-performance cell and another one with lower performance (Li et al., 2016).

8.2 POWER CONVERSION EFFICIENCY

The main application of a solar cell is to serve as a power supply unit that produces electricity from sunlight, or from ambient light in indoor applications. The central feature to assess the solar cell performance is the electrical power that can be extracted from the available radiation level. The electrical power at a given voltage operation point of the solar cell, P_{el}, has the value

$$P_{el} = jV \tag{8.6}$$

Figure 8.1b shows the power supplied by the solar cell as a function of the voltage. The power is zero at both open- and short-circuit conditions. In between the extreme cases of low power lies the *maximum power point* (mpp) at which voltage (V_{mp}) the solar cell should be operated for electricity production. The maximum power provided by the photovoltaic device is

$$P_{el,\max} = j_{mp} V_{mp} \tag{8.7}$$

For the characterization of energy converter devices, the main figure of merit is the maximum conversion efficiency of the solar cell, that is, the PCE, that consists of the electrical power supplied at the mpp with respect to the incoming photon energy, as stated in Equation PSC.6.9,

$$\eta_{PCE} = \frac{j_{mp} V_{mp}}{\Phi_{E,tot}^{source}} \tag{8.8}$$

The PCE depends on the operation conditions that need to be defined. The PCE of a solar cell is usually reported under simulated standard terrestrial spectrum $AM1.5G$, which bears an integrated power of $\Phi_E^{AM1.5G} = 1\,\mathrm{kW\,m^{-2}} = 100\,\mathrm{mW\,cm^{-2}}$. This type of illumination is usually denominated "1 sun," see Section PSC.1.8.

The efficiency increases when the short-circuit photocurrent j_{ph} and open-circuit voltage V_{oc} increase, but η_{PCE} also depends critically on the mpp. The shape of the j–V curve determines at which voltage we extract the electrons as electric current. As observed in Figure 8.1, there is a tradeoff between current and voltage. At low voltage, extraction is easy, and the current is determined by the quantum yield (conversion of photons to electron carriers) of the absorber. However, a high voltage produces a current that is contrary to the photocurrent, and eventually the power decreases. In general, if we extract the electrons at high useful energy (voltage), there is a price in the lowering of the number we can extract. If this point is close to V_{oc}, then the operational voltage and current are much larger than if V_{mpp} occurs at low voltage close to $V_{oc}/2$. The parameter that tracks this property is the *fill factor* (FF) defined as

$$\mathrm{FF} = \frac{j_{mp} V_{mp}}{j_{ph} V_{oc}} \tag{8.9}$$

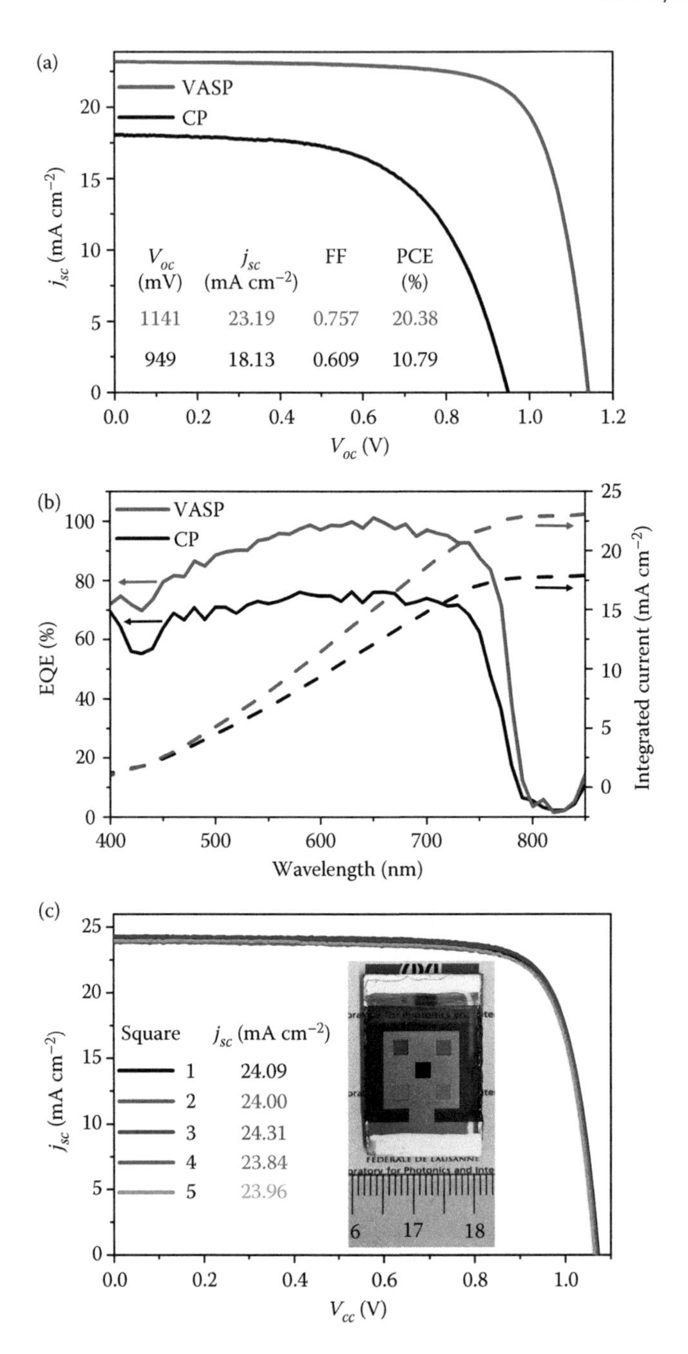

FIGURE 8.2 (a) Current–voltage curves for $FA_{0.81}MA_{0.15}PbI_{2.51}Br_{0.45}$ (FA: formamidinium, MA: methylammonium) perovskite solar cell devices using perovskite films prepared by the single-step solution deposition process (CP) and vacuum-flash solution processing (VASP) methods measured under standard *AM*1.5 solar radiation. (b) *Solid line*: EQE_{PV} curves of cells fabricated by the CP and VASP method. Measurements were taken with chopped monochromatic light under a white light bias corresponding to 5% solar intensity. *Dashed lines*: Short-circuit current density calculated from the overlap integral of the EQE_{PV} spectra with the standard *AM*1.5 solar emission. (c) Photovoltaic parameters for a representative $FA_{0.81}MA_{0.15}PbI_{2.51}Br_{0.45}$-based perovskite device fabricated by VASP measured from five different spots with an aperture area of 0.16 cm^2 selected from the total active area of 1.2 × 1.2 cm^2 under standard *AM*1.5*G* illumination. All *j*–*V* curves were recorded at a scanning rate of 50 mV s^{-1} in reverse direction. (Reproduced with permission from Li, X. et al. A vacuum flash-assisted solution process for high-efficiency large-area perovskite solar cells. *Science* 2016, 10.1126/science.aaf8060.)

We obtain the convenient expressions

$$P_{el,\max} = j_{ph} \times \mathrm{FF} \times V_{oc} \tag{8.10}$$

$$\eta_{PCE} = \frac{j_{ph} \times \mathrm{FF} \times V_{oc}}{\Phi_{E,tot}^{source}} \tag{8.11}$$

8.3 ANALYSIS OF FF

The FF depends on the form of the j–V curve, determining the mpp and exerting a large influence on the PCE of the solar cell. If the FF is high, the drop in current at high voltage is delayed, and we can extract the electrons at high voltage while the current is still close to j_{ph}. For a good diode characteristic, the power P increases linearly at low voltage, Figure 8.3b, and decreases abruptly at $V > V_{mp}$. Figure 8.3a compares the j–V characteristic of an ideal diode described by Equation 8.3, which has FF = 0.85, with a j–V "ohmic" characteristic that consists of the straight line that represents a resistor in the j–V plane, displaced upward with the addition of a photocurrent. In the last case, the photocurrent at maximum power is $j_{mp} = j_{ph}/2$, $V_{mp} = V_{oc}/2$, thus FF = 0.25, and the

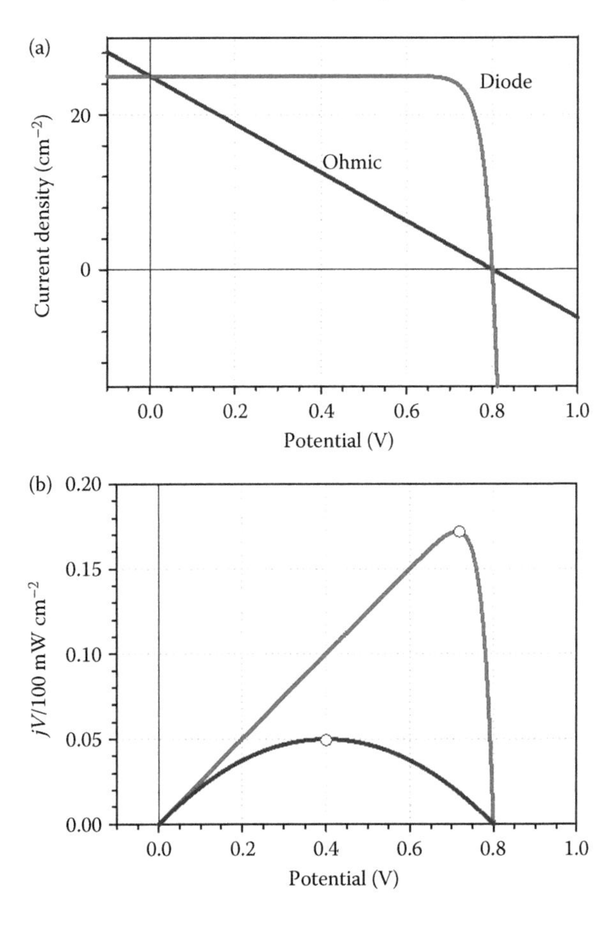

FIGURE 8.3 (a) Current density–potential characteristics of two solar cells: an ideal diode with quality factor $m = 1$ and a straight characteristic with added photocurrent denoted ohmic. (b) The electrical power supplied by the solar cells as a function of voltage. The power is normalized to 100 mW cm^{-2}. The mpp is indicated with a circle.

efficiency is very low. A low FF is frequently found in solar cells made with a poorly conducting material (Ji et al., 2010; Lopez-Varo et al., 2016).

While the measurement of the open-circuit photovoltage and short-circuit photocurrent is straightforward, the determination of the mpp requires a tracking of the point of maximum energy conversion (Christians et al., 2015; Zimmermann et al., 2016). Assuming that a perfectly stable characteristic has been achieved, the voltage of maximum power V_{mp} is obtained by the solution of the equation $d(jV)/dV = 0$. For the ideal diode model, the equation can be expressed as

$$V_{mp} = V_{oc} - \frac{mk_BT}{q}\ln\left(1+\frac{qV_{mp}}{mk_BT}\right) \tag{8.12}$$

Obtaining the mpp requires to solve numerically this transcendental equation for V_{mp}. Some useful approximated solutions have been developed. By an iteration of Equation 8.12, we find

$$V_{mp} = V_{oc} - \frac{mk_BT}{q}\ln\left(1+\frac{qV_{oc}}{mk_BT}-\ln\left[1+\frac{qV_{mp}}{mk_BT}\right]\right) \tag{8.13}$$

We still have the unknown at the right-hand side, but this term can be neglected with respect to the second term in the parenthesis. Thus, we obtain

$$V_{mp} = V_{oc} - \frac{mk_BT}{q}\ln\left(1+\frac{qV_{oc}}{mk_BT}\right) \tag{8.14}$$

We can calculate the current from Equations 8.2 and 8.3

$$j_{mp} = j_{ph}\frac{1}{1+\dfrac{mk_BT}{qV_{oc}}} \tag{8.15}$$

Hence,

$$\text{FF} = \frac{\dfrac{qV_{oc}}{mk_BT}-\ln\left(1+\dfrac{qV_{oc}}{mk_BT}\right)}{\dfrac{qV_{oc}}{mk_BT}+1} \tag{8.16}$$

A variety of approximations of this type have been proposed, and the most accurate formula is the following (Green, 1982):

$$\text{FF} = \frac{\dfrac{qV_{oc}}{mk_BT}-\ln\left(0.72+\dfrac{qV_{oc}}{mk_BT}\right)}{\dfrac{qV_{oc}}{mk_BT}+1} \tag{8.17}$$

In conclusion, in the diode model, the FF is completely determined by the values of V_{oc} and m. The lines in Figure 8.4 indicate the FF value depending on the diode quality factor, and the values

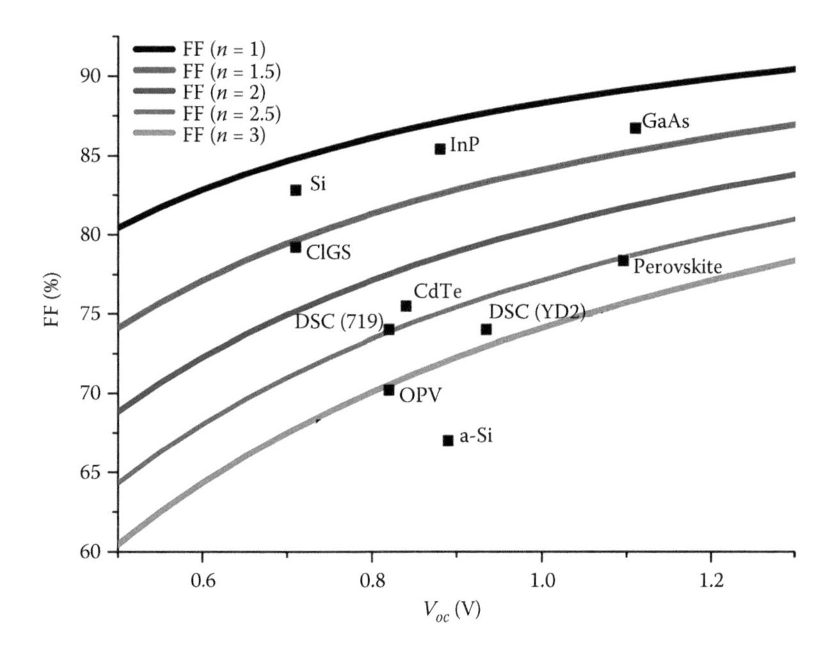

FIGURE 8.4 FFs of best laboratory cells of different categories. The solid lines represent the expected FF value as a function of V_{oc} for a given value of the diode ideality factor, Equation 8.17. CIGS stands for $Cu(In,Ga)Se_2$; OPV stands for organic photovoltaic cell; DSC indicates dye-sensitized solar cell with the dye used in parentheses. (Adapted from Nayak, P. K. et al. *Energy & Environmental Science* 2012, 5, 6022–6039.)

for top performing cells in different photovoltaic technologies are also indicated. A more advanced analytical solution for j–V characteristics including the series resistance is described by Banwell and Jayakumar (2000).

Let us discuss some examples that illustrate the j–V characteristics in a solar cell. We compare in Figures 8.5 and 8.6 two different solar cells A and B. The photovoltaic parameters are indicated in Table 8.1. As observed in Figure 8.5, as the illumination power is reduced, the photocurrent decreases but the recombination characteristics (in this simple model) are unchanged, so the curve is

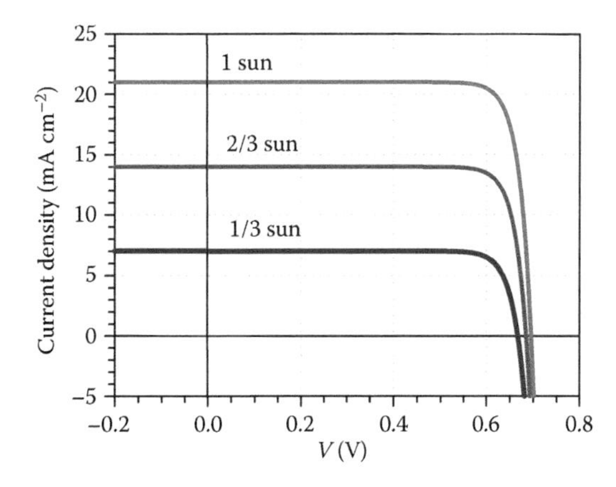

FIGURE 8.5 Current density–potential characteristics of a solar cell (A) that consists of a fundamental diode with dark saturation current $j_0 = 4.34 \times 10^{-11}$ mA cm^{-2} and diode quality factor $m = 1$, under different illumination intensities.

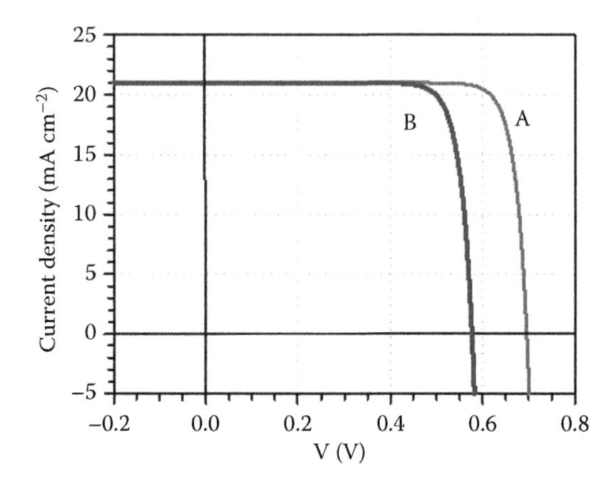

FIGURE 8.6 Current density–potential characteristics of two solar cells that consist of fundamental diodes with dark saturation current $j_0 = 4.34 \times 10^{-11}$ mA cm^{-2} (cell A), $j_0 = 4.34 \times 10^{-9}$ mA cm^{-2} (cell B) and diode quality factor $m = 1$ in both cases, under illumination of 1 sun.

TABLE 8.1
Performance Parameters of Solar Cells (Figures 8.5 and 8.6)

Cell	Illumination Intensity Φ_E (mW cm^{-2})	Short-Circuit Photocurrent j_{ph} (mA cm^{-2})	Open-Circuit Voltage V_{oc} (V)	Fill Factor *FF*	Efficiency η_{PCE} (%)
A	(1/3) × 100	7.0	0.671	0.840	11.84
A	(2/3) × 100	14.0	0.690	0.844	12.21
A	100	21.0	0.700	0.845	12.42
B	100	21.0	0.580	0.820	10.00

displaced downward. As the collection efficiency of this exercise is $EQE_{PV} = 1$, the photocurrent is simply proportional to the illumination intensity. However, the photovoltage varies slowly with illumination as indicated in Equation PSC.7.10, and the FF consequently changes very slowly as implied by Equation 8.17. Therefore, the PCE shows only a slight decrease as the light intensity decreases. Normally in experiments, the PCE increases as the light intensity decreases, and increases for cells with smaller area, as discussed in Section PSC.10.10. The cell B in Figure 8.6 has larger recombination rate, as determined by the reverse saturation current density j_0, which causes the bending down of the curve at a lower forward voltage and consequently, the open-circuit voltage decreases.

One effect that produces a serious modification of the FF is the series resistance R_s. The series resistance may correspond to the external resistance of the selective contact layers or to internal transport effects. For simplicity, we now model a constant series resistance (per area) due to the ohmic transport in one contact layer that takes a potential drop jR_s. The external voltage is distributed as a combination of the ohmic drop and the "Fermi level voltage" associated with the splitting of the Fermi levels at the outer edges of the absorber layer, V_F, as explained in Equation ECK.8.18. Therefore,

$$V = V_F - jR_s \tag{8.18}$$

Since the solar cell continues to work internally as in Equation 8.3, the current density depends on voltage as

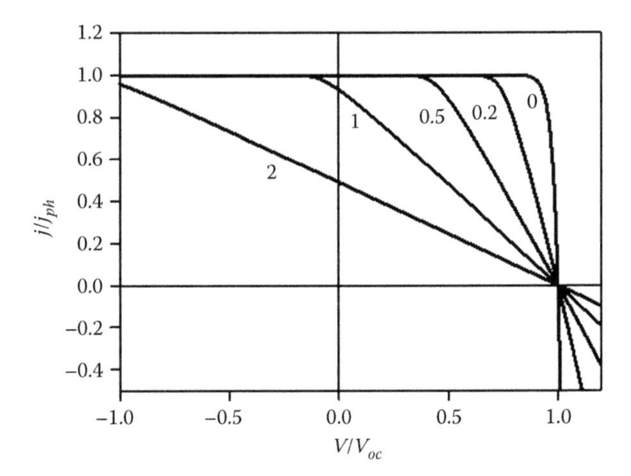

FIGURE 8.7 Current–voltage characteristics of the fundamental diode model including a series resistance. The value indicated in each curve is $R_s j_{ph}/V_{oc}$.

$$j = j_{ph} - j_0(e^{q(V+jR_s)/mk_BT} - 1) \tag{8.19}$$

The resulting current–voltage characteristics are shown in Figure 8.7 for different values of R_s. Note that all the curves have the same shape in terms of V_F, but when plotted against the external voltage, the FF becomes progressively degraded. The value $R_s = V_{oc}/j_{ph}$ causes the characteristic to be ohmic in the first quadrant as discussed earlier in Figure 8.3, which decreases the FF to a value 0.25.

8.4 SOLAR CELL EFFICIENCY LIMITS

The final goal of this chapter is to summarize the fundamental limitations to solar energy conversion and quantitatively establish the fraction of the incoming energy that can be transformed into electrical power by the use of a given absorber material. The central method to provide these results was derived by Shockley and Queisser (1961) (SQ). The rationale of the method is to describe a solar cell in terms of the absorber with optically sharp bandgap E_g and temperature T_A.

The fundamental model of the solar cell has been studied in the previous chapters. The minimal possible losses allowed by detailed balance are given by the exclusive radiative recombination. As a matter of fact, we have already derived expressions for the photocurrent, Equation PSC.6.8, the photovoltage, PSC.7.11, and the FF, Equation 8.17, in this model, so that we can apply Equation 8.11. The result of this calculation when the incoming flux is described by blackbody radiation at the temperature of the sun is the dashed curve given in Figure PSC.6.2c. The record efficiency obtained in dominant photovoltaic technologies is shown for comparison.

The SQ method is universally acknowledged as the central figure of merit to characterize the target PCE of a given class of photovoltaic materials. In addition, the method can be improved to a more realistic description of the theoretical efficiency using the desired source of radiation as *AM*1.5*G* solar irradiance (shown in Figure PSC.6.2c) and by describing the specific light absorption and emission properties of the materials by means of the actual EQE_{PV} of the device, as in Equation PSC.6.1. A method for a better estimation of photovoltaic properties in the numerical screening of materials is discussed at the end of this section.

Having described the SQ method and results, we now consider the different factors θ associated with fundamental effects that reduce the energy conversion capabilities of a single absorber solar cell. To describe each effect that limits the conversion of photons to electrical energy, we calculate the corresponding power loss with respect to the incoming solar power $\Phi_{E,tot}(T_S)$ (see also Hirst

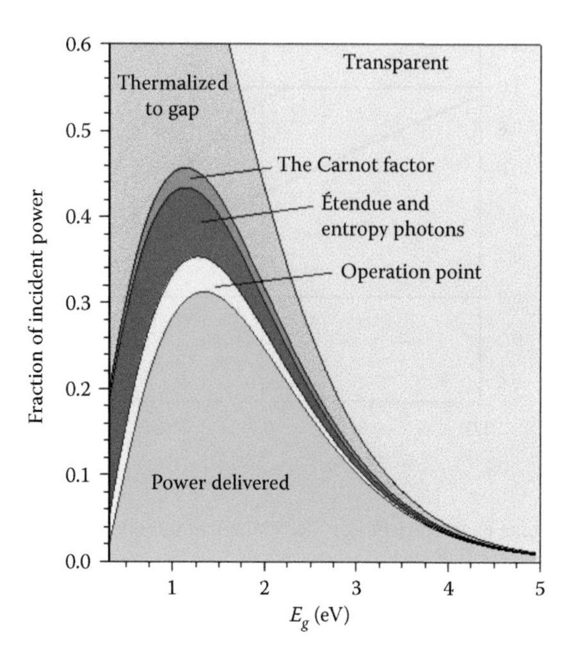

FIGURE 8.8 Different fundamental intrinsic losses of power with respect to that of incident radiation and the delivered electrical power dependence on E_g.

and Ekins-Daukes, 2011). We will use the modeling of sunlight by blackbody radiation at T_S, as discussed in Section PSC.6.1, and the approximation $E_g \gg k_B T_S$ ($\Gamma \approx E_g^2$). Each of the factors of the following list is represented in a different color in Figure 8.8:

1. The transparency of the cell below E_g, with the expression

$$\theta_{E,trans}(E_g, T_S) = \frac{\Phi_E(0, E_g)}{\Phi_{E,tot}(T_S)} \tag{8.20}$$

where the quantity in the denominator is given in Equation PSC.6.15. The effect of the factor $\theta_{E,trans}$ has already been described in Figure PSC.6.1b. The fraction of power lost by the unabsorbed photons increases with the bandgap energy.

2. The thermalization of photons. The fraction of power lost by the thermalization to the bandgap is

$$\theta_{E,ther}(E_g, T_S) = \frac{\Phi_E(E_g, \infty)}{\Phi_{E,tot}} - \frac{E_g \Phi_{ph}(E_g, \infty)}{\Phi_{E,tot}} \tag{8.21}$$

This factor is also shown in Figure PSC.6.1b, and together with transparency, they are the major losses of energy in photovoltaic conversion, giving rise to the Trivich–Flinn efficiency in Equation PSC.6.16.

In the next set of losses, we will discuss the reduction of electrical energy from E_g to V_{oc}, using the fundamental expression of the radiative voltage in Equation PSC.7.13. To each loss of voltage, there corresponds a loss of power given by the product of the voltage in units of energy and the absorbed photon flux, $qV_{oc}\Phi_{ph}$.

3. The Carnot losses

$$\theta_{E,C}(E_g,T_S) = \frac{T_A}{T_S}\frac{E_g\Phi_{ph}(E_g,\infty)}{\Phi_{E,tot}} \tag{8.22}$$

4. The étendue and photon entropy losses

$$\theta_{E,ee}(E_g,T_S) = \left[k_B T_A \ln\left(\frac{\varepsilon_S}{\pi}\right) + k_B T_A \ln\left(\frac{T_S}{T_A}\right)\right]\frac{\Phi_{ph}(E_g,\infty)}{\Phi_{E,tot}} \tag{8.23}$$

The previous factors complete the reduction of energy of electrical carriers so that the resulting efficiency is

$$\eta = \frac{qV_{oc}^{rad}\Phi_{ph}(E_g,\infty)}{\Phi_{E,tot}} \tag{8.24}$$

However, we have already noted that the carriers are extracted at current and voltage of the mpp, so that a further reduction given by the FF occurs in passing to Equation 8.11. The FF of this model has already been calculated in Equation 8.17. Therefore, SQ efficiency is

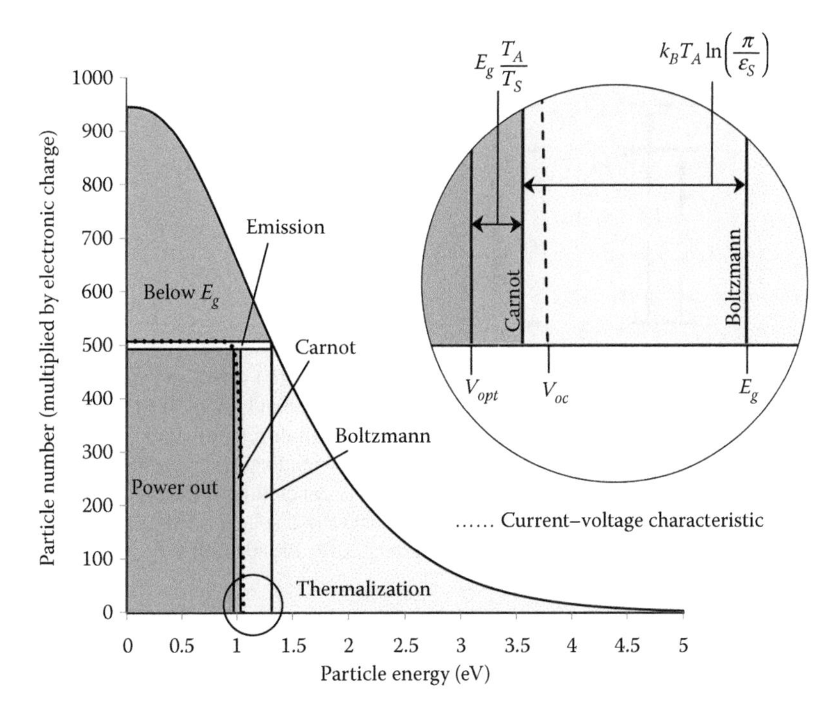

FIGURE 8.9 Intrinsic losses occurring in a device with optimal E_g (1.31 eV) under 1 sun illumination, indicated by the colored areas in the surface below the curve of incident-collected photons versus their energy. The Carnot, Boltzmann, and thermalization losses reduce the optimal operating voltage, while below-E_g radiation and emission losses reduce the current. The combination of both types of effects dictates the form of the current–voltage characteristic. Analytical descriptions of the Carnot and Boltzmann voltage drops are given. (Reproduced with permission from Hirst, L. C.; Ekins-Daukes, N. J. *Progress in Photovoltaics: Research and Applications* 2011, 19, 286–293.)

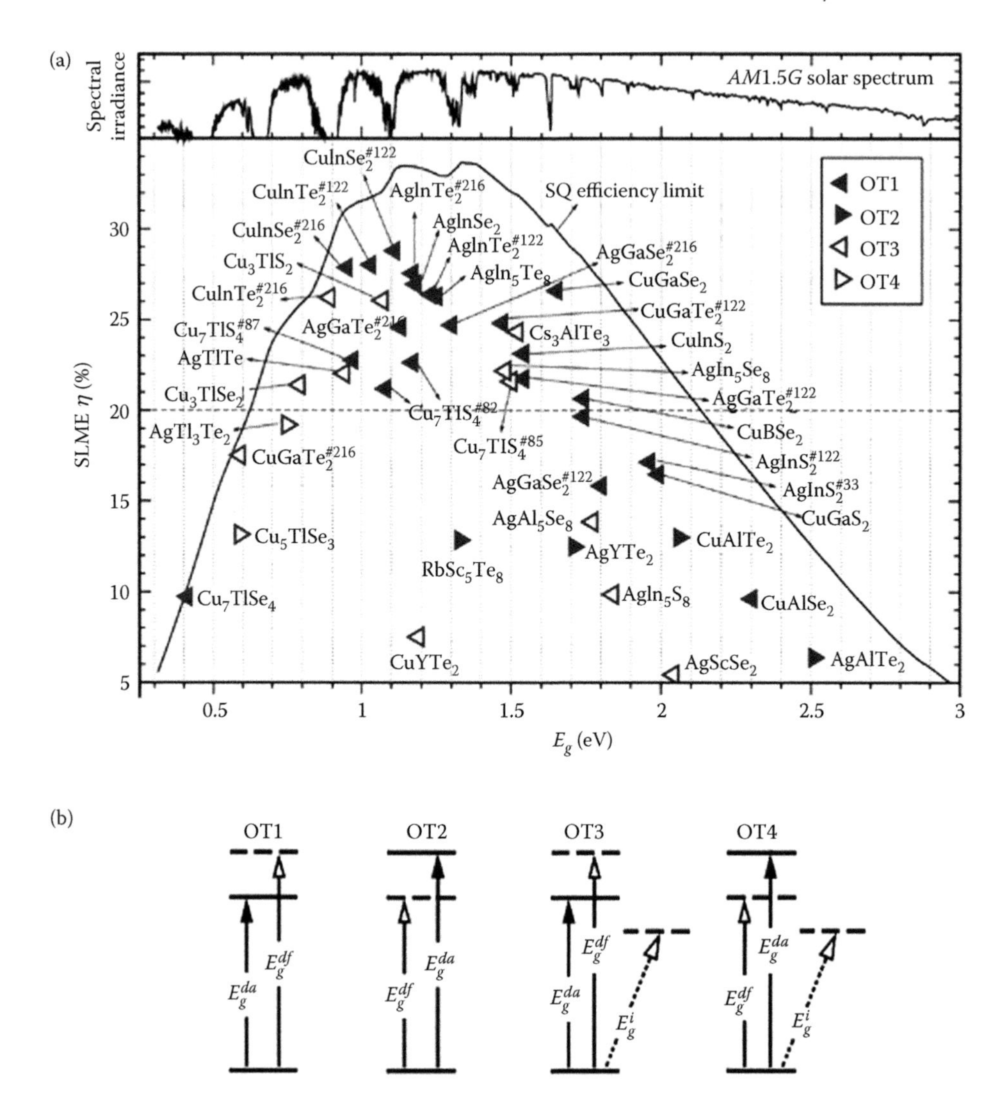

FIGURE 8.10 (a) Spectroscopic limited maximum efficiency (SLME) versus the minimum gap E_g for generalized I–III–VI chalcopyrite materials at $d = 0.5$ μm. The compounds with SLME < 5% are not shown. The shown space group number (*superscript*) is used to distinguish different materials with the same chemical formula. (b) Schematic diagrams of different types of semiconductors according to optical transitions. Electric-dipole-allowed (*forbidden*) direct optical transition is denoted by a line with an arrow pointing to solid (*dashed*) horizontal line. Indirect states are shown as laterally displaced *dashed lines*. (Reproduced with permission from Yu, L.; Zunger, A. *Physical Review Letters* 2012, 108, 068701.)

$$\eta = \frac{q\Phi_{ph}(E_g,\infty)\times \text{FF}\left(V_{oc}^{rad}\right)\times V_{oc}^{rad}}{\Phi_{E,tot}} \tag{8.25}$$

and it consists of the combination of the features shown in Figure 8.8.

Another way to represent the losses is shown in Figure 8.9. Here, a loss in power from the incident spectrum for each mechanism is associated with the product of current density and energy of the photons shown by a colored area. The final output power obtained by subtraction of all the losses is shown by the blue square.

As mentioned earlier, the SQ methods give a first approximation to photovoltaic efficiency by assuming direct absorption and a sharp bandgap. However, the shape of the absorption coefficient near the onset of absorption and nonradiative recombination channels play a dominant role in the actual efficiency that may be achieved. A more advanced method of calculation for realistic materials screening by Yu and Zunger (2012) takes into account the bandgap, the shape of absorption spectra, and the material-dependent nonradiative recombination losses. Figure 8.10a shows the screening of 256 materials belonging to I–III–VI chalcopyrite group. The panel (b) shows the type of materials according to dipole-allowed or dipole-forbidden transitions, allowing to quantify the nonradiative recombination, which is dominant when the gap is smaller than the dipole-allowed transition. The effect of different recombination pathways on the photovoltaic efficiency has been analyzed by Kirchartz and Rau (2017).

GENERAL REFERENCES

Analysis of the fill factor: Green (1982), Banwell and Jayakumar (2000), Maa and Abdel-Motaleb (2002), De Soto et al. (2006), and Bashahu and Nkundabakura (2007).

Solar cell operation: Sze (1981), Partain (1995), Nelson (2003), Würfel (2009), Dittrich (2014), Fonash (2010), Tress (2014), and Schmidt-Mende and Weickert (2016).

Solar cell efficiency limits: Ross (1967), Henry (1980), Ruppel and Wurfel (1980), Araújo and Martí (1994), Baruch et al. (1995), Hanna and Nozik (2006), Markvart (2008), Würfel (2009), Hirst and Ekins-Daukes (2011), Yu and Zunger (2012), Rau et al. (2014), and Vossier et al. (2015).

REFERENCES

Araújo, G. L.; Martí, A. Absolute limiting efficiencies for photovoltaic energy conversion. *Solar Energy Materials and Solar Cells* 1994, 33, 213–240.

Banwell, T. C.; Jayakumar, A. Exact analytical solution for current flow through diode with series resistance. *Electronics Letters* 2000, 36, 291–292.

Baruch, P.; De Vos, A.; Landsberg, P. T.; Parrott, J. E. On some thermodynamic aspects of photovoltaic solar energy conversion. *Solar Energy Materials and Solar Cells* 1995, 36, 201–222.

Bashahu, M.; Nkundabakura, P. Review and tests of methods for the determination of the solar cell junction ideality factors. *Solar Energy* 2007, 81, 856–863.

Christians, J. A.; Manser, J. S.; Kamat, P. V. Best practices in perovskite solar cell efficiency measurements. Avoiding the error of making bad cells look good. *The Journal of Physical Chemistry Letters* 2015, 6, 852–857.

De Soto, W.; Klein, S. A.; Beckman, W. A. Improvement and validation of a model for photovoltaic array performance. *Solar Energy* 2006, 80, 78–88.

Dittrich, T. *Materials Concepts for Solar Cells*; Imperial College Press: London, 2014.

Fonash, S. J., *Solar Cell Device Physics*, 2nd Edition; Academic Press: New York, 2010.

Green, M. A. Accuracy of analytical expressions for solar cell fill factors. *Solar Cells* 1982, 7, 337–340.

Hanna, M. C.; Nozik, A. J. Solar conversion efficiency of photovoltaic and photoelectrolysis cells with carrier multiplication absorbers. *Journal of Applied Physics* 2006, 100, 074510.

Henry, C. H. Limiting efficiencies of ideal single and multiple energy gap terrestrial solar cells. *Journal of Applied Physics* 1980, 51, 4494.

Hirst, L. C.; Ekins-Daukes, N. J. Fundamental losses in solar cells. *Progress in Photovoltaics: Research and Applications* 2011, 19, 286–293.

Ji, W.; Yao, K.; Liang, Y. C. Bulk photovoltaic effect at visible wavelength in epitaxial ferroelectric $BiFeO_3$ thin films. *Advanced Materials* 2010, 22, 1763–1766.

Kirchartz, T.; Rau, U. Decreasing radiative recombination coefficients via an indirect band gap in lead halide perovskites. *The Journal of Physical Chemistry Letters* 2017, 8, 1265–1271.

Li, X.; Bi, D.; Yi, C.; Décoppet, J.-D.; Luo, J.; Zakeeruddin, S. M.; Hagfeldt, A.; Grätzel, M. A vacuum flash-assisted solution process for high-efficiency large-area perovskite solar cells. *Science* 2016, 10.1126/science.aaf8060.

Lopez-Varo, P.; Bertoluzzi, L.; Bisquert, J.; Alexe, M.; Coll, M.; Huang, J.; Jimenez-Tejada, J. A. et al. Physical aspects of ferroelectric semiconductors for photovoltaic solar energy conversion. *Physics Reports* 2016, 653, 1–40.

Maa, Y. J.; Abdel-Motaleb, I. M. Analysis of the diode characteristics using the thermodynamic theories. *Solid-State Electronics* 2002, 46, 735–742.

Markvart, T. Solar cell as a heat engine: Energy–entropy analysis of photovoltaic conversion. *Physica Status Solidi A* 2008, 205, 2752–2756.

Nayak, P. K.; Garcia-Belmonte, G.; Kahn, A.; Bisquert, J.; Cahen, D. Photovoltaic efficiency limits and material disorder. *Energy & Environmental Science* 2012, 5, 6022–6039.

Nelson, J. *The Physics of Solar Cells*; Imperial College Press: London, 2003.

Partain, L. D. *Solar Cells and Their Applications*; Wiley: Weinheim, 1995.

Rau, U.; Paetzold, U. W.; Kirchartz, T. Thermodynamics of light management in photovoltaic devices. *Physical Review B* 2014, 90, 035211.

Ross, R. T. Some thermodynamics of photochemical systems. *The Journal of Chemical Physics* 1967, 46, 4590–4593.

Ruppel, W.; Wurfel, P. Upper limit for the conversion of solar energy. *IEEE Transactions on Electron Devices* 1980, 27, 877–882.

Schmidt-Mende, L.; Weickert, J. *Organic and Hybrid Solar Cells. An Introduction*; de Gruyter: Berlin, 2016.

Shockley, W.; Queisser, H. J. Detailed balance limit of efficiency of p-n junction solar cells. *Journal of Applied Physics* 1961, 32, 510–520.

Sze, S. M. *Physics of Semiconductor Devizes*; 2nd edition; John Wiley and Sons: New York, 1981.

Tress, W. *Organic Solar Cells. Theory, Experiment, and Device Simulation*; Springer: Switzerland, 2014.

Vossier, A.; Gualdi, F.; Dollet, A.; Ares, R.; Aimez, V. Approaching the Shockley-Queisser limit: General assessment of the main limiting mechanisms in photovoltaic cells. *Journal of Applied Physics* 2015, 117, 015102.

Würfel, P. *Physics of Solar Cells. From Principles to New Concepts*, 2nd edition; Wiley: Weinheim, 2009.

Yu, L.; Zunger, A. Identification of potential photovoltaic absorbers based on first-principles spectroscopic screening of materials. *Physical Review Letters* 2012, 108, 068701.

Zimmermann, E.; Wong, K. K.; Müller, M.; Hu, H.; Ehrenreich, P.; Kohlstädt, M.; Würfel, U. et al. Characterization of perovskite solar cells: Toward a reliable measurement protocol. *APL Mater.* 2016, 4, 091901.

9 Charge Separation in Solar Cells

In the photovoltaic conversion process, the excitation created by a photon in the light absorber is converted into separate electrons and holes, which produce current and voltage. In this chapter, the separation of the excitation into distinct carriers is analyzed, with particular emphasis on those devices composed of different nanostructured phases that realize the charge separation by energetic gradients close to the generation point, such as DSCs and organic bulk heterojunction cells. Realistic material limitations to the photovoltage beyond the radiative limit are summarized. In particular, the use of separate phases for electron and hole transport imposes constraints on the open-circuit voltage value due to the fact that the Fermi level is limited by the density of states of the transport materials.

9.1 LIGHT ABSORPTION

The operation of a solar cell can be analyzed in terms of a sequential set of processes, which eventually cause the conversion of the free energy of the incoming photons into a voltage and current extracted at the outer contacts of the device. We will discuss in turn the processes of light absorption and charge separation. The final step of charge extraction is treated in Chapter 10.

The first process is the light absorption that converts the absorbed photon into an excitation in the semiconductor such as a bound electron–hole pair or exciton. The effective harvesting of solar photons is an important requirement for effective solar energy conversion, as discussed in Chapter PSC.2. One primordial property is the thickness of the absorber, d. Since the low-energy photons usually have the longest pathway for absorption, the required layer thickness will be determined by the value of absorption coefficient in the photon frequency range of the absorption onset. The required range of values of d is also determined by light reflection/scattering properties of the device that increase the optical path length of weakly absorbed photons in the absorber. The thickness of the absorber layer is much smaller for strongly absorbing semiconductors, which typically necessitate only $d \approx 1$ μm, and these are termed thin film solar cells. A thinner absorber produces a significant advantage from the point of view of manufacturing and cost.

9.2 CHARGE SEPARATION

When an electron–hole pair is created by the absorption of a photon, bound states like excitons may be formed. Charge separation means that initially formed electron–hole pairs after photogeneration dissociate to form two separate unbound charge carriers of opposite polarity. Effective local charge separation is an essential feature to avoid immediate geminate recombination of individually photogenerated electron–hole pairs in low permittivity absorbers and/or at low temperatures (that decrease the thermal energy available to overcome the binding state). After the initial separation of the geminate pair, the so-called nongeminate recombination may occur between electrons in the conduction and holes in the valence band.

Specific mechanisms of charge separation may differ broadly depending on the type of absorber and its relative permittivity. In homogeneous semiconductors, such as standard inorganic thin films or crystalline silicon (c-Si) with relatively high-dielectric permittivities and low-exciton binding energies, local charge separation is very efficient without the assistance of electrical fields, at least at room temperature. Thus, each photogenerated carrier rapidly forms part of the respective ensemble of free carriers in the conduction or valence band, after a carrier thermalization time, see Section

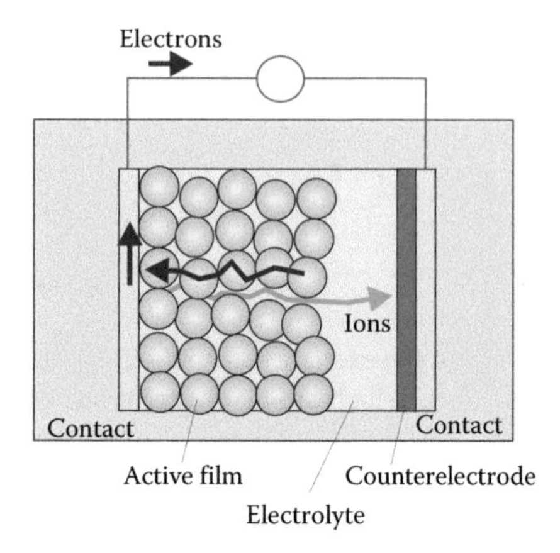

FIGURE 9.1 Scheme of a DSC indicating electron transport in TiO_2 metal oxide nanoparticulate network and hole transport by ion carriers in the liquid electrolyte.

ECK.5.6. In lead halide perovskite solar cells, the exciton population at room temperature seems to be negligible (Manser and Kamat, 2014; Yamada et al., 2014). The photogenerated carriers are separated on ps time scale and the radiative recombination occurs between uncorrelated electron–hole carriers rather than geminate pairs (Chen et al., 2014).

Many types of solar cells rely on a homogeneous semiconductor absorber layer that realizes both the functions of charge separation and carrier transport for their collection at the selective contacts. However, several types of nanostructured solar cells that use organic materials as absorbers establish a combination of materials to realize effective charge separation and subsequently, the transport of carriers occurs in separated phases. The morphology of a DSC, based on a TiO_2 nanostructure to which the dye is attached and filled with redox electrolyte that actuates as the hole conductor, is shown in Figure 9.1. The operation of the DSC has been commented upon in chapters ECK.8, ECK.9 and PSC.5, see Figures ECK.8.12 and PSC.5.8. The structure of a fully organic, bulk heterojunction (BHJ) organic solar cell is shown in Figure 9.2. Organic blends consist of a mixture of

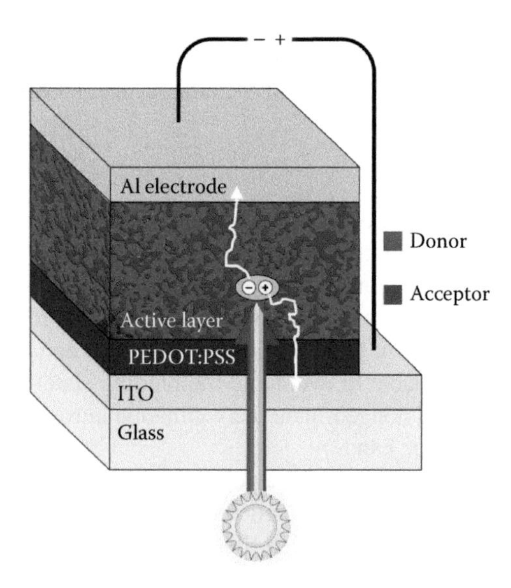

FIGURE 9.2 Structure of a BHJ organic solar cell. (Reproduced with permission from Verploegen, E. et al. *Advanced Functional Materials* 2010, 20, 3519–3529.)

electron donor and electron acceptor, as discussed previously in Figures ECK.6.5, PSC.3.26, and PSC.7.10. Light is absorbed at a polymer material, although small molecules or oligomers also give very good results as absorbers (Mishra and Bäuerle, 2012). The electron donor functions as the hole-transport material (HTM). Microstructure and morphology of the BHJ are key issues for the blend operational properties, and there is enormous variability of reported structures, but ideally, each phase should be formed by aggregates or crystallites a few nm thick, continuously connected along the film thickness.

All these types of solar cells employ the structure of a blend of nanomaterials and/or molecules to achieve the efficient separation of photogenerated electrons and holes, as indicated in Figure 9.3. The junctions are arranged to facilitate charge separation immediately after absorption of a photon. The interfaces that exist very close to the generation point offer a downhill gradient of energy to at least one of the carriers. The relaxation of carriers takes place in different materials and the relevant energy gap for determining the photovoltage, the *transport gap* (Bredas, 2014), is formed by the difference between the electron energy level in the electron-transport material (ETM) and the hole energy level in the HTM.

The detailed steps for charge generation and charge separation in a DSC and BHJ are indicated in Figure 9.4a and b. In the DSC, the photoexcited electrons are preferentially injected to TiO_2 and the hole is thus transferred to the redox carrier or a solid hole conductor such as Spiro-OMeTAD, while

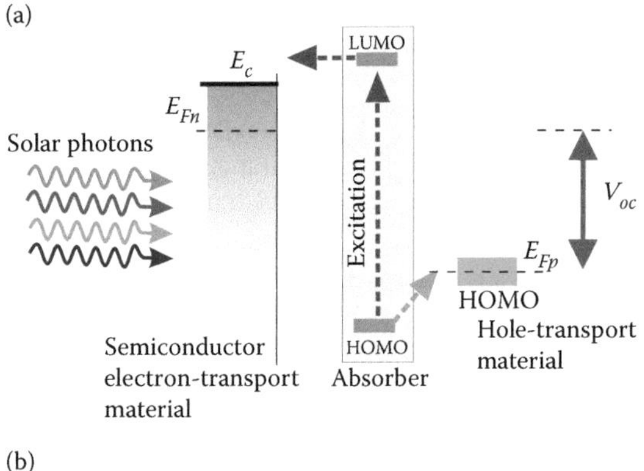

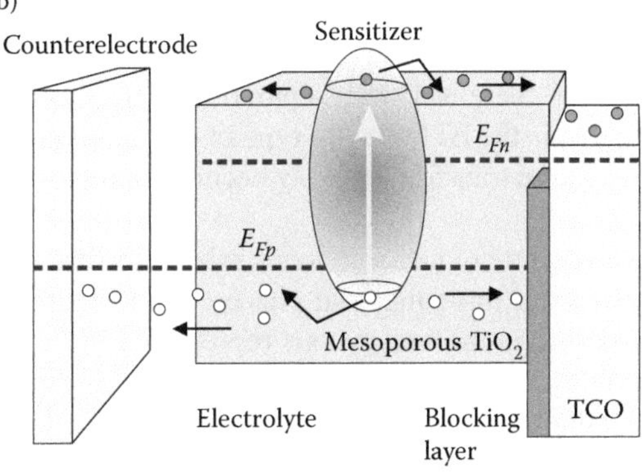

FIGURE 9.3 (a) Schematic steps of charge separation after the excitation of the dye in a DSC. (b) Charge separation and transport in a DSC. Electrons are injected from the excited state of a sensitizer to the TiO_2 framework, and they travel by diffusion across the semiconductor until collected at the transparent conducting oxide (TCO) substrate. Holes in the ground state of the absorber are ejected into a hole-transport medium. Holes are prevented to enter the electron collecting TCO by a blocking layer, and they are accepted at the counterelectrode (CE). The separation of the Fermi levels of electrons (E_{Fn}) and holes (E_{Fp}) produces a photovoltage. (Reproduced with permission from Kirchartz, T. et al. *Physical Chemistry Chemical Physics* 2015, 17, 4007–4014.)

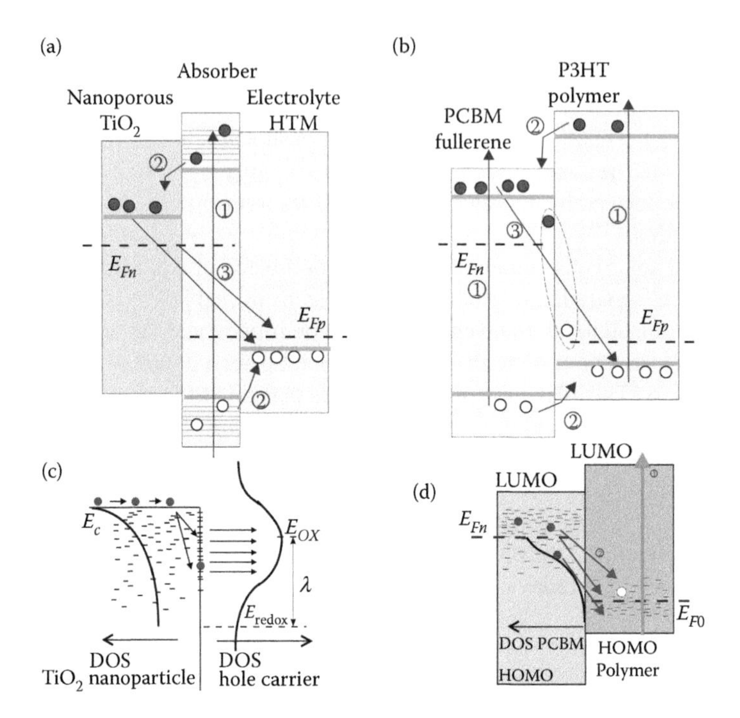

FIGURE 9.4 Energetic scheme of the different phases in (a) DSC and (b) organic BHJ indicating the different steps of (1) photogeneration of carriers, (2) charge separation from the absorber to the acceptor(s), and (3) recombination. (c, d) Represent the energy levels and charge transfer steps in the disordered media.

the opposite pathway is kinetically and energetically forbidden for each carrier. The diode functionality is achieved by adequate energy level matching and kinetic selectivity, as commented in Section PSC.5.5 and suggested in Figure 9.3. The photovoltaic operation of the standard DSCs relies to a large extent on unique properties of I^-/I_3^- redox species, the redox hole carrier that ensures excellent kinetically induced rectification of electron transfer both at TiO_2/dye and liquid/counterelectrode contacts (O'Regan and Grätzel, 1991; Peter, 2007). Kinetic preference for charge extraction can switch the photovoltage sign, as shown in a layer of quantum dots in Figure PSC.4.10b (Mora-Sero et al., 2013). In an organic BHJ, there is a rapid separation of the electron–hole pair by transfer of an electron to the fullerene that is the ETM in this type of cells, although it has been suggested that dissociation of an excitonic charge transfer state may become an important step for efficient conversion (Clarke and Durrant, 2010).

So far, we have obtained a picture of heterogeneous solar cells that is composed of a fine grained morphology that combines two or three material phases, as in Figures 9.1 and 9.2, which form a local charge-separation structure in the energy axis represented in Figure 9.4. The first reason for the nanostructured morphology in a BHJ is to realize charge separation in the polymer absorber phase in close vicinity of an interface. The active layer of the DSC is formed by a monolayer of dye that is locally sandwiched between nanostructured electron and hole conductors.

The small size of the constituents of the two transport phases facilitates effective electrical charge shielding, avoiding the space charge formation in long-range transport. If the sizes of the units of the phases are smaller than the Debye length, then the overall electroneutrality of positive and negative carriers can be maintained and transport can be realized by diffusion. These issues have been commented on in detail in Chapter FCT.3. Charge shielding in a DSC is supported by the electrolyte that causes the effective compensation of electronic charge injected to the metal oxide framework, as indicated in Figure ECK.8.12b. Organic blends for BHJ contain a significant amount of doping, usually p-type, that also facilitate the existence of a minority carrier. For example, P3HT

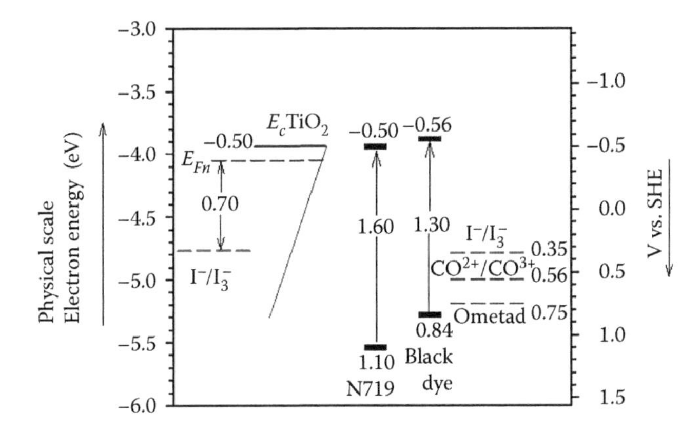

FIGURE 9.5 Energetic scheme of the components of a DSC. On the left is the DOS of TiO_2 and the position of the conduction band, indicating also the photovoltage by difference of the Fermi level of electrons and the redox potential in the electrolyte. On the right are shown the redox potentials of conventional hole conductors. In the center, the ground and excited state of standard dyes are shown.

forms a defect complex with oxygen that promotes p-doping in P3HT-PCBM molecular blends, where the doping is in the range 10^{14}–10^{16} cm^{-3} (Abdou et al., 1997).

Another important issue for the organic nanostructured cells is the electronic disorder indicated in the bottom row of Figure 9.4. Facile low-temperature deposition techniques adopted for the preparation of nanostructured materials and blends often lead to components that contain a large amount of structural defects and irregular morphologies. Therefore, charge transport and charge transfer determining the recombination rates adopt the properties characteristic of disordered materials, which have been amply described in Chapter FCT.4. By following the methods of chemical capacitance explained in Chapters ECK.7 and ECK.8, it is possible to establish the density of states (DOS) in the ETM and HTM that form organic nanostructured cells, as suggested in Figure 9.4. The exponential DOS usually found in nanostructured TiO_2 has been discussed in Figures ECK.7.7, ECK.7.8, and ECK.9.34. The result of a quantitative determination of the TiO_2 energy levels in a DSC in reference to dye and redox electrolyte levels are shown in Figure 9.5. Figure 9.6 shows an example of the determination of the DOS of BHJ cells that use two different fullerene acceptors. Figure 9.7 shows a comparison of the dominant energy levels for photovoltaic operation in a DSC and in a BHJ.

9.3 MATERIALS LIMITS TO THE PHOTOVOLTAGE

When the charge separation step described in the previous section has been realized, we obtain an ensemble of free charge carriers in the semiconductor, or in separate phases, where they are mobile. Since these are excess populations of electrons and holes with respect to the equilibrium situation, we have formed a situation of separation of the Fermi levels and the recombination process starts immediately, while the continuous absorption of photons produces a renewed supply of free electronic carriers. If the device is at open-circuit condition, the carriers accumulate inside causing a photovoltage V_{oc} but they cannot leave the device to produce electrical work. Nevertheless, as discussed in Chapter PSC.8, the photovoltage is a central feature of the solar cell determining to a great extent the power conversion efficiency. In Section PSC.7.2, we have characterized the ideal properties and fundamental limitations of the photovoltage. We now examine different types of nonidealities and limitations imposed by materials properties on the V_{oc}.

First of all, we note that the number of electrons and holes that determine the separation of the Fermi levels is established by an equilibrium of the processes of generation and recombination.

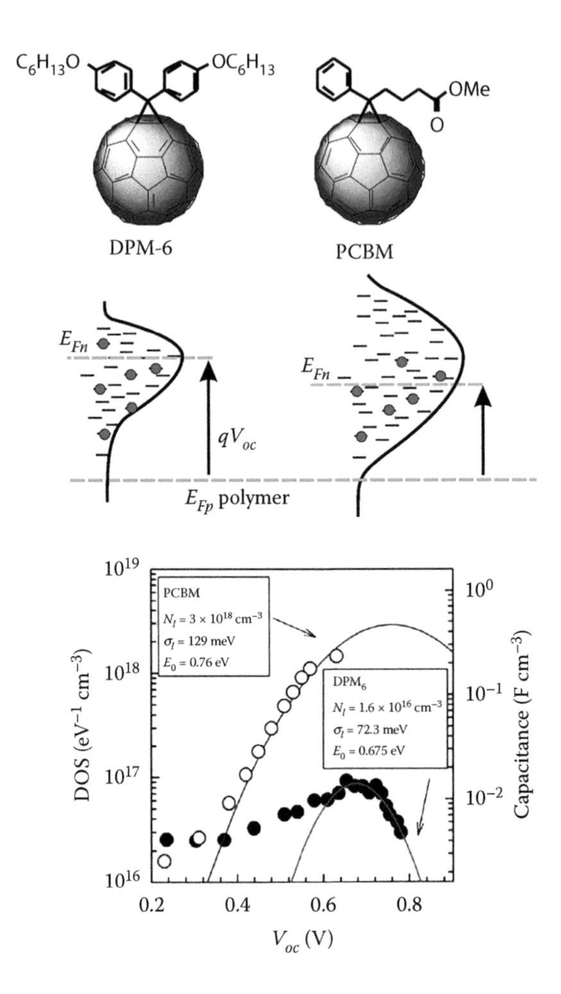

FIGURE 9.6 Capacitance values measured from low-frequency values of impedance spectroscopy as a function of V_{oc} reached under varying illumination levels. White dots correspond to PCBM-based solar cells and black dots to DPM_6-based solar cells. Gaussian DOS (*solid lines*) and distribution parameters resulting from fits. The top drawing shows the fullerene molecules and a pictorial representation of the DOS in each case. (Reproduced with permission from Garcia-Belmonte, G. et al. *Journal of Physical Chemistry Letters* 2010, 1, 2566–2571.)

From the expression that relates the Fermi level separation to the product densities of electrons and holes, Equation ECK.5.37, we obtain

$$qV_{oc} = E_g + k_B T \ln\left(\frac{np}{N_c N_v}\right) \tag{9.1}$$

As observed in Chapter PSC.7, the photovoltage depends critically on the rate of recombination that determines the product density np. When we consider a recombination process in the bulk absorber, any deviation of the recombination from an ideal bimolecular law will have a signature on the V_{oc} of the solar cell. Previously, we have stated two central models for recombination current under the assumptions of minority carrier and electroneutrality-dominated recombination in Section PSC.5.2. We now describe more generally the bimolecular recombination rate as a function of local carrier densities as follows (Correa-Baena et al., 2017):

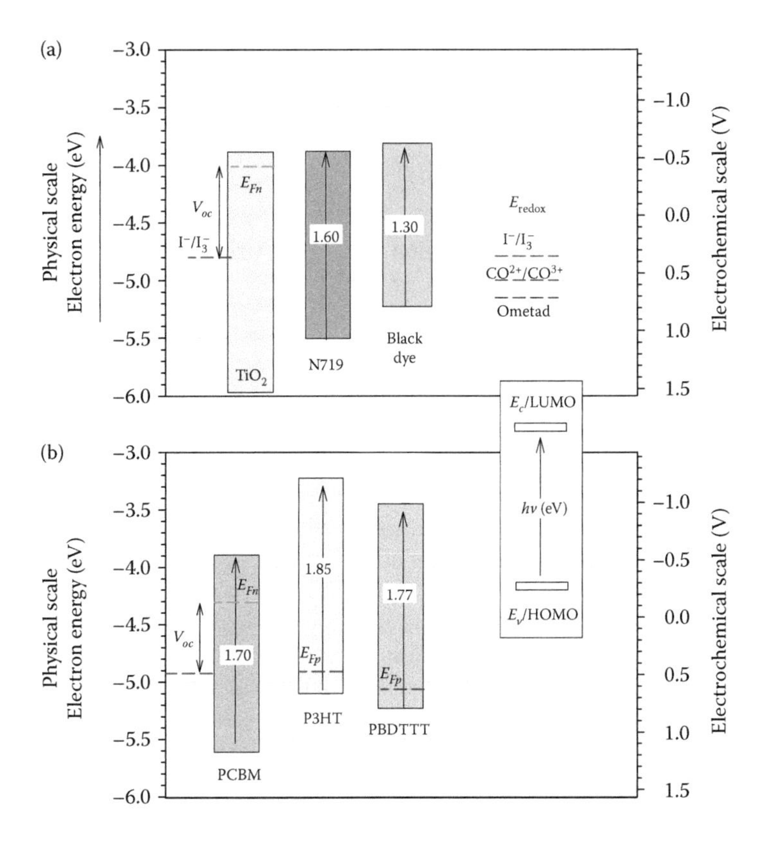

FIGURE 9.7 Schematic energy diagrams of the materials components of (a) dye-sensitized and (b) organic BHJ solar cells. Standard energy levels are given on the solid state, one electron energy scale, while the origin of the electrochemical scale is taken at −4.44 eV. E_{Fn} and E_{Fp} are the Fermi levels of electrons and holes (the I^-/I_3^- redox potential in most DS cells). Their difference relates to V_{oc} as indicated. TiO_2 and dye energy levels depend on the solution components; dye levels may also depend on criteria for the absorption onset. (Reproduced with permission from Nayak, P. K. et al. *Advanced Materials* 2011, 23, 2870–2876.)

$$U_n = B_{rec}\left[(np)^{1/m_d} - n_i^{2/m_d}\right] \tag{9.2}$$

Here, the parameter B_{rec} is a kinetic recombination rate and m_d accounts for the nonideal behavior of recombination that departs from the strict bimolecular law corresponding to Equation PSC.2.42, and the product is given in Equation PSC.5.13. The ideal value for band-to-band recombination is $m_d = 1$ by Equation PSC.5.15, but any source of trap-assisted recombination leads to $m_d > 1$ (Ansari-Rad et al., 2012). The local generation rate by absorbed light flux Φ_{ph} is approximated as

$$G \propto \frac{1}{d}\Phi_{ph} \tag{9.3}$$

Under intense illumination, the number of photogenerated carriers is larger than the native doping density. Using electroneutrality condition (PSC.5.14) and neglecting the constant term in Equation 9.2, the recombination rate takes the form

$$U_n = B_{rec} n^{2/m_d} \tag{9.4}$$

The open-circuit condition $G = U_n$ implies that the carrier density depends on photon flux as

$$n = \left(\frac{1}{d\,B_{rec}} \Phi_{ph} \right)^{m_d/2} \tag{9.5}$$

The photovoltage dependence on illumination flux can be written as

$$\frac{dV_{oc}}{d\ln(\Phi_{ph})} = \frac{m_d k_B T}{q} \tag{9.6}$$

Therefore, the nonideality of the recombination model is reflected in the diode parameter and produces a change of V_{oc} in agreement with Equations PSC.7.10 and PSC.8.3. Consequently, the measurement of the slope of the open-circuit voltage versus light intensity is a useful method to determine the recombination mechanisms in a solar cell device (Huang et al., 1997; Gouda et al., 2015). Figure 9.8 shows the characteristic curve of lead halide perovskite solar cell with metal oxide nanostructures of different composition. The low-light intensity region is sensitive to the contact properties, while at high-generation rates, both cells show the same behavior, which allows to identify the bulk recombination properties (Correa-Baena et al., 2017).

As suggested by Figure 9.8, besides the details of bulk recombination, layers that are used as selective contacts may impose important restrictions to the V_{oc}. Recombination at defects in the contact layers is a major source of decrease of performance in crystalline solar cells. In crystalline silicon solar cells, electrons and holes are separated, accumulated, and transported in the absorber medium itself. The selective contacts consist of extremely localized regions of strong electrical fields at the edges of a thick absorber, as further discussed in the final paragraphs of this section. To prevent the carrier loss in silicon solar cells, a back surface field is formed, which reflects the majority carrier, or a passivation layer is introduced that may be made up of materials like SiO_2, amorphous Si, or Al_2O_3 (Taguchi et al., 2000; Schmidt et al., 2008). In lead halide perovskite solar cells, it has been observed that surface recombination at the outer contact is a major source of loss of photogenerated carriers (Zarazua et al., 2016).

In thin film inorganic CdTe solar cells, a $CdCl_2$ "activation" treatment considerably enhances the performance by improving the properties of CdTe/CdS heterointerfaces (Jaegermann et al., 2009). As indicated in Figure 9.9, in the initial abrupt interface, a large density of interfacial electronic defects causes the Fermi level pinning and large surface recombination rates. The $CdCl_2$ activation

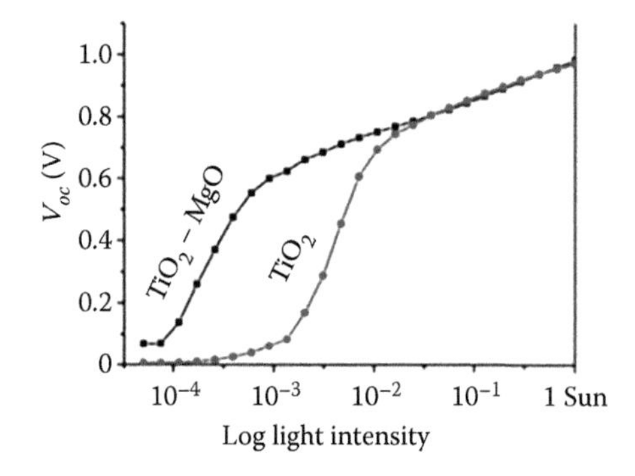

FIGURE 9.8 V_{oc} versus log I_0 (light intensity) curves. Comparison between a TiO_2 perovskite solar cell and a TiO_2–MgO perovskite solar cell. (Reproduced with permission from Gouda, L. et al. *The Journal of Physical Chemistry Letters* 2015, 6, 4640–4645.)

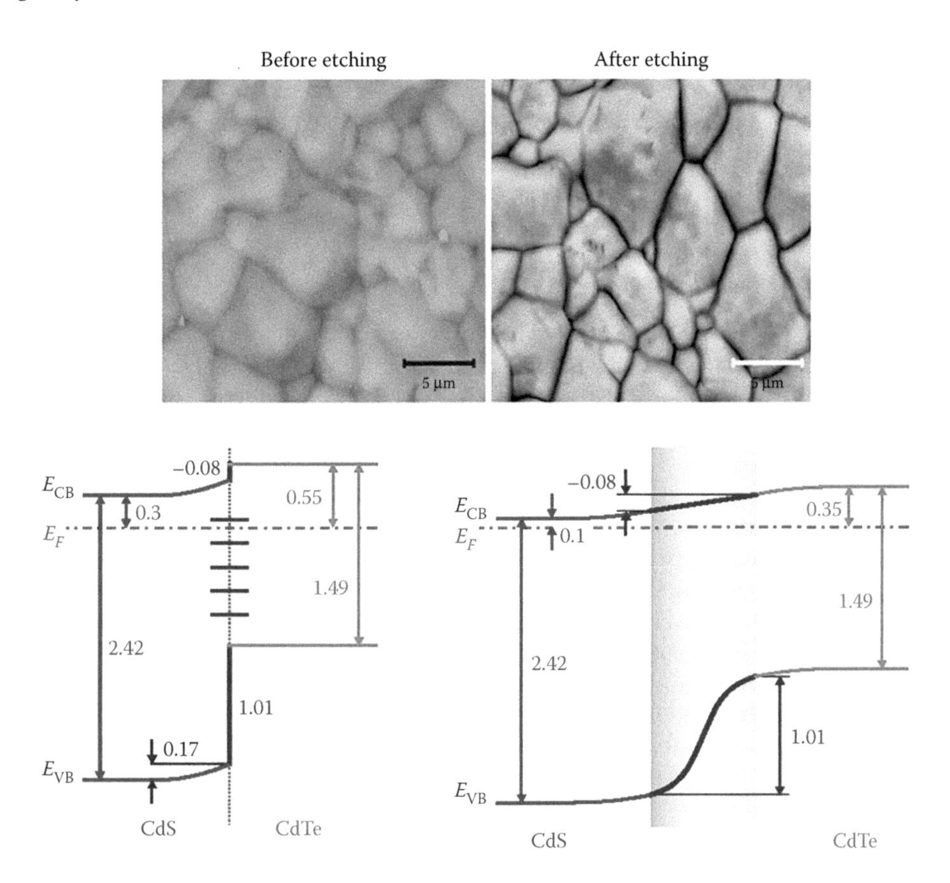

FIGURE 9.9 Grain structure and band energy diagram of CdS/CdTe interfaces prepared by PVD before activation (*left*) and after activation (*right*). (Reproduced with permission from Jaegermann, W. et al. *Advanced Materials* 2009, 21, 4196–4206.)

favors interdiffusion of CdS and CdTe at the phase boundary, which leads to a reduction of the interface density of states. In general, a proper passivation of the interfacial electronic defect levels is an essential precondition for the realization of efficient thin film solar cells.

We now consider the limitations to the photovoltage imposed by the materials properties of the ETM and HTM in nanostructured solar cells. In the DSC (composed by three materials: absorber, ETM, and HTM), rapid charge extraction from the dye absorber requires a downhill driving force, as emphasized in Figure 9.3. Therefore, the conduction band in the ETM and the transport level in the HTM (such as the redox energy of the electrolyte) need to be lower and higher, respectively, than the energy level of the excited carriers in the dye. Since the V_{oc} is determined by the difference of Fermi levels in the ETM and HTM, the limits of the Fermi level variation in these materials impose the constraint to the maximal V_{oc} that may be obtained. For the energetically favorable electron transfer, the energy of the excited state of the dye has to be about 0.3 eV higher than that of the conduction band (CB) of the electron transporter, that is, TiO_2. It is possible that E_{Fn} raises up to the CB edge, but here, a very large density of states prevents the Fermi level from further increase. On the hole transporter side, a driving force for the regeneration of the oxidized dye is required as well. This is up to 500 mV with the best carrier, an I_3^-/I^- redox couple in an organic solvent. Unfortunately, the redox energy of this vital element of the DSC lies high in the energy scale, at $+0.35\ V_{NHE}$, see Figure 9.5. The V_{oc} will be higher if the HTM has a lower-energy level in the energy scale, as indicated in several examples in Figures 9.5 and 9.7, such as a Co-based redox couple. However, these alternative HTMs present lower kinetic selectivity than the I_3^-/I^- redox couple.

In summary, in a DSC, the primary available voltage is related to the separation of chemical potential of electrons and holes *in the dye*, but the photovoltage measured from outer contacts is produced by the difference of the Fermi level of electrons in TiO_2 and the redox level of the ionic carriers. The combination of the two energy steps for charge separation produces a limitation of the V_{oc}, as follows:

$$qV_{oc,\max} = E_{c,\mathrm{ETM}} - E_{F,\mathrm{HTM}} \tag{9.7}$$

Thus, maximal V_{oc} becomes unrelated to the ideal model of recombination in the absorber, which is manifested by the very low-external luminescence, as already commented on in Section PSC.7.3.

We have already indicated that the organic BHJ is a two-material solar cell with a distributed donor–acceptor interface. In Sections PSC.3.6 and PSC.7.5, we have remarked on the optical transitions that may occur in a donor–acceptor blend. The photon absorption across the intrinsic gap E_D is the main light absorption channel in an organic BHJ. For achieving the effective charge separation in the blend, the injection state of the organic absorber needs to be about 0.3 eV higher than that of the CB of the electron transporter that is the LUMO level of the fullerene (Veldman et al., 2009). The transport gap is defined as $E_0 = E_{LUMO(A)} - E_{HOMO(D)}$. Since $E_{LUMO(A)}$ and $E_{HOMO(D)}$ pose a limitation to the separation of the Fermi levels in the blend, the quantity E_0 is the maximal possible value of V_{oc}.

The existence of a broad DOS in the electron and HTMs has important implications for the photovoltage, as already commented in Figure ECK.8.11 (Nayak et al., 2012). The electron carriers generated will pile up at the bottom of the available states of the DOS, as shown in Figure ECK.7.10. Hence, a material with a small density of states in the bandgap will favor a higher-electron Fermi level and consequently a larger V_{oc}, and a similar argument applies for holes. Similarly, a correlation between the DOS and the photovoltage can be observed in organic BHJ solar cells (Garcia-Belmonte and Bisquert, 2010). In the comparison shown in Figure 9.6, it is found that DPM_6 (4,4′-dihexyloxy-diphenylmethano[60]fullerene) has a smaller DOS in the gap than PCBM, so that a full Gaussian DOS is observed in the former case. If the recombination rate is similar in both cases, then the Fermi level of electrons raises higher in the fullerene with the smaller DOS, producing a larger photovoltage as indicated in the scheme of Figure 9.6 and observed experimentally (Garcia-Belmonte et al., 2010).

The effect of accumulation of carriers in the Gaussian DOS upon obtaining a photovoltage can be quantitatively described using the properties of the distributions that have been discussed in Chapter ECK.8.6 in terms of the disorder parameters σ_n, σ_p that give the width of the distributions. For low-occupancy conditions, the carriers thermalize in the bottom of the DOS and the formula in Equation 9.1 is modified as follows (Garcia-Belmonte, 2010):

$$qV_{oc} = E_g - \frac{\sigma_n^2 + \sigma_p^2}{2k_BT} + k_BT \ln\left(\frac{np}{N_cN_v}\right) \tag{9.8}$$

Similar to inorganic thin film solar cells, in hybrid and organic nanostructured solar cells, the macroscopic contacts play an important role in selecting just one carrier from the mixture of phases upon arrival to the contact. This function is usually assisted by the use of blocking layers described in Section PSC.4.6. Barrier layers at the contacts of a BHJ organic cell improve the selectivity of contacts by repelling one of the carriers, as shown in Figure 9.10. In solid DSCs, a thin TiO_2 blocking layer protecting the TCO contact has a crucial effect to prevent a shorting current from the hole transporting phase. The same type of blocking layer is usually adopted in lead halide perovskite solar cells.

The distribution of electric potential in a nanostructured cell in relation to the explanation of the origin of the photovoltage has been often discussed in the area of nanostructured cells (Pichot and

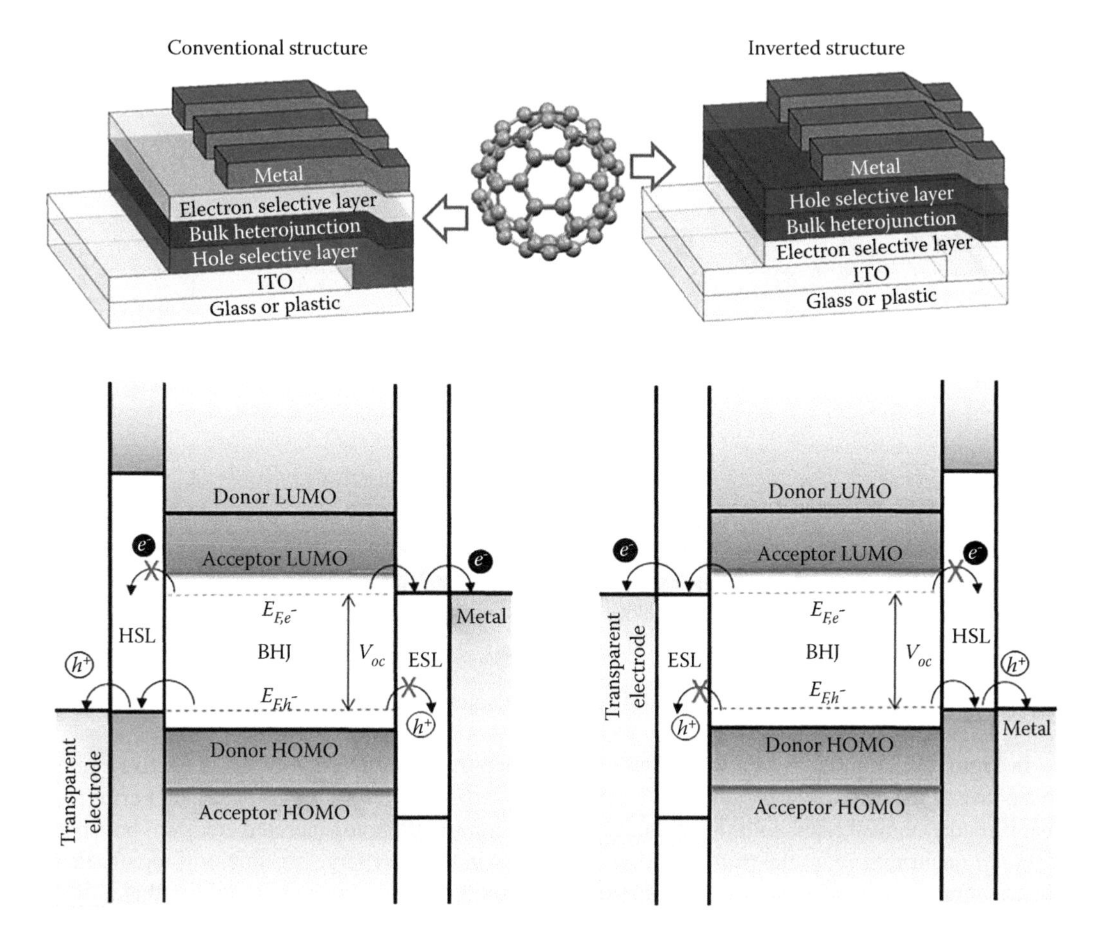

FIGURE 9.10 Schematic illustration of the structure of organic blend of fullerene/polymer layer and electron selective layer of conventional and inverted polymer solar cells, and their energy diagram. (Reproduced with permission from Li, C.-Z. et al. *Journal of Materials Chemistry* 2012, 22, 4161–4177.)

Gregg, 2000). When the applied voltage is modified in a DSC device, a change of the vacuum level will necessarily occur, and similarly, when a photovoltage is generated under illumination. Since the Debye length in the electrolyte has the value of a few nm, establishing a macroscopic field in the direction normal to the contacts is unlikely, see Figure ECK.9.26, which implies that the variation of local potential must be absorbed in a size of several nm, close to the outer interface with the TCO, as shown in Figure 9.11. To absorb the whole modification of the external potential difference in a region close to the interface, a change of band bending occurs in a highly doped medium, as in the TCO, or a change of voltage of the Helmholtz layer takes place at the substrate surface. Effectively, the interface dipole at the semiconductor–substrate interface is modified in equal magnitude to the change of the Fermi level, as remarked earlier in Figure PSC.4.8c.

The operation of a p–n junction as a selective contact has been discussed in Figure ECK.9.21. In Figure 9.12a, we indicate an operation of the p–n junction as a charge separating electrical field reminiscent of the model of Figure 4.11. This view is very typical in older texts on solar cells. It should be emphasized that silicon is an indirect semiconductor and poor light absorber; thus a thick absorber layer of about 200–300 μm (depending on light trapping schemes) is necessary to capture all the solar photons. For characteristic doping density of 10^{17} cm^{-3}, the size of the depletion region in the p–n junction is about 50 nm (Figure ECK.9.4). This means that in contrast to the drawing of

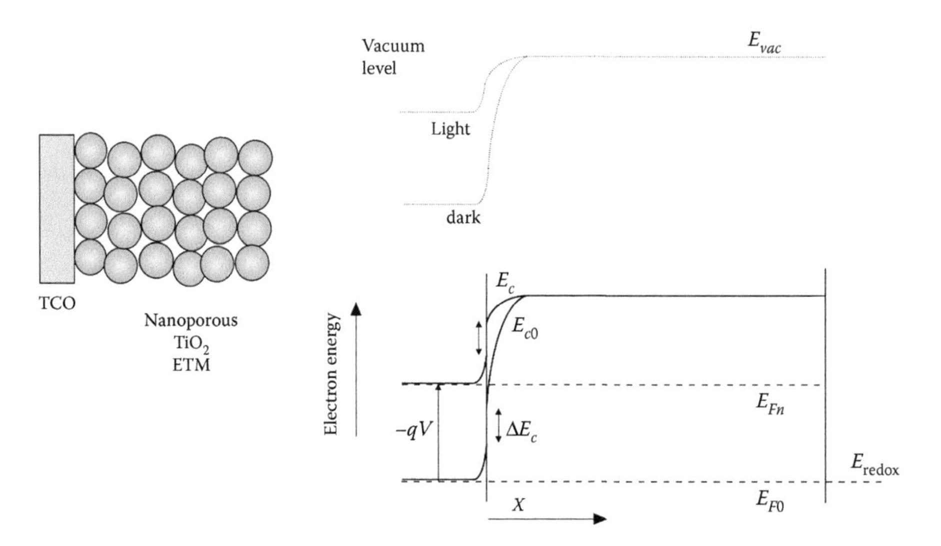

FIGURE 9.11 Changes of the position of the conduction band in a DSC when the voltage is increased, showing also the changes of the VL across the device. The left diagram shows the morphology of the active layer and the contact.

Figure 9.12a, the p–n junction in a crystalline silicon solar cell occupies an extremely thin region and is located on the boundary.

In Figure 9.12b and c, we show the operation of the p–n junction as a very thin selective contact that converts the minority carrier Fermi level in the absorber side into a majority carrier Fermi level that is readily equilibrated with the metal. Charge separation of photogenerated electron–hole pairs *does not* occur mainly in the thin depletion region, but in the very large quasineutral region. From the conceptual point of view, the p–n junction is just another mechanism of selectivity that absorbs

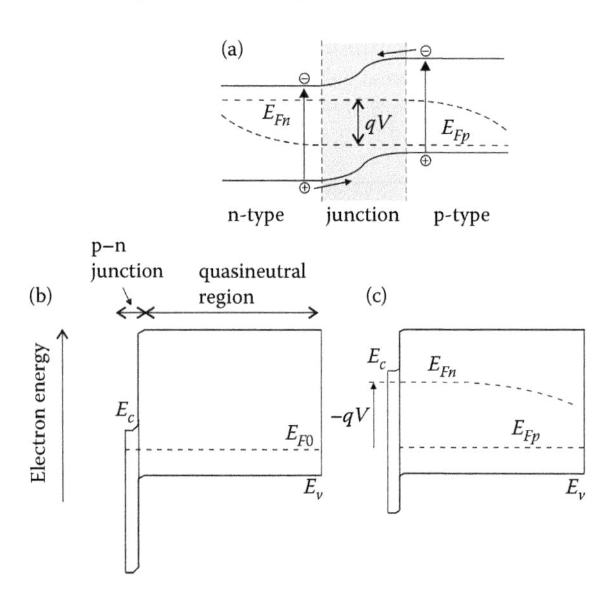

FIGURE 9.12 (a) Energy diagram of p–n junction. Depletion layers existing in the junction region are shown, as well as the Fermi levels of electrons and holes whose separation is the voltage across the junction. (b) Zero voltage and (c) forward voltage (or photovoltage under illumination) scheme of a p-type crystalline silicon solar cell. The p–n junction is a thin region at the left contact, where minority carrier electrons from the p-side are converted to majority carriers at the n-side of the junction.

the variation of voltage in a rather thin interfacial layer, very similar to the TCO/TiO_2 contact in a DSC previously discussed in Figure 9.11.

GENERAL REFERENCES

Modification of potential at the contacts and origin of photovoltage: Cahen et al. (2000), Pichot and Gregg (2000), Kron et al. (2003), Turrión et al. (2003), Bisquert et al. (2004), and Bisquert and Garcia-Belmonte (2011).

Morphology of organic solar cells: Pfannmöller et al. (2011), Creddington and Durrant (2012), and Guerrero and Garcia-Belmonte (2016).

Effect of disorder on open-circuit voltage: Tiedje (1982), Yablonovitch et al. (1982), Garcia-Belmonte (2010), Garcia-Belmonte and Bisquert (2010), Nayak et al. (2011), Nayak et al. (2012), and Garcia-Belmonte (2013).

REFERENCES

Abdou, M. S. A.; Orfino, F. P.; Son, Y.; Holdcroft, S. Interaction of oxygen with conjugated polymers: Charge transfer complex formation with poly(3-alkylthiophenes). *Journal of the American Chemical Society* 1997, 119, 4518–4524.

Ansari-Rad, M.; Abdi, Y.; Arzi, E. Reaction order and ideality factor in dye-sensitized nanocrystalline solar cells: A theoretical investigation. *The Journal of Physical Chemistry C* 2012, 116, 10867.

Bisquert, J.; Cahen, D.; Rühle, S.; Hodes, G.; Zaban, A. Physical chemical principles of photovoltaic conversion with nanoparticulate, mesoporous dye-sensitized solar cells. *The Journal of Physical Chemistry B* 2004, 108, 8106–8118.

Bisquert, J.; Garcia-Belmonte, G. On voltage, photovoltage, and photocurrent in bulk heterojunction organic solar cells. *Journal of Physical Chemistry Letters* 2011, 2, 1950–1964.

Bredas, J.-L. Mind the gap! *Materials Horizons* 2014, 1, 17–19.

Cahen, D.; Hodes, G.; Grätzel, M.; Guillemoles, J. F.; Riess, I. Nature of photovoltaic action in dye-sensitized solar cells. *The Journal of Physical Chemistry B* 2000, 104, 2053–2059.

Chen, K.; Barker, A. J.; Morgan, F. L. C.; Halpert, J. E.; Hodgkiss, J. M. Effect of carrier thermalization dynamics on light emission and amplification in organometal halide perovskites. *The Journal of Physical Chemistry Letters* 2014, 6, 153–158.

Clarke, T. M.; Durrant, J. R. Charge photogeneration in organic solar cells. *Chemical Reviews* 2010, 110, 6736–6767.

Correa-Baena, J.-P.; Turren-Cruz, S.-H.; Tress, W.; Hagfeldt, A.; Aranda, C.; Shooshtari, L.; Bisquert, J.; Guerrero, A. Changes from bulk to surface recombination mechanisms between pristine and cycled perovskite solar cells. *ACS Energy Letters* 2017, 2, 681–688.

Creddington, D.; Durrant, J. R. Insights from transient optoelectronic analyses on the open-circuit voltage of organic solar cells. *Journal of Physical Chemistry Letters* 2012, 3, 1465–1478.

Garcia-Belmonte, G. Temperature dependence of open-circuit voltage inorganic solar cells from generation–recombination kinetic balance. *Solar Energy Materials and Solar Cells* 2010, 94, 2166–2169.

Garcia-Belmonte, G. Carrier recombination flux in bulk heterojunction polymer: Fullerene solar cells: Effect of energy disorder on ideality factor. *Solid-State Electronics* 2013, 79, 201–205.

Garcia-Belmonte, G.; Bisquert, J. Open-circuit voltage limit caused by recombination through tail states in bulk heterojunction polymer-fullerene solar cells. *Applied Physics Letters* 2010, 96, 113301.

Garcia-Belmonte, G.; Boix, P. P.; Bisquert, J.; Lenes, M.; Bolink, H. J.; La Rosa, A.; Filippone, S.; Martín, N. Influence of the intermediate density-of-states occupancy on open-circuit voltage of bulk heterojunction solar cells with different fullerene acceptors. *Journal of Physical Chemistry Letters* 2010, 1, 2566–2571.

Gouda, L.; Gottesman, R.; Ginsburg, A.; Keller, D. A.; Haltzi, E.; Hu, J.; Tirosh, S.; Anderson, A. Y.; Zaban, A.; Boix, P. P. Open circuit potential build-up in perovskite solar cells from dark conditions to 1 sun. *The Journal of Physical Chemistry Letters* 2015, 6, 4640–4645.

Guerrero, A.; Garcia-Belmonte, G. Recent advances to understand morphology stability of organic photovoltaics. *Nano-Micro Letters* 2016, 9, 10.

Huang, S. Y.; Schlichthörl, G.; Nozik, A. J.; Grätzel, M.; Frank, A. J. Charge recombination in dye-sensitized nanocrystallyne TiO_2 solar cells. *The Journal of Physical Chemistry B* 1997, 101, 2576–2582.

Jaegermann, W.; Klein, A.; Mayer, T. Interface engineering of inorganic thin-film solar cells—Materials-science challenges for advanced physical concepts. *Advanced Materials* 2009, 21, 4196–4206.

Kirchartz, T.; Bisquert, J.; Mora-Sero, I.; Garcia-Belmonte, G. Classification of solar cells according to mechanisms of charge separation and charge collection. *Physical Chemistry Chemical Physics* 2015, 17, 4007–4014.

Kron, G.; Egerter, T.; Werner, J. H.; Rau, W. Electronic transport in dye-sensitized nanoporous TiO_2 solar cells—Comparison of electrolyte and solid-state devices. *The Journal of Physical Chemistry B* 2003, 107, 3556.

Li, C.-Z.; Yip, H.-L.; Jen, A. K. Y. Functional fullerenes for organic photovoltaics. *Journal of Materials Chemistry* 2012, 22, 4161–4177.

Manser, J. S.; Kamat, P. V. Band filling with free charge carriers in organometal halide perovskites. *Nature Photonics* 2014, 8, 737–743.

Mishra, A.; Bäuerle, P. Small molecule organic semiconductors on the move: Promises for future solar energy technology. *Angewandte Chemie International Edition* 2012, 51, 2020–2067.

Mora-Sero, I.; Bertoluzzi, L.; Gonzalez-Pedro, V.; Gimenez, S.; Fabregat-Santiago, F.; Kemp, K. W.; Sargent, E. H.; Bisquert, J. Selective contacts drive charge extraction in quantum dot solids via asymmetry in carrier transfer kinetics. *Nature Communications* 2013, 2, 2272.

Nayak, P. K.; Bisquert, J.; Cahen, D. Assessing possibilities and limits for solar cells. *Advanced Materials* 2011, 23, 2870–2876.

Nayak, P. K.; Garcia-Belmonte, G.; Kahn, A.; Bisquert, J.; Cahen, D. Photovoltaic efficiency limits and material disorder. *Energy & Environmental Science* 2012, 5, 6022–6039.

O'Regan, B.; Grätzel, M. A low-cost high-efficiency solar cell based on dye-sensitized colloidal TiO_2 films. *Nature* 1991, 353, 737–740.

Peter, L. M. Transport, trapping and interfacial transfer of electrons in dye-sensitized nanocrystalline solar cells. *Journal of Electroanalytical Chemistry* 2007, 599, 233–240.

Pfannmöller, M.; Flügge, H.; Benner, G.; Wacker, I.; Sommer, C.; Hanselmann, M.; Schmale, S. et al. Visualizing a homogeneous blend in bulk heterojunction polymer solar cells by analytical electron microscopy. *Nano Letters* 2011, 11, 3099–3107.

Pichot, F.; Gregg, B. A. The photovoltage-determining mechanism in dye-sensitized solar cells. *The Journal of Physical Chemistry B* 2000, 104, 6–10.

Schmidt, J.; Merkle, A.; Brendel, R.; Hoex, B.; de Sanden, M. C. M. v.; Kessels, W. M. M. Surface passivation of high-efficiency silicon solar cells by atomic-layer-deposited Al_2O_3. *Progress in Photovoltaics: Research and Applications* 2008, 16, 461–466.

Taguchi, M.; Kawamoto, K.; Tsuge, S.; Baba, T.; Sakata, H.; Morizane, M.; Uchihashi, K.; Nakamura, N.; Kiyama, S.; Oota, O. HITTM cells—High-efficiency crystalline Si cells with novel structure. *Progress in Photovoltaics: Research and Applications* 2000, 8, 503–513.

Tiedje, T. Band tail recombination limit to the output voltage of amorphous silicon solar cells. *Applied Physics Letters* 1982, 40, 627–629.

Turrión, M.; Bisquert, J.; Salvador, P. Flatband potential of $F{:}SnO_2$ in a TiO_2 dye-sensitized solar cell: An interference reflection study. *The Journal of Physical Chemistry B* 2003, 107, 9397–9403.

Veldman, D.; Meskers, S. C. J.; Janssen, R. A. J. The energy of charge-transfer states in electron donor-acceptor blends: Insights into the energy losses in organic solar cells. *Advanced Functional Materials* 2009, 19, 1939–1948.

Verploegen, E.; Mondal, R.; Bettinger, C. J.; Sok, S.; Toney, M. F.; Bao, Z. Effects of thermal annealing upon the morphology of polymer–fullerene blends. *Advanced Functional Materials* 2010, 20, 3519–3529.

Yablonovitch, E.; Tiedje, T.; Witzke, H. Meaning of the photovoltaic band gap for amorphous semiconductors. *Applied Physics Letters* 1982, 41, 953–955.

Yamada, Y.; Nakamura, T.; Endo, M.; Wakamiya, A.; Kanemitsu, Y. Photocarrier recombination dynamics in perovskite $CH_3NH_3PbI_3$ for solar cell applications. *Journal of the American Chemical Society* 2014, 136, 11610–11613.

Zarazua, I.; Han, G.; Boix, P. P.; Mhaisalkar, S.; Fabregat-Santiago, F.; Mora-Seró, I.; Bisquert, J.; Garcia-Belmonte, G. Surface recombination and collection efficiency in perovskite solar cells from impedance analysis. *The Journal of Physical Chemistry Letters* 2016, 7, 5105–5113.

10 Charge Collection in Solar Cells

This chapter completes the analysis of the operation of solar cells by describing the mechanisms of charge collection, composed of the combination of carrier transport, recombination, and extraction at the selective contacts. We establish the main intuitive ideas concerning the charge collection by either diffusion, or by drift in strong electrical fields associated with space-charge regions. The modeling of the solar cell by transport–recombination equations, as well as the boundary conditions, is analyzed in detail. Increasingly complex analytical models are reviewed, and then we discuss the numerical simulation that allows to treat any type of complex morphology of charge and current distribution. We form a basic classification of the main operation modes for charge collection, depending on the size of the depletion region with respect to the thickness of the absorber. We finish with some practical considerations for the measurement and reporting of solar cell performance.

10.1 INTRODUCTION TO CHARGE COLLECTION PROPERTIES

The photovoltaic operation leading to a photocurrent can be analyzed in terms of the functions of charge generation, charge separation, and charge collection, as commented in Chapter PSC.9. Once the initial charge separation step is achieved, which in many inorganic and hybrid perovskite solar cells is an extremely fast process, it is necessary to establish a flux of both types of carriers toward separate contacts. The carrier collection across the active layer in competition with the recombination losses that take place either in the bulk or at interfaces of the material will be analyzed in the forthcoming sections of this chapter.

In the previous analysis of Chapter PSC.4, we have presented two archetype models for the operation of a solar cell with respect to charge collection. First is the model of Figures PSC.4.8 and PSC.5.1, where the bands are flat in all circumstances. Here, the problem of transport is suppressed by declaring that the mobilities are infinite, in other words, an internal quantum efficiency (IQE_{PV}) of 1 is assumed. This model represents an "ideal" solar cell where charge collection is optimal without the need for fields that produce drift transport, and the current–voltage characteristic is completely determined by recombination, which modulates the internal loss current depending on the applied voltage.

However, in Section PSC.4.5 and Figure PSC.4.11, we have discussed a solar cell formed by an intrinsic absorber semiconductor with slanted bands, in which the performance is limited by charge collection determined by the drift current in the electrical field.

We now start a broader investigation of the mechanisms of generation, recombination, and transport in solar cells, based on the extensive discussion of the carrier transport properties in the volume FCT of this collection. Chapter FCT.3 describes the drift–diffusion and conservation equations and is especially relevant for the general problem of solar cell operation. We reached the conclusion that the transport is governed by the gradient of each Fermi level, so that calculating the distribution of electron and hole Fermi levels is one important goal of the simulation and modeling of solar cells. We will treat this problem with increasing degrees of complexity, showing a set of representative and important models, and applying simulation tools to obtain a general description of the Fermi levels in a solar cell device under illumination. For most applications, it can be assumed that the excess carrier transport occurs by carriers thermalized to the band edges, so that the relationships $n(E_{Fn})$ and $p(E_{Fp})$ in Equations PSC.5.11 and PSC.5.12 can be used to change between carrier density and the Fermi level representation of the charge distribution.

In organic and nanostructured solar cells, we find the features of disorder that have been commented upon in Chapter PSC.9. The presence of energy or structural disorder affects the photovoltage and implies that transport characteristics require the specific methods and concepts of disordered systems such as the hopping conductivity. These tools have been already reviewed in Chapter FCT.4 and will not be extensively considered here.

The combination of light harvesting, charge generation, and charge collection phenomena, including recombination effects, determine the photovoltaic EQE, which has been described in Equation PSC.2.28. The EQE_{PV} is the central function that establishes the practical value of the photocurrent. It also determines the fundamental operation properties of the solar cell, as discussed in Chapter PSC.7. The total photocurrent j_{ph} can be directly measured, Equation PSC.8.4, but additional insight is obtained by the composition of EQE_{PV} and the spectral photon number, as indicated in Equation PSC.6.1.

To illustrate real characteristics of charge collection, we show the typical behavior of EQE_{PV} and total photocurrent in a number of solar cell technologies. Figure 10.1 shows the features of the organic absorber molecules in a DSC. The figure compares the performance of the reference ruthenium dye N719 already discussed in Figure PSC.5.12 with two porphyrin dyes coded YD0 and YD2 (Barea et al., 2011). Porphyrins are molecules that contain a heterocyclic macrocycle with a π-aromatic core, showing an intense Soret band at 400–450 nm and moderate Q bands at

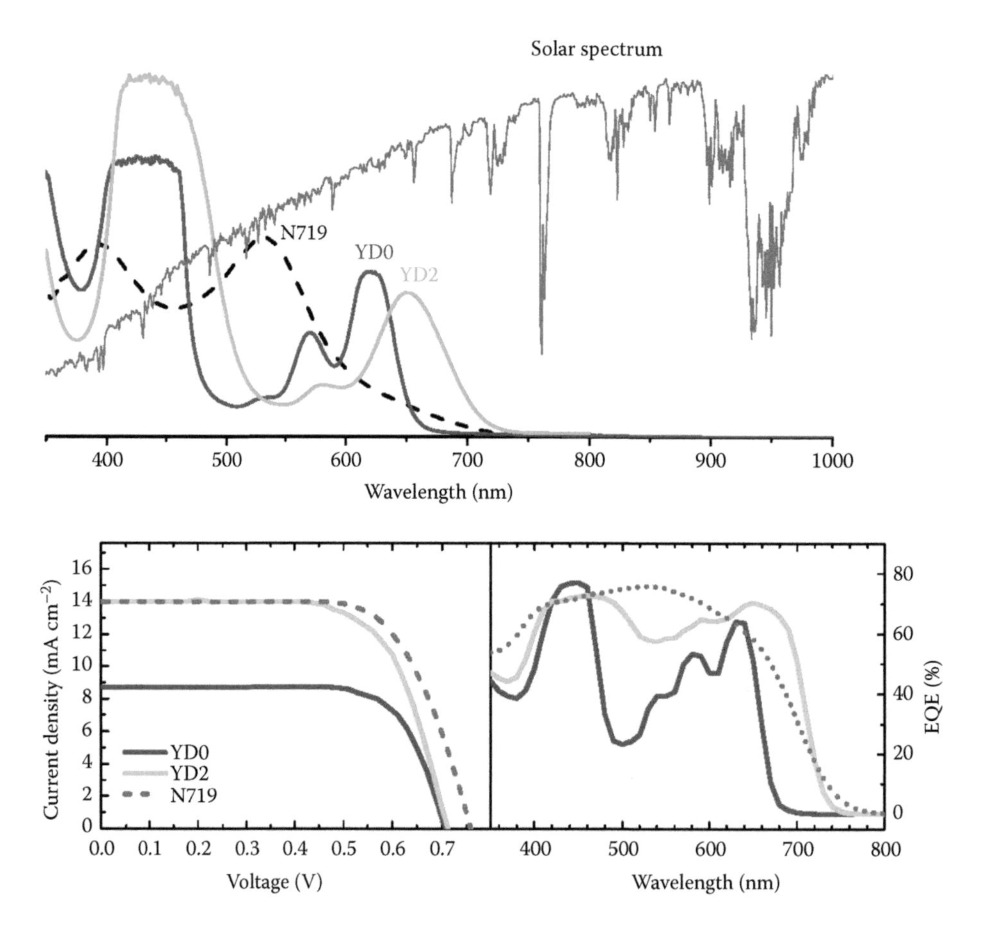

FIGURE 10.1 Absorption coefficient of N719 dye and two porphyrin dyes, in comparison with the spectral photon flux, and the corresponding photovoltaic performance and EQE_{PV}. (Adapted from Li, L.-L.; Diau, E. W.-G. *Chemical Society Reviews* 2012, 42, 291–304.)

500–650 nm. The shift of the Q band toward longer wavelength by ligand modification in the case of YD2 results in a much better match of the absorption coefficient with the solar spectrum region, where the photon density is large. This implies much larger photocurrent as shown in Figure 10.1. However, since N719 has larger light absorption and EQE_{PV} in the region around 550 nm, the current resulting from both YD2 and N719 is similar.

A range of characteristics for different solar cell technologies are shown in Figure 10.2, with the normalized EQE_{PV}, and in Figure 10.3, with current–voltage characteristics. Figure PSC.8.2 already showed the match of the photocurrent with the integrated EQE_{PV} in perovskite solar cells.

10.2 CHARGE COLLECTION DISTANCE

In a solar cell under intense photogeneration, there occurs a competition of transport and recombination that determines whether the carriers generated at an internal point will reach the external contacts, which is a necessary condition for the efficient operation of the device under sunlight. It was discussed in the chapters FCT.1 to FCT.3 that the transport in semiconductors can often be separated into two distinct mechanisms, provided that interactions between carriers can be neglected. The diffusion transport is governed by the gradient of the carrier concentration, while the drift transport of the carriers occurs in an electrical field. In a cell with flat and horizontal bands, no electrical fields assist the transport, which occurs entirely by diffusion. The band bending or band slanting indicates the presence of electrical fields that assist charge separation and transport by directing oppositely charged carriers in contrary directions. This distinction is useful for establishing simple models of solar cell operation. It is often assumed that the recombination of carriers can be drastically reduced by intense drift in space-charge regions. More generally, diffusion and drift can cooperate in a device as explained in Section 10.6 in the analysis of the Gärtner model.

Let us assume that no electrical fields exist along the absorber thickness. The central parameter that establishes the charge collection features of the solar cell device is the diffusion length, explained in Section FCT.2.3. We recall that the diffusion length is the most likely distance that a carrier can travel before recombination. For electrons, it is denoted

$$L_n = \sqrt{D_n \tau_n} \tag{10.1}$$

in terms of the electron diffusion coefficient D_n and recombination lifetime τ_n. By Equation PSC.2.36, we also have

$$L_n = \left(\frac{D_n}{B_{rec} p_0} \right)^{1/2} \tag{10.2}$$

We consider first diffusion transport in the dark in a semiconductor layer of thickness d. Minority carrier electrons can be injected in the layer by forward bias applied to the electron selective contact (ESC). The resulting carrier distribution dramatically depends on the ratio L_n/d, as shown in Figure 10.4 (the graph is plotted with expressions described in Section 10.5). When the diffusion length is short, the electrons do not penetrate all the way to the back contact. When the diffusion length is long, the concentration of carriers is homogeneous in the layer. This last case corresponds to our previous assumptions about flat Fermi levels and homogeneous minority carrier density in Figure PSC.5.1.

In the case of a short diffusion length, when the diode operates as a solar cell, only the carriers generated close to the contact are collected, as indicated in Figure 10.5, which is a schematic picture that will be developed more fully in Sections 10.5 and 10.8. Nevertheless, we obtain a valid and

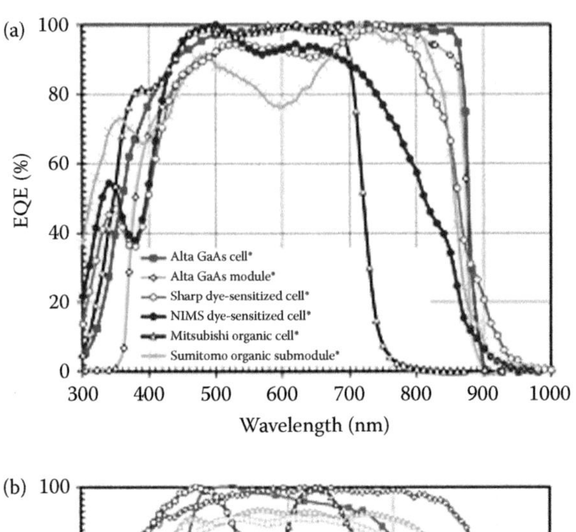

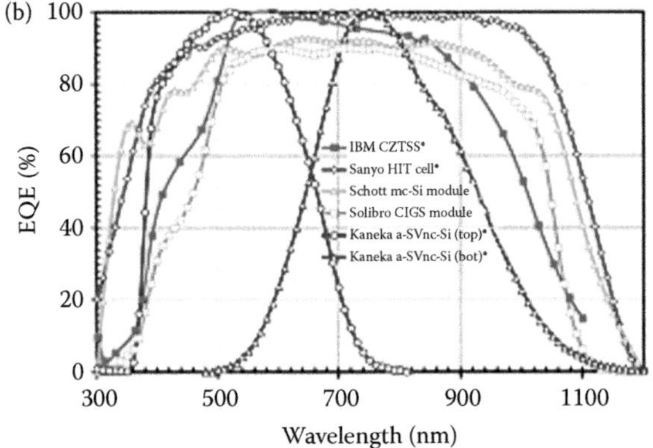

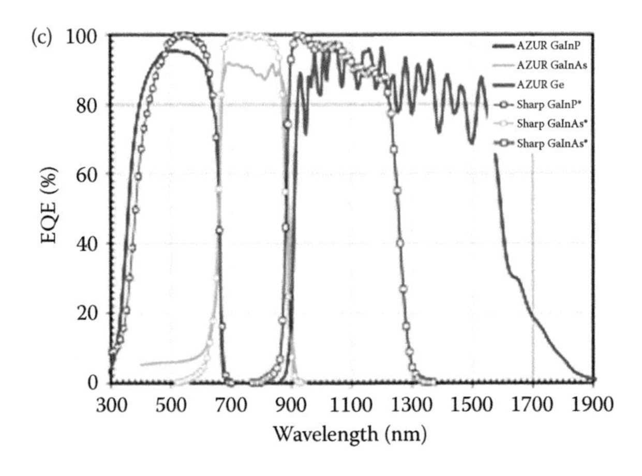

FIGURE 10.2 Normalized external quantum efficiency (EQE_{PV}) for solar cells and modules. (a) GaAs cell and module, dye-sensitized cell, organic cell, and module; (b) Silicon cell and module and copper zinc tin sulfide-selenide (CZTSe) cell; (c) Triple junction multijunction cells. (Reproduced with permission from Green, M. A. et al. Solar cell efficiency tables (version 39). *Progress in Photovoltaics: Research and Applications* 2012, 20, 12–20.)

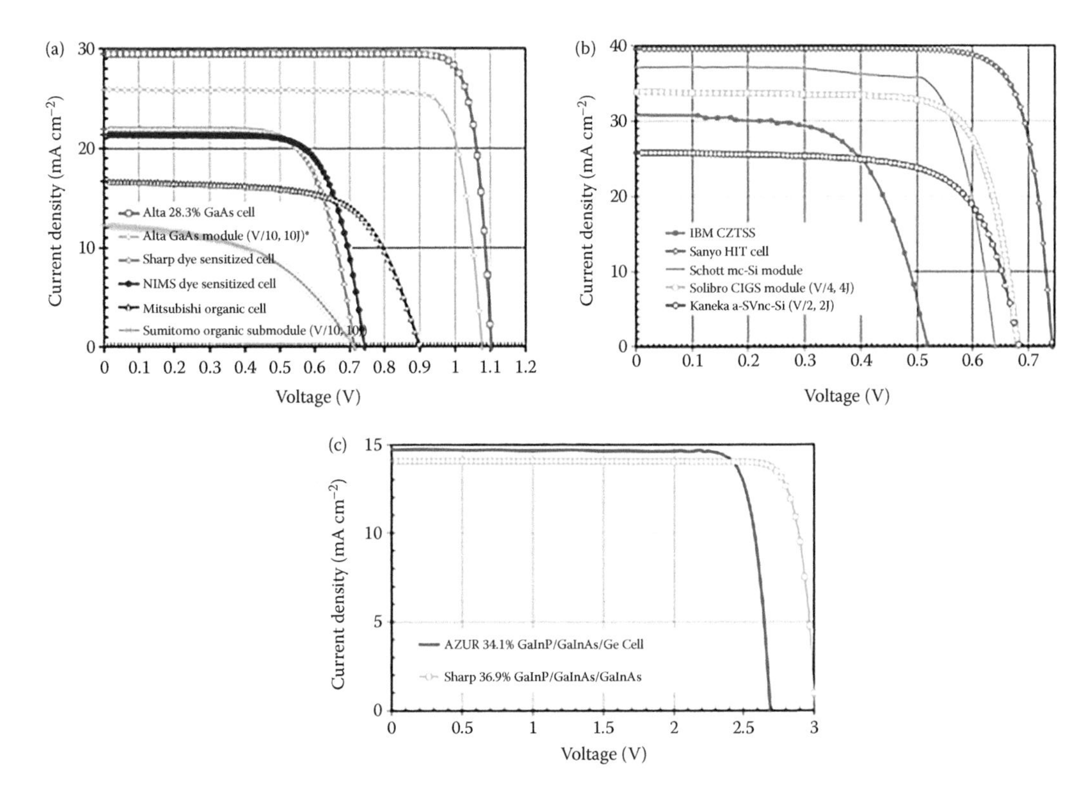

FIGURE 10.3 Current density–voltage (j–V) curves for solar cells and modules. (a) GaAs cell and module, dye-sensitized cell, organic cell, and module; (b) silicon cell and module and CZTSe cell; (c) triple junction multijunction cells. (Reproduced with permission from Green, M. A. et al. Solar cell efficiency tables (version 39). *Progress in Photovoltaics: Research and Applications* 2012, 20, 12–20.)

informative expression simply observing that the saturation current density in Equation PSC.7.27 is modified as follows:

$$j_0 = q\int_0^{L_n} U_{rec}\,dx = qL_nB_{rec}n_0p_0 \tag{10.3}$$

That is, only the carriers generated in dark at $x \le L_n$ can be extracted and contribute to the saturation current. Using Equation 10.2, we obtain the recombination parameter

$$j_0 = q\frac{D_n}{L_n}n_0 \tag{10.4}$$

The current density–voltage curve of the solar cell is given by Equation PSC.8.2, where j_0 in Equation 10.4 is independent of the thickness of the sample. Using the density of acceptors N_A as described in Section ECK. 5.4, we obtain with respect to the measured photocurrent j_{ph}

$$V_{oc} = \frac{k_BT}{q}\ln\left(\frac{j_{ph}L_nN_A}{qD_nn_i^2}\right) \tag{10.5}$$

This relationship is nearly quantitatively obeyed for silicon solar cells with excellent interface contacts (Maldonado et al., 2008).

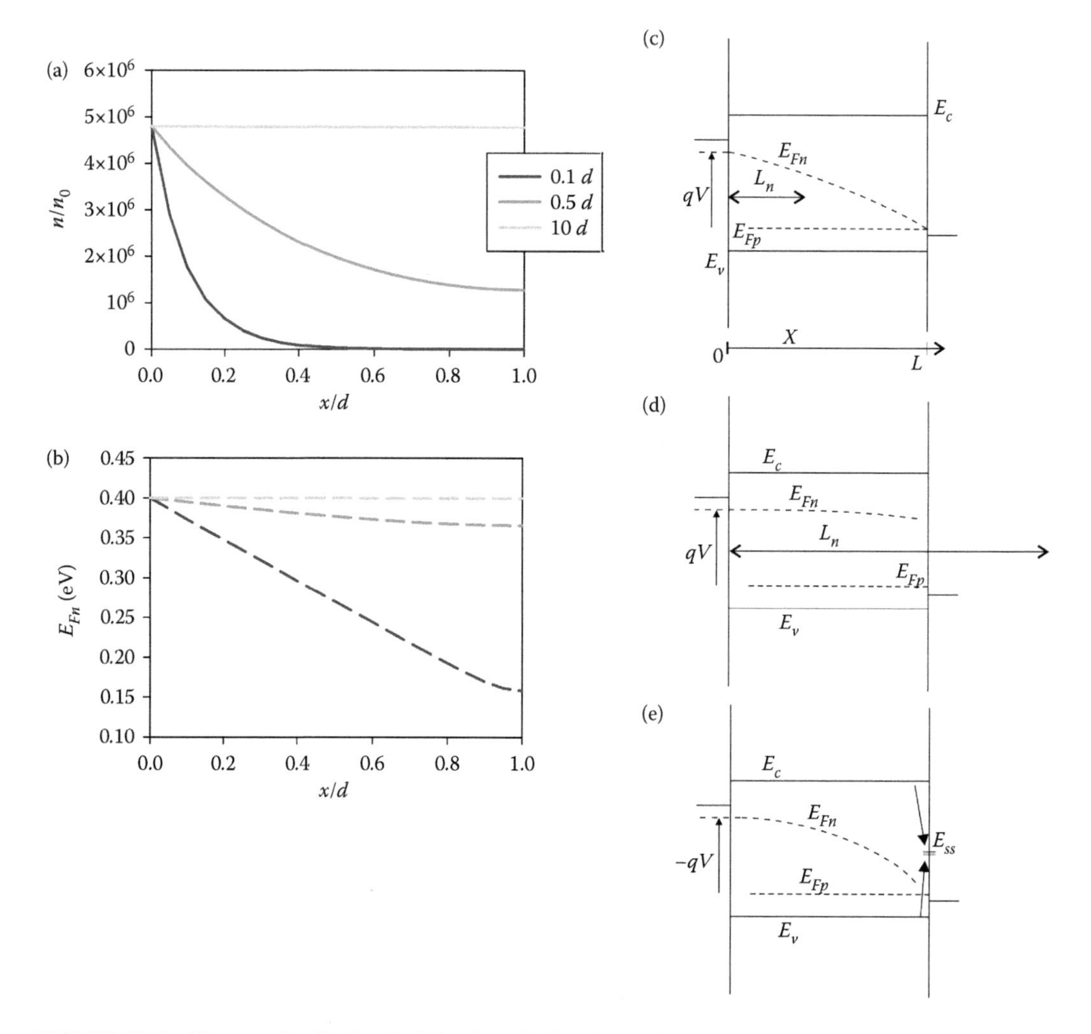

FIGURE 10.4 Electron distribution for injection of minority carrier electrons in the dark for a forward bias of 0.4 V, by diffusion and recombination, in a layer of thickness d for different values of the diffusion length as indicated. (a) electron density and (b) electron Fermi level. The right column shows the schemes of electrons and holes Fermi level under forward bias for the following cases: (c) Short diffusion length, (d) long diffusion length and reflecting boundary condition, and (e) long diffusion length and absorbing boundary condition by strong SRH recombination at the contact.

If the current is governed by drift in an electrical field, as in Figure PSC.4.11, then the mean distance covered by a carrier before recombination is obtained by the product of the drift velocity and the recombination lifetime

$$L_{drift} = v\tau_n \tag{10.6}$$

Using Equation FCT.1.9, we find the expression

$$L_{drift} = u_n\tau_n\mathbf{E} \tag{10.7}$$

in terms of the mobility and the electrical field. At given field strength, the mobility-lifetime product is the quantity determining the drift length.

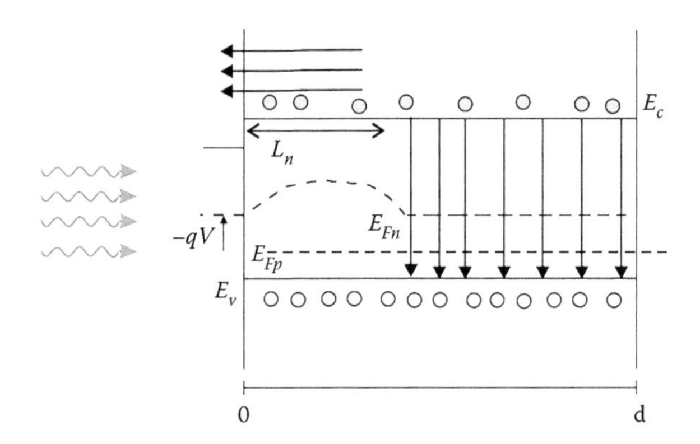

FIGURE 10.5 Schematic model indicating the photogeneration of electrons in a solar cell with flat bands where the diffusion length L_n is shorter than layer thickness. The minority carriers situated at $x \leq L_n$ are collected at the contact, while those created away from the contact recombine with majority carrier holes.

10.3 GENERAL MODELING EQUATIONS

In order to analyze the dominant processes of operation of a solar cell for charge transport and charge extraction, we will adopt a model in which the carriers obey the general drift–diffusion equations for electrons and holes in a layer $0 \leq x \leq d$, combined with the continuity and the Poisson equations that have been described in Section FCT.3.4. We have the equations

$$j_n = -qJ_n = qnu_n\mathbf{E} + qD_n\frac{\partial n}{\partial x} \tag{10.8}$$

$$j_p = +qJ_p = qpu_p\mathbf{E} - qD_p\frac{\partial p}{\partial x} \tag{10.9}$$

$$j = j_n + j_p \tag{10.10}$$

$$\frac{1}{q}\frac{\partial j_n}{\partial x} + G_\Phi - U_{rec} = 0 \tag{10.11}$$

$$-\frac{1}{q}\frac{\partial j_p}{\partial x} + G_\Phi - U_{rec} = 0 \tag{10.12}$$

$$\frac{\partial \mathbf{E}}{\partial x} = \frac{q}{\varepsilon_0\varepsilon_r}(p - n - N_A) \tag{10.13}$$

$$\mathbf{E} = -\frac{\partial \phi}{\partial x} \tag{10.14}$$

If not otherwise stated, we adopt the bimolecular model for band-to-band recombination defined in Equation PSC.2.34.

$$U_{rec} = B_{rec}\left(np - n_i^2\right) \tag{10.15}$$

The carrier generation rate at the position x is given by the attenuation of the photon number, $G_\Phi(x) = d\Phi_{ph}(x)/dx$. Using the Beer–Lambert law, and integrating over the wavelength of the incident radiation λ,

$$G_\Phi(x) = \int (1-R)\alpha(\lambda)\Phi_{ph}(0,\lambda)e^{-\alpha(\lambda)x}d\lambda \tag{10.16}$$

where $\Phi_{ph}(0,\lambda)$ is the incident photon number and R is the reflection coefficient. For monochromatic radiation it is $G_\Phi(x) = \alpha(\lambda)\Phi_{ph}(0)e^{-\alpha(\lambda)x}$.

10.4 THE BOUNDARY CONDITIONS

We have remarked the central significance of selective contacts in the basic solar cell structure in Chapter PSC.4. For a one-dimensional solar cell model, we use the differential equations summarized in Section 10.3 to establish the interdependence of the charge distribution, electrical field, and carrier currents in the active layer $0 \le x \le d$. To complete the problem mathematically, it is necessary to establish a set of boundary conditions at the edges that express the operation of contacts. These conditions often play a determinant role in the physical properties of the model and require a judicious selection. The selective contacts must have the property of extracting one carrier and blocking the oppositely charged carrier, as emphasized in Figure 10.6. In real cases, the property of charge blocking occurs to a total or partial extent, combined with some amount of surface recombination. We now discuss separately the main types of boundary conditions for extraction and blocking of the carriers.

10.4.1 CHARGE EXTRACTION BOUNDARY CONDITION

Let us assume that the Fermi level of the majority carrier is generally fixed in the absorber layer while an increase of the minority carrier occurs under photogeneration, that is, we remain under the constraint $n \ll p \approx p_0$ or $p \ll n \approx n_0$. We can impose boundary conditions on the carrier density. One common assumption is to fix the carrier density at a constant value, the equilibrium value n_0 or p_0, as explained in Section FCT.1.3. In Figures PSC.4.5, ECK.9.2, and ECK.9.3, we have shown an n-doped semiconductor with a contact to the majority carrier electrons at the right side. Correspondingly, we write the condition for the number density at the surface as

$$n_s = n_0 \tag{10.17}$$

This is called in mathematics as the Dirichlet boundary condition, and the absorbing boundary condition in the theory of diffusion. For a majority carrier in low-injection conditions, it is a natural condition implying that the carrier density is maintained at the contact with zero impedance for

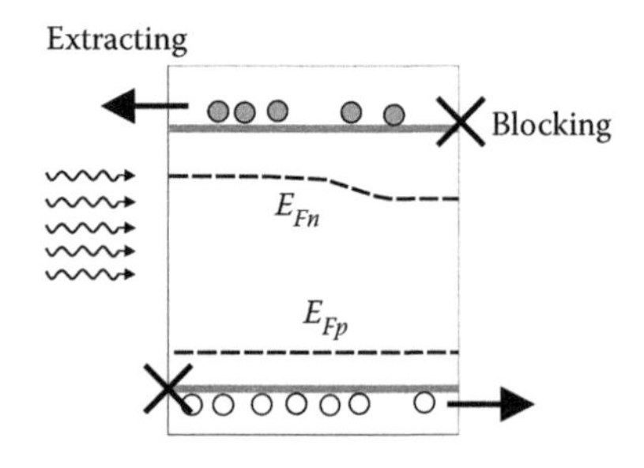

FIGURE 10.6 Fundamental solar cell model indicating the extracting and blocking properties of the contacts, which determine the carrier selectivity.

transport across the contact. It is called an "ohmic" boundary condition. However, for the minority carrier, the implication of a constant density is different, as discussed below.

The most important contact of the solar cell structure is that of the extraction of minority carriers. In the simple model of Figure PSC.4.8, the variation caused by the voltage is located completely at the minority carrier extraction interface. The change of the VL occurs at the surface dipole layer. At this contact, the surface concentration in the semiconductor relates directly to the voltage in the device as

$$n_s = n_0 e^{qV/k_BT} \tag{10.18}$$

This model boundary is realized in the cases of Figures PSC.9.11 and PSC.9.12.

However, as discussed in Figure FCT.1.9, the original built-in voltage may be distributed either partly or fully inside the absorber layer. This situation has been analyzed in the model SB diode discussed in Chapter PSC.4. The changes of the SB are indicated in detail in Figure 10.7. The height of the barrier, which corresponds to the built-in potential, is given by the difference of the Fermi level of the semiconductor, $\Phi_s = E_{F0}$ and the cathode work function

$$V_{bi} = (E_{F0} - \Phi_c)/q \tag{10.19}$$

Polarization of the junction first causes the decrease of the space-charge region and size of the barrier, keeping the density of minority carrier constant as the Fermi level raises upward, as indicated in Figure 10.7c. At further forward bias, the density of electrons rises at the surface according to the expression

$$n_s = n_0 e^{q(V - V_{bi})/k_BT} \tag{10.20}$$

The formation and operation of the contact including depletion in the absorber layer is illustrated in Figure 10.8 for organic BHJ solar cells. In "regular" configuration, the HSC is the transparent combination of ITO/PEDOT, and in the "inverted" cell, the transparent contact is ZnO/ITO. A difference of size of the depletion region occurs due to the fact that the blend is spin coated on top of the anode for regular cells, whereas, for inverted cells, it is deposited on top of the cathode, forming different degrees of doping in the two situations.

In the initial contact formation, the original difference of work functions indicated in Equation 10.19 can be distributed in two parts, a depletion region of barrier height corresponding to the built-in voltage V_{bi}, and the interfacial dipole Δ_i that accommodates part of the potential drop, as discussed in Section ECK.4.6. Then Equation 10.19 is modified to the more general expression

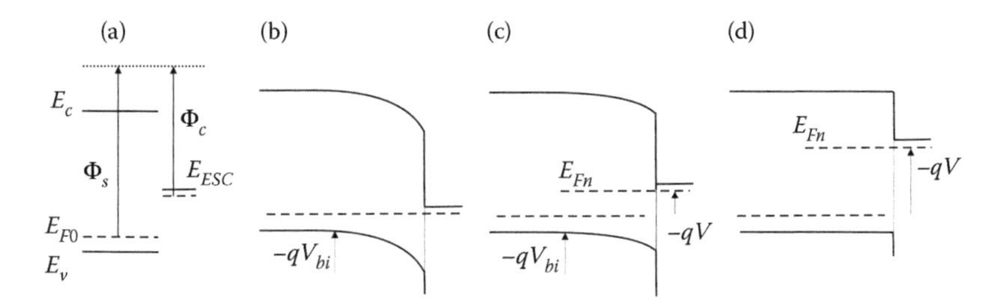

FIGURE 10.7 Schematic representative energy diagram of a p-type organic layer and ESC as contact for the device. (a) The separate materials. (b) In contact, the original difference of work functions forms a depletion zone with barrier qV_{bi}. (c) At forward bias, the SB decreases. (d) When the built-in potential is overcome by negative potential or by photogeneration of electrons, the electron density at the surface increases.

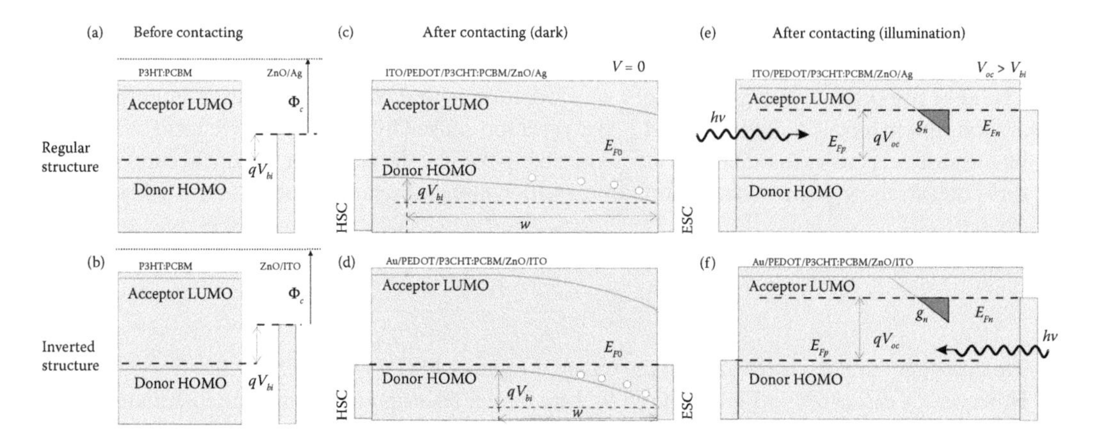

FIGURE 10.8 Energy diagram of a P3HT:PCBM organic bulk heterojunction with a contact for holes in the left side (PEDOT) and a contact to electrons at the right side, in regular (ZnO/Ag) or inverted (ZnO/ITO) configurations. (a, b) Separate representation of the blend and the cathode. (c, d) Equilibrium after contact ($V_{app} = 0$). Band bending appears near the cathode. (e, f) Forward voltage larger than flat band condition at the cathode. The Fermi level of minority electrons scans the density of states g_n in the bandgap associated with disorder. (Reproduced with permission from Boix, P. P. et al. *Journal of Physical Chemistry Letters* 2011, 2, 407–411.)

$$V_{bi} = (E_{F0} - \Phi_c - \Delta_i)/q \tag{10.21}$$

10.4.2 Blocking Boundary Condition

At the blocking contact, we can find different situations depending on the assumed properties of the interface, as shown in Figure 10.4. The best case for applications is a reflecting boundary condition for the minority carrier at the majority extraction contact. The diffusion flux is zero:

$$\left.\frac{\partial n}{\partial x}\right|_{x=d} = 0 \tag{10.22}$$

This condition is called the von Neumann boundary condition or a reflecting boundary condition. If the electrical field is intense at the boundary, the flux in Equation 10.22 must include the drift component (Beaumont and Jacobs, 1967).

We have remarked in the fundamental solar cell model of Figure PSC.5.1 that the minority carrier density must increase under illumination, causing the splitting of the Fermi levels and hence generating the photovoltage. For enhanced solar cell performance, the minority carrier density should increase everywhere in the absorber including at the majority extraction contact, which is well realized by Equation 10.22, which imposes a horizontal Fermi level at the blocking boundary. In Figure PSC.4.5b, the minority carrier density increases at the majority extraction contact. In contrast to this, in Figure PSC.4.6, the minority carrier density is fixed at the dark equilibrium level, which has the consequence that its density cannot increase:

$$p_s = p_0 \tag{10.23}$$

This condition was remarked in Equation 10.17 for the extraction of majorities, but for the blocking of minorities, the interpretation is different. If the rate of surface recombination is very large, then the population of minority carrier at the contact cannot rise (Figure 10.4e). Thus, according to

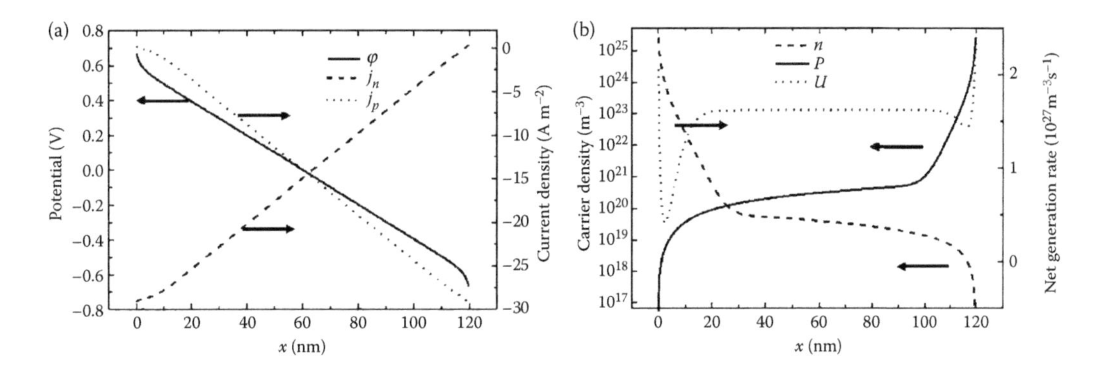

FIGURE 10.9 Device model with the assumption of constant density boundary conditions at short circuit: (a) shows the potential and current densities, (b) shows the carrier densities and the net recombination rate. (Reproduced with permission from Koster, L. J. A. et al. *Physical Review B* 2005, 72, 085205.)

Equation 10.23, the Fermi levels of electrons and holes may separate in the bulk but are forced to coincide at this contact under the assumption of constant density at the boundary. A similar result is obtained with another approach that consists in fixing the energy barriers $\Phi_{B,n}$ and $\Phi_{B,p}$ at the contacts.

The use of the boundary conditions that fix the charge is a common procedure for p–i–n type solar cells discussed in Figure PSC.4.11, in which transport layers form the contacts to an insulator layer. This assumption has the effect that the charge density of electrons and holes at either edge of the active layer becomes fixed, and it is very large at the respective selective contact (due to the initial match of energy levels), as remarked in Equations FCT.1.22 and FCT.1.23. The resulting characteristic energy diagrams are shown in Figure 10.9, which correspond to short-circuit condition as in Figure PSC.4.11a. The Fermi level of the carrier cannot change at the contact and the variation of the potential is necessarily absorbed across the active layer, with similar results to those of the MIM model of Figure FCT.1.10. In order to allow a variation of the carrier density at the edge of the active layer, one can introduce additional contact layers that serve as a buffer from the highly doped transport layer, and then variations across the contact are permitted, as shown in Figure 10.10.

10.4.3 Generalized Boundary Conditions

A more general condition relates the carrier flux to the excess concentration at the contact, similar to Equation PSC.4.3. This condition is stated in terms of a "surface recombination velocity" parameter S_n

$$J_n(0) = S_n(n_s - n_0) \tag{10.24}$$

Depending on the strength of S_n, either Equation 10.22 or 10.23 can be obtained in extreme cases of Equation 10.24, respectively $S_n = 0$ and $S_n \to \infty$. The evolution of the shape of the Fermi level from blocking to absorbing at the blocking contact is shown in Figure 10.11.

The condition of recombination at the surface can be more realistically described by SRH recombination, with the rate given in Equation ECK.6.35. For a blocking boundary

$$J_n(0) = -N_t \frac{\beta_n \beta_p (n(0)p(0) - n_0 p_0)}{\beta_n n(0) + \varepsilon_n + \beta_p p(0) + \varepsilon_p} \Delta d \tag{10.25}$$

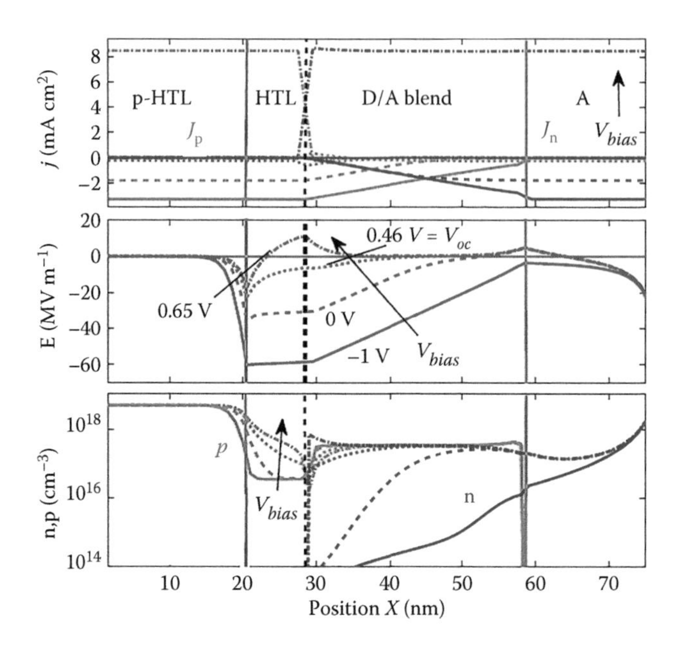

FIGURE 10.10 Model of a bulk heterojunction solar cell with contact layers. Profiles of current density (j), electric field (**E**), and charge-carrier densities (n,p) with applied bias (V_{bias}) as parameter, green lines (high values on p-side) show hole currents and densities, blue lines (high values in acceptor [A]) show values for electrons. Field and charge-carrier density profiles indicate the applied boundary conditions: constant doping concentration on p-side, thermal injection on n-side with a work function of the cathode of 4.2 eV. $V_{bias} = -1$, 0, 0.46, 0.65 V. (Reproduced with permission from Tress, W.; Leo, K.; Riede, M. *Advanced Functional Materials* 2011, 21, 2140–2149.)

Δd is the thickness close to the contact where the SRH recombination takes place. For an extracting boundary

$$J_p(0) = N_t \frac{\beta_n \beta_p (n(0)p(0) - n_0 p_0)}{\beta_n n(0) + \varepsilon_n + \beta_p p(0) + \varepsilon_p} \Delta d + S_p(p(0) - p_0) \tag{10.26}$$

The diagrams in Figure 10.12 (further discussed in Section 10.8) indicate the presence or absence of surface states according to the passivation of the back contact of the solar cell. Further discussion of charge transfer models at metal–semiconductor contact, which can actuate as boundary conditions, is presented in Section ECK.6.7.

Charge transfer at semiconductor–electrolyte interface is a fundamental step in solar fuel production that will be described in Section PSC.11.3. As discussed previously, in Figures ECK.6.24 and FCT.7.3, the electron or hole transfer may originate either from extended states in the band edges, or by an intermediate step of capturing at a surface state. The capture of an electron at the surface states increases the probability of recombination with a photogenerated hole in the valence band. The competition between charge transfer and recombination at the surface states is a major factor in the operation of metal oxide photoelectrodes for water splitting reactions (Klahr et al., 2012), and the boundary condition at the extraction contact needs to incorporate the mechanism of surface states capture (Gimenez and Bisquert, 2016).

10.5 A PHOTOVOLTAIC MODEL WITH DIFFUSION AND RECOMBINATION

We now establish a complete photovoltaic model taking into account the detailed features of generation, transport, and recombination in a solar cell with perfect selective contacts. Our main objective

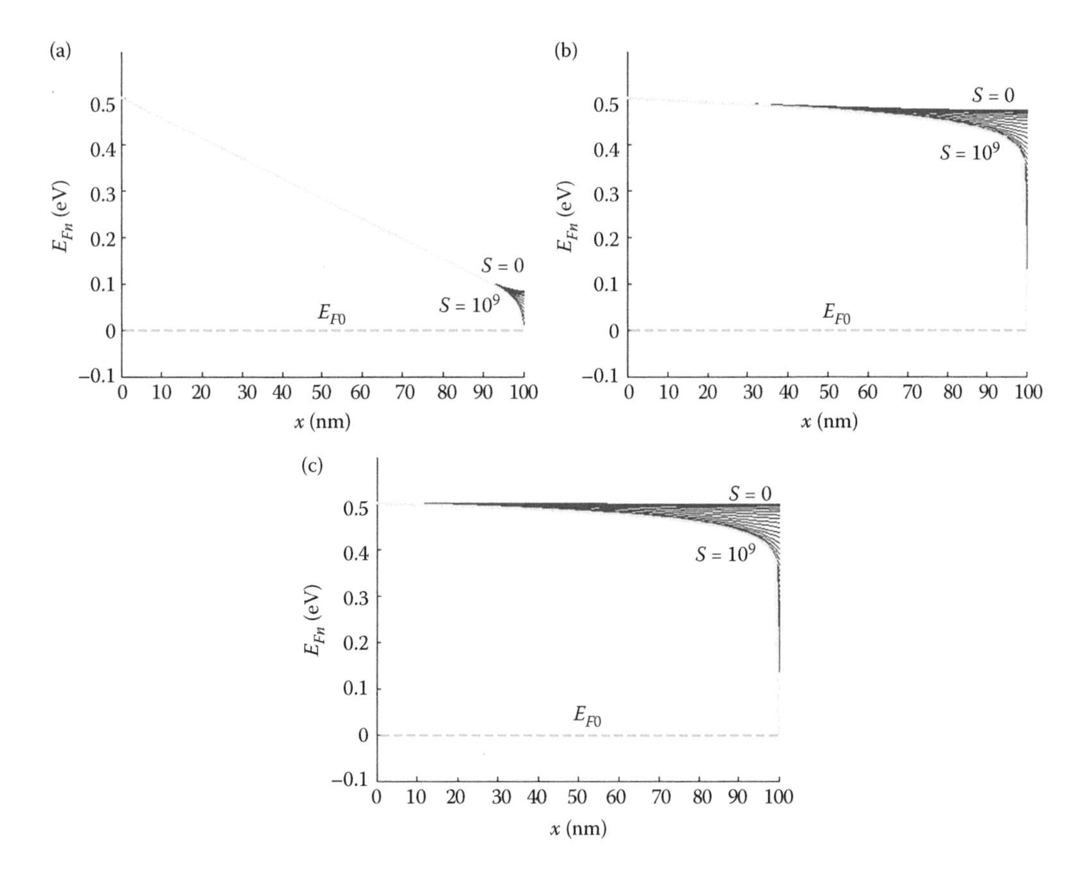

FIGURE 10.11 Injection of electrons in the dark in a semiconductor layer of thickness 100 nm. Electron Fermi level is shown for applied voltage at electron contact of 0.5 V, and diffusion lengths of (a) 6, (b) 60, and (c) 600 nm. At the right contact, surface recombination velocity S ranging from 0 (blocking boundary) to 10^9 cm s^{-1} (absorbing boundary condition), $n(d) = n_0$.

is to obtain the current dependence on voltage and illumination intensity, and to discuss the features of IQE_{PV} and EQE_{PV} for monochromatic illumination. The situation we consider is shown in the scheme of Figure 10.5. We assume a large majority carrier density and nearly flat bands. We discuss the case of illumination from the contact of minority carrier extraction at $x = 0$. In Section 10.6, this model will be extended with the collection at the space-charge region, which as of now we assume is negligible.

The total electrical current density produced by the transport of carriers, see Equation FCT.1.3, is

$$j = \sum_i j_i \tag{10.27}$$

The sum includes electrons, holes, and also other carriers, such as ions in a DSC. If we consider a semiconductor where ions are immobile, then we have the drift–diffusion problem indicated in Section 10.3, where the current corresponds to Equation 10.10.

Let us discuss the implications of the different conductivities of the two species, electrons and holes, which lead to the conclusion that diffusion of minority carriers is the dominant effect we need to describe analytically in our model.

We note that the Poisson equation is not needed in this model, as the compensation of charge already present in dark equilibrium is not disturbed by the presence of additional generated

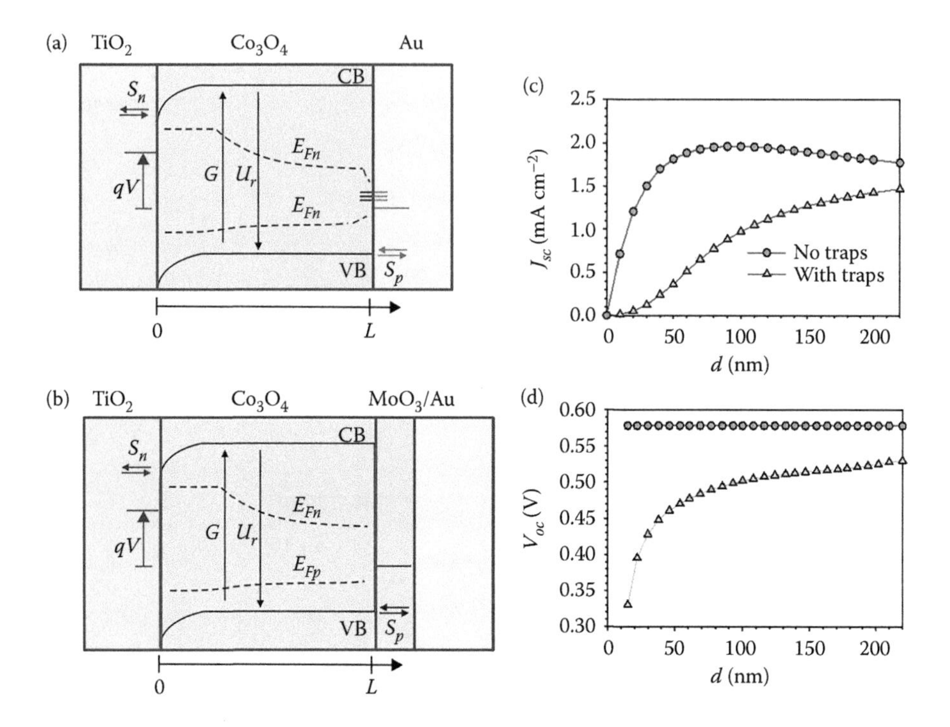

FIGURE 10.12 (a, b) Schemes of all-oxide solar cells with different structure of recombination at the back contact. (c, d) Model calculation of photovoltage and photocurrent as function of absorber thickness. (Reproduced with permission from Majhi, K.; et al. *Advanced Materials Interfaces* 2016, 3, 10.1002/admi.201500405.)

minority carrier electrons. The majority carrier concentration is orders of magnitude larger, $p_0 \gg n$. According to the dominant electrical conductivity in the bulk semiconductor layer, nearly all the electrical current is taken by the majority carrier. An imperceptible band slant, associated with a small electrical field, is enough to drive the required electrical current inside the device. However, this is not true at the point of the ESC, as here the majority species holes cannot conduct the current. The problem of finding the outgoing current can then be reduced to a calculation of the minority carrier current just at this point. We have mentioned that the electrical field is negligible. Then the carrier transport is driven by diffusion.

Based on the previous remarks, the electrons operate in a medium where charge compensation is granted. Thus, the solar cell operation is controlled by diffusion and recombination, as described in Section FCT.3.6, which produces the total diffusion current at the extraction ESC point. We require the solution of generation–diffusion–recombination in a quasineutral region, Equation FCT.2.25, which is obtained by the combination of Equations 10.8 and 10.12:

$$D_n \frac{\partial^2 n}{\partial x^2} - U_n + \alpha \Phi_{ph} e^{-\alpha x} = 0 \tag{10.28}$$

Using the linear recombination model of Equation PSC.2.35 and the diffusion length of Equation 10.1, the conservation equation can be expressed as

$$\frac{\partial^2 n}{\partial x^2} - \frac{n - n_0}{L_n^2} + \frac{\alpha \Phi_{ph}}{D_n} e^{-\alpha x} = 0 \tag{10.29}$$

This equation is an extension of Equation FCT.2.31, and it can be solved analytically with the solution

$$n = n_0 + A_1 \cosh\left(\frac{x}{L_n}\right) + A_2 \sinh\left(\frac{x}{L_n}\right) + \gamma_n e^{-\alpha x} \tag{10.30}$$

where

$$\gamma_n = \frac{L_n^2 \alpha \Phi_{ph}}{D_n\left(1 - L_n^2 \alpha^2\right)} \tag{10.31}$$

To determine the constants A_1 and A_2, we fix the boundary conditions. In the extraction contact, we take Equation 10.18, and we select the most favorable Equation 10.22, which will impose a flat Fermi level at the blocking contact. We obtain the following results:

$$A_1 = n_0(e^{qV/k_BT} - 1) - \gamma_n \tag{10.32}$$

$$A_2 = \frac{\alpha L_n \gamma_n e^{-\alpha d}}{\cosh(d/L_n)} - A_1 \tanh(d/L_n) \tag{10.33}$$

The carrier distribution in the dark has been shown earlier in Figure 10.4, and for the more general boundary condition (10.24), the solution is shown in Figure 10.11.

The diffusion flux at the extraction point is

$$\begin{aligned} J_n(0) &= -D_n \left.\frac{\partial n}{\partial x}\right|_{x=0} \\ &= \frac{D_n}{L_n}\left(-\frac{\alpha L_n \gamma_n e^{-\alpha d}}{\cosh(d/L_n)} + A_1 \tanh(d/L_n)\right) + \alpha D_n \gamma_n \end{aligned} \tag{10.34}$$

The flux is the combination of the photocurrent and recombination flux

$$J_n(0,V) = -J_{ph} + J_{rec}(V) \tag{10.35}$$

The photocurrent has the expression (Södergren et al., 1994; Jennings et al., 2010)

$$j_{ph} = j_n(V=0) = qJ_{ph} \tag{10.36}$$

$$J_{ph} = \frac{L_n^2 \alpha^2}{\left(1 - L_n^2 \alpha^2\right)\cosh(d/L_n)}\left(e^{-\alpha d} + \frac{1}{L_n \alpha}\sinh(d/L_n) - \cosh(d/L_n)\right)\Phi_{ph} \tag{10.37}$$

and the recombination current is given by

$$j_{rec} = qJ_{rec}(V) = j_0(e^{qV/k_BT} - 1) \tag{10.38}$$

$$j_0 = \frac{qD_n n_0}{L_n} \tanh(d/L_n) \tag{10.39}$$

The ratio D_n/L_n has units of cm s^{-1} and this quantity is often referred to as the diffusion velocity v_d. The current–voltage characteristic is given by the equation

$$j = j_{ph} - j_0(e^{qV/k_BT} - 1) \tag{10.40}$$

The quantum efficiencies have the form

$$EQE_{PV} = \frac{J_{ph}}{\Phi_{ph}} \tag{10.41}$$

$$IQE_{PV} = \frac{J_{ph}}{\Phi_{ph}(1-e^{-\alpha d})} \tag{10.42}$$

If the diffusion length is long with respect to the film thickness, then

$$J_{ph} = \frac{L_n^2 \alpha^2 \Phi_{ph}}{L_n^2 \alpha^2 - 1}(1 - e^{-\alpha d}) \tag{10.43}$$

and we obtain

$$EQE_{PV} = \frac{1 - e^{-\alpha d}}{1 - (L_n \alpha)^{-2}} \tag{10.44}$$

$$IQE_{PV} = \frac{1}{1 - (L_n \alpha)^{-2}} \tag{10.45}$$

If $L_n \gg \alpha^{-1}$, we have total collection by diffusion and $IQE_{PV} \approx 1$. For $L_n \gg d$, the recombination parameter in Equation 10.39 reduces to that of Equation PSC.7.27 corresponding to homogeneous (dark) recombination in the whole semiconductor layer

$$j_0 \approx \frac{qD_n d n_0}{L_n^2} = qdB_{rec} n_0 p_0 \tag{10.46}$$

However, if the diffusion length is short, then Equation 10.39 reduces to Equation 10.4. Furthermore, we get

$$EQE_{PV} = \frac{L_n \alpha}{1 + L_n \alpha} \tag{10.47}$$

Figure 10.13 shows the distribution of electrons (indicated by the Fermi level) in the diffusion–recombination model under illumination. A number of examples covering different cases of the absorption depth and diffusion length are shown, with parameters indicated in Table 10.1. When

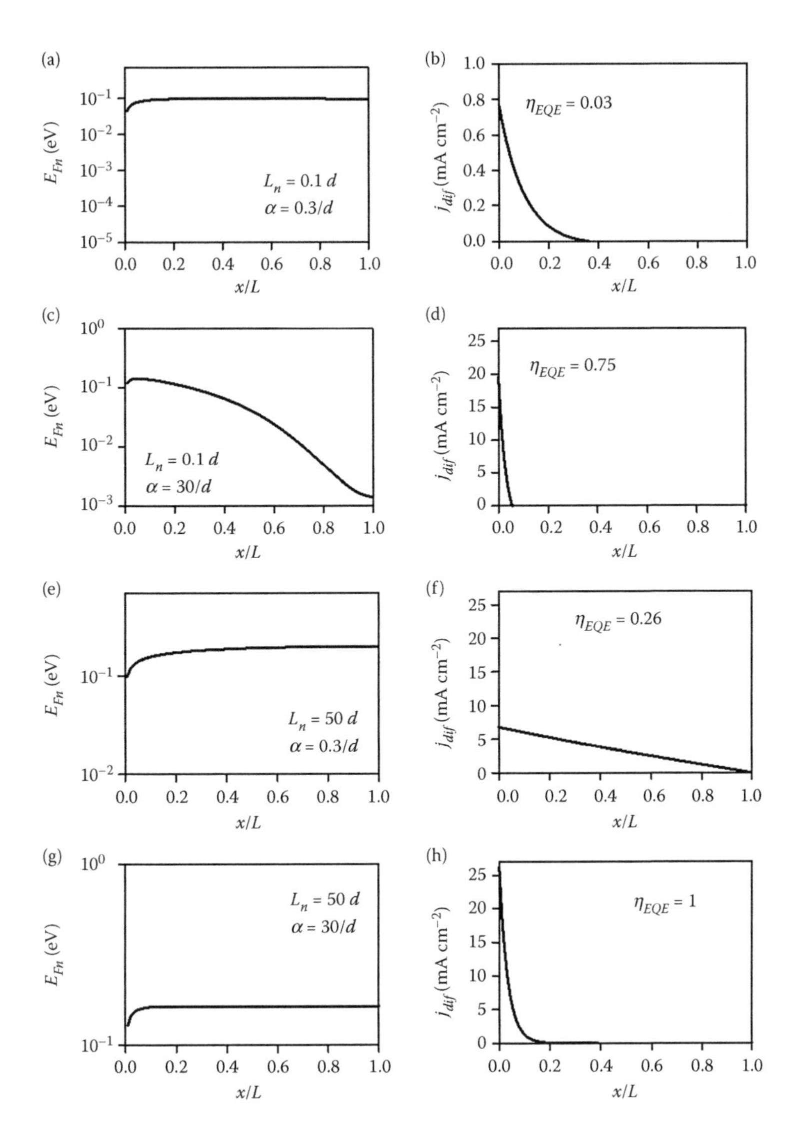

FIGURE 10.13 Electron Fermi level (*left column*) and local diffusion current density (*right column*) in a solar cell controlled by the diffusion–recombination model, covering different cases of the absorption depth and diffusion length, as indicated in Table 10.1. The simulations correspond to short-circuit situation. (Courtesy of Luca Bertoluzzi.)

$\alpha^{-1} > d$, the electrons are generated homogeneously across the layer, and $EQE_{PV} \ll IQE_{PV}$ due to the fact that many photons leave the layer through the back contact. Thus, in panel (a) and (e), it is observed that the distribution of electrons is rather homogeneous. The gradient increases close to the extraction contact as here, the total current is carried exclusively by diffusion of the minority carrier, while in the rest of the active layer, the current is sustained by the majority carrier, as explained earlier. The diffusion currents for a more general situation of combined electronic and ionic transport in a DSC are shown in Figure FCT.3.5. In the case of strong absorption $\alpha^{-1} \ll d$, electrons are generated only close to the transparent contact. Nevertheless, if the diffusion length is long, $L_n \gg d$, panel e, the electrons do not remain at the generation point but spread across the layer again creating a nearly homogeneous distribution. However, when both $\alpha^{-1} \ll d$ and $L_n \ll d$, the electrons are localized close to the generation point and a rather inhomogeneous profile is obtained as shown in panel (c).

TABLE 10.1
Characteristics of Simulated Solar Cells in Figure 10.13

	(a)	(c)	(e)	(g)
L_n/d	0.10	0.10	50	50
αd	0.30	30	0.30	30
j_{ph} (A cm^{-2})	0.763	19.6	6.79	26.2
V_{oc} (V)	0.101	0.185	0.420	0.456
EQE_{PV}	0.0291	0.750	0.259	1.00
IQE_{PV}	0.112	0.750	1.00	1.00
j_0 (A cm^{-2})	0.0160	0.0160	6.40×10^{-7}	6.40×10^{-7}
$k_{rec} = B_{rec}p_0$ (s^{-1})	1000	1000	4.00×10^{-3}	4.00×10^{-3}

Parameters: $d = 10$ μm, $T = 300$ K, $n_0 = 10^{15}$ cm^{-3}, $D_n = 10^{-5}$ cm^2 s^{-1}, $\Phi_{ph} = 10^{16}$ cm^{-2} s^{-1}.

Based on these observations, we consider the EQE_{PV} dependence on wavelength of a solar cell driven by electron diffusion. These characteristics are shown in Figure 10.14 from the landmark article by Södergren et al. (1994), which showed quantitatively the diffusive transport of electrons in a DSC. If the solar cell is illuminated from the TCO substrate side, it will always collect the electrons, even if the diffusion length is short. The EQE_{PV} will be unity at the shorter, strongly absorbed wavelengths and will decrease at longer wavelength due to the transparency of the device. But for illumination from the electrolyte side, the behavior of $EQE_{PV}(\lambda)$ is different. For strongly absorbed wavelengths and short diffusion lengths, the electrons cannot reach the collecting contact. At longer wavelengths, the behavior from illumination of either side must be rather similar. By fitting the spectra to the diffusion–recombination model, the diffusion length can be obtained. However,

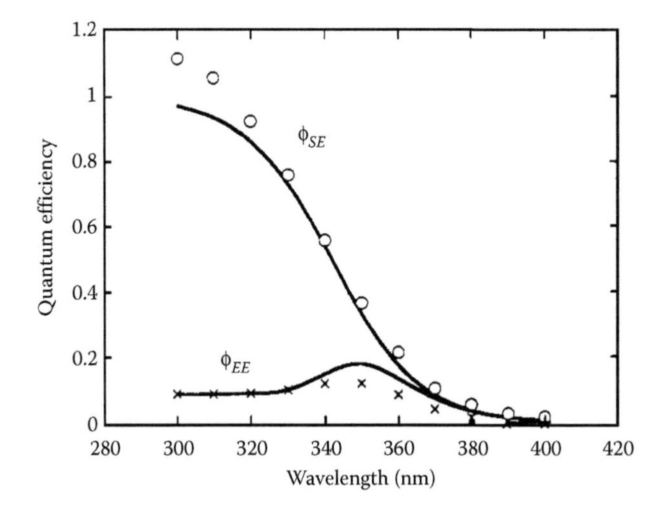

FIGURE 10.14 Experimental data for the EQE_{PV} of a 2.5 μm thick colloidal TiO_2 film electrode in 0.1 M KSCN in ethanol with a diffusion length of $L_n = 0.8$ μm, and fit to the diffusion–recombination model. ϕ_{SE} corresponds to the illumination from the substrate side and ϕ_{EE} to the illumination from the side of the electrolyte. The absorption coefficient as a function of wavelength was approximated by $\ln\alpha(\mu m) = 29–85\lambda$ (μm). (Reproduced with permission from Södergren, S. et al. *The Journal of Physical Chemistry* 1994, 98, 5552–5556.)

this methodology can be applied only to cells that show a low transport rate (Halme et al., 2008). As we remarked in Figure 10.13, if the diffusion length in the device becomes sufficiently long, such analysis will not be possible, as $IQE_{PV} = 1$ and all the electrons will be collected wherever they are generated.

Another way to calculate the external quantum efficiency is to apply a reciprocity theorem derived by Donolato (1985, 1989). It states that the probability of collection of a carrier generated at position x, $f_c(x)$, is related to the dark carrier distribution value, $n(x)$, divided by its thermal equilibrium value $n_0(x)$,

$$f_c(x) = \frac{n(x)}{n_0} \tag{10.48}$$

For a generation profile $G(x)$, the collected carrier flux at $x = 0$ is

$$J_{ph} = \int_0^d G(x) f_c(x) \tag{10.49}$$

Expressions for EQE_{PV} under general boundary conditions for surface recombination are given in Green (2009).

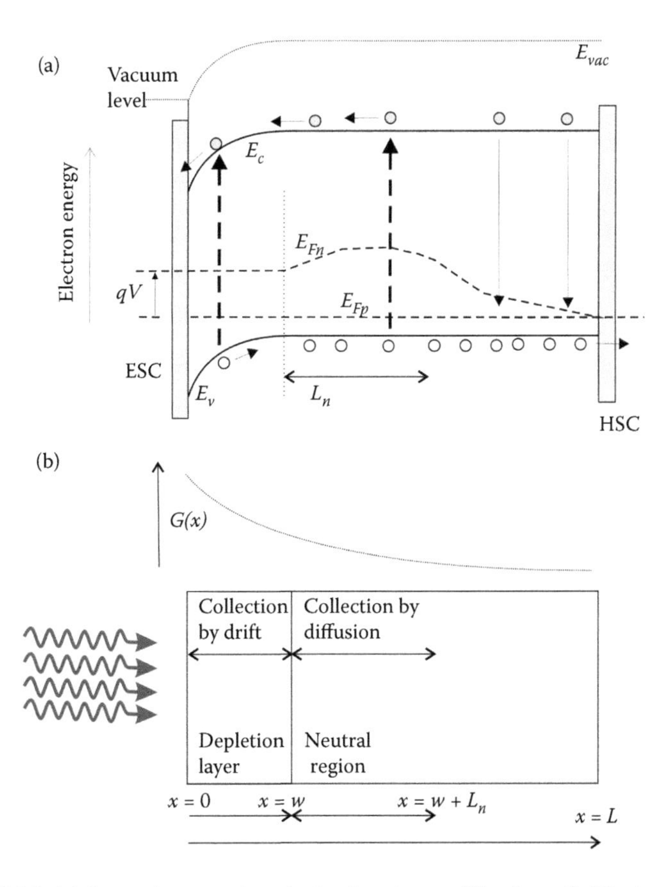

FIGURE 10.15 (a) Model for a p-type semiconductor forming an SB solar cell. Electronic processes under illumination are indicated. (b) Photogeneration gradient G in the two parts of the n-type semiconductor material, the depletion layer and the neutral region, indicating the collection in the region of a diffusion length.

10.6 THE GÄRTNER MODEL

The solar cell with the band-bending region at one contact associated with an SB has been analyzed in Figures ECK.9.3, PSC.4.1, and PSC.4.5. When the diffusion process is insufficient to extract the charge efficiently, the formation of an SB creates a strong electrical field that assists charge separation and facilitates charge collection in situations of low mobility and strong recombination, as in the application of some metal oxide electrodes to solar water splitting (Klahr and Hamann, 2011). In Figure 10.15a is shown the operation of this model under photogeneration, which is usually associated with the work of Gärtner (1959). The original model assumed a short diffusion length with respect to film thickness. It is also assumed that the layer is sufficiently thick to absorb all incoming photon flux, $\alpha^{-1} \ll d$.

The charge collection process is composed of two elements, namely the space-charge zone and the diffusion length zone, as indicated in Figure 10.15b. In the depletion layer formed by the SB, there are two concerted factors that reduce recombination. First, the majority carrier population is depleted which enhances the lifetime of minorities. And second the electrical field moves the carriers in opposite directions favoring rapid charge separation. In consequence, in a first approximation recombination can be neglected. The current of minority carriers generated in the depletion region of thickness w is

$$j_{ph,dl} = q\int_0^w G_\Phi \, dx = q\Phi_{ph}(1-e^{-\alpha w}) \tag{10.50}$$

The second part of the collection process is that of carriers that are generated in the neutral region. These do not have a preferred direction for the displacement, as suggested in Figure 10.15a. Only the minority carriers generated at a distance less than L_n from the edge of the depletion region are collected, as discussed before. The diffusion current arriving at $x = w$ is obtained by the solution of the diffusion–recombination model, Equation 10.29, with the boundary conditions correspondent to Figure 10.15a. The result derived in Equation 10.47 gives

$$j_{ph,diff} = q\Phi_{ph}e^{-\alpha w}\frac{\alpha L_n}{1+\alpha L_n} \tag{10.51}$$

Summing the two terms, we obtain the photocurrent

$$j_{ph} = q\Phi_{ph}\left[1-\frac{1}{1+\alpha L_n}e^{-\alpha w}\right] \tag{10.52}$$

From Equation 10.52, the quantum efficiency is given by

$$IQE_{PV}(\lambda) = \frac{j_{ph}(\lambda)}{q\Phi_{ph}(\lambda)} = \left[1-\frac{1}{1+\alpha(\lambda)L_n}e^{-\alpha(\lambda)w}\right] \tag{10.53}$$

Due to the reflectivity R of the front surface, the EQE_{PV} relates to IQE_{PV} as

$$EQE_{PV} = (1-R)\eta_{IQE} \tag{10.54}$$

Equation 10.53 can be used to determine the minority carrier diffusion length, if the absorption coefficient $\alpha(\lambda)$ is known and depletion layer thickness is calculated independently by capacitance

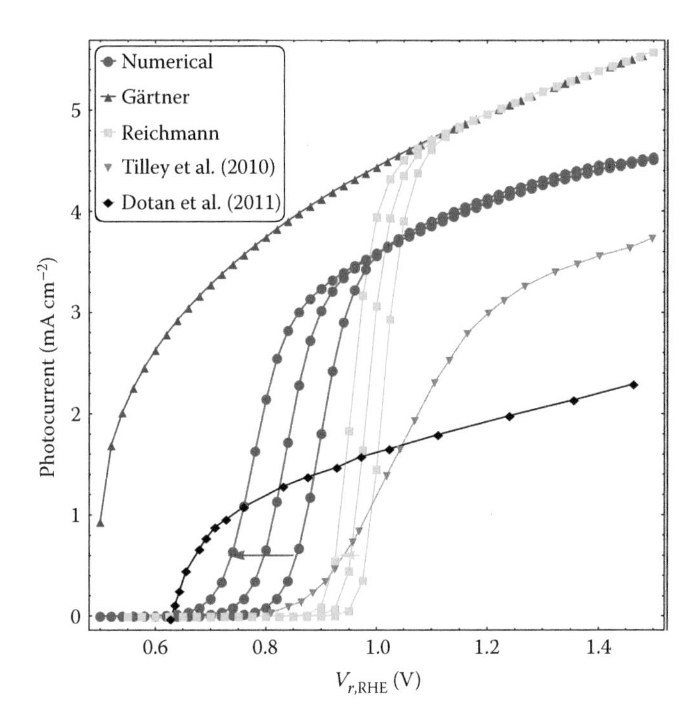

FIGURE 10.16 Comparison of photocurrent–voltage curves from: a numerical model, the Gärtner model, the Reichmann model, and the measured data from Tilley et al. (2010) and Dotan et al. (2011) for n-doped hematite photoelectrodes. The different simulated curves in each case are for different values of charge transfer constant. (Reproduced with permission from Cendula, P.; Tilley, S. D. et al. *The Journal of Physical Chemistry C* 2014, 118, 29599–29607.)

measurements (Ritenour et al., 2012). If $w \ll \alpha^{-1}$, then the approximate relation holds (Werner et al., 1993):

$$IQE_{PV}^{-1} = 1 + \frac{1}{\alpha L_n} \tag{10.55}$$

The Gärtner model has been amply used in the field of water splitting with metal oxide semiconductors, since these photoelectrodes normally rely on the space-charge region for the efficient transfer of holes in the water decomposition reaction (Gimenez and Bisquert, 2016). The assumption of full collection in the space-charge region is usually not well satisfied due to recombination in the case of slow transport, such as in typical hematite electrodes (Klahr and Hamann, 2011). This physical effect is suitably described by the model of Reichman (1980). Figure 10.16 shows the exact numerical calculation of the current–voltage curve using the methods indicated in Section 10.3, compared with the models of Gärtner and Reichmann and showing also characteristic experimental results (Cendula et al., 2014).

10.7 DIFFUSION–RECOMBINATION AND COLLECTION IN THE SPACE-CHARGE REGION

We derive the general analysis of the model of Figure 10.15, consisting of the combination of unity collection in the space-charge region of width w and the diffusion–recombination in the neutral region with the reflecting boundary condition at $x = d$. The general solution (without a restriction to the case of short diffusion length) is

$$n = n_0 + R\cosh\left(\frac{x-w}{L_n}\right) + S\sinh\left(\frac{x-w}{L_n}\right) + \gamma_n e^{-\alpha x} \tag{10.56}$$

where

$$\gamma_n = \frac{L_n^2 \alpha \Phi_{ph}}{D_n\left(1 - L_n^2 \alpha^2\right)} \tag{10.57}$$

$$R = n(w) - n_0 - \gamma_n e^{-\alpha w} \tag{10.58}$$

$$S = \frac{\alpha L_n \gamma_n e^{-\alpha d}}{\cosh y_d} - R\tanh y_d \tag{10.59}$$

$$y_d = \frac{d-w}{L_n} \tag{10.60}$$

The electron flux at the edge of the SCR is

$$\begin{aligned} J_n(w) &= -D_n \left.\frac{\partial n}{\partial x}\right|_{x=w} \\ &= \frac{D_n}{L_n}\left(-\frac{\alpha L_n \gamma_n e^{-\alpha d}}{\cosh y_d} + R\tanh y_d\right) + \alpha D_n \gamma_n e^{-\alpha w} \end{aligned} \tag{10.61}$$

In the SCR ($0 \le x \le w$), we have just the generation term

$$\frac{\partial J_n}{\partial x} = \alpha \Phi_{ph} e^{-\alpha x} \tag{10.62}$$

Integrating Equation 10.62, we obtain

$$J_n(0) = J_n(w) + \Phi_{ph}(e^{-\alpha w} - 1) \tag{10.63}$$

The concentration of electrons at both edges of the space-charge region is related by

$$n(w) = n(0)e^{-qV_{sc}/k_B T} \tag{10.64}$$

where V_{sc} is the voltage across the SCR. Inserting Equation 10.64 in Equation 10.58, we obtain the relation between electron flux and concentration at the edge of the semiconductor layer:

$$J_n(0) = -J_g + \frac{D_n}{L_n}\tanh y_d (n(0)e^{-qV_{sc}/k_B T} - n_0) \tag{10.65}$$

where

$$J_g = \Phi_{ph}\left[1 - e^{-\alpha w} + \frac{L_n^2\alpha^2}{\left(1 - L_n^2\alpha^2\right)\cosh y_d}\left(e^{-\alpha d} + \frac{e^{-\alpha w}}{L_n\alpha}\sinh(y_d) - e^{-\alpha w}\cosh(y_d)\right)\right] \tag{10.66}$$

The extracted flux in Equation 10.65 is a function of the electron density at the edge of the semiconductor layer. This allows for more complex boundary conditions to be added, such as the effect of a dipole layer, as shown in Figure PSC.4.6 (Bisquert et al., 2008). If the electron concentration is simply controlled by the external voltage as indicated in Equation 10.18, then Equation 10.65 gives the diode model

$$j = qJ_g - \frac{qD_n n_0}{L_n}\tanh y_d\left(e^{qV/k_BT} - 1\right) \tag{10.67}$$

In the particular case in which $d \gg L_n$, we obtain

$$j = j_{ph} - \frac{qD_n n_0}{L_n}\left(e^{qV/k_BT} - 1\right) \tag{10.68}$$

where j_{ph} is the value of the Gärtner model given in Equation 10.52.

10.8 SOLAR CELL SIMULATION

We describe the photovoltaic operation of a generic solar cell with selective contacts of different extraction properties. Instead of using a simplified approach as in the previous sections, here we adopt the general drift–diffusion equations, combined with the continuity and the Poisson equations for both electrons and holes that have been described in Sections FCT.3.4 and 10.3. We can add the specific recombination model and boundary conditions that apply to the given problem according to the interface properties and then run a numerical simulation, which will provide the band-bending, electron and hole concentrations, the Fermi levels, and currents.

We use a p-doped semiconductor with the ESC at $x = 0$, where an SB is formed as shown in Figure 10.17. The boundary conditions are

$$J_n(0) = qS_n\left(n(0) - n_0 e^{qV_{bi}/kT}\right) \tag{10.69}$$

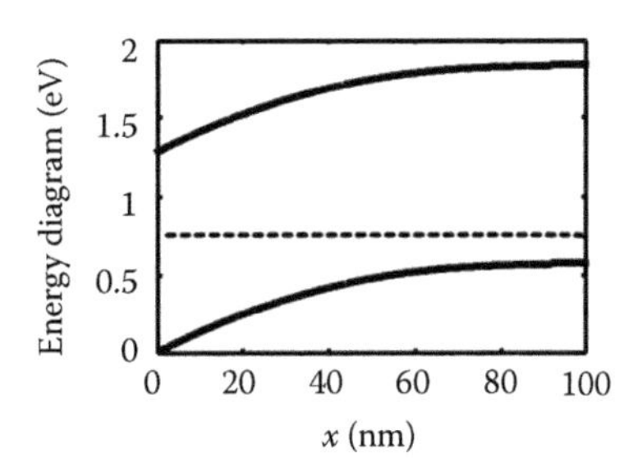

FIGURE 10.17 Band diagram of an SB solar cell in dark. The ESC is at $x = 0$ and the HSC is at $x = d$. The physical parameters used in the simulation are: $u_{n,p} = 0.1$ cm^2 V^{-1} s^{-1}, $\varepsilon_r = 10$, $d = 100$ nm, $E_g = 1.2$ eV, N_C, $N_V = 10^{19}$ cm^{-3}, energy barrier for holes at the HSC $\Phi_a = 0.18$ eV, energy barrier for electrons at the ESC $\Phi_b = 0.45$ eV, $\alpha d = 1$, $\tau_n = (5d)^2/D_n$, and $N_A = 10^{17}$ cm^{-3}.

$$J_p(0) = 0 \tag{10.70}$$

The origin of the electrostatic potential is taken at the ESC at $x = 0$:

$$\varphi(0) = 0 \tag{10.71}$$

$$\varphi(d) = V - V_{bi} \tag{10.72}$$

At the HSC ($x = d$), we consider different cases of the properties of the interface:

1. A reflecting contact for electron and hole transfer limited by surface recombination of holes

$$J_p(d) = qS_p(p(d) - p_0(d)) \tag{10.73}$$

$$J_n(d) = 0 \tag{10.74}$$

2. SRH recombination and hole transfer limited by surface recombination, Equations 10.25 and 10.26:

$$J_n(d) = -qN_t \frac{\beta_n \beta_p (n(d)p(d) - n_0 p_0)}{\beta_n n(d) + \varepsilon_n + \beta_p p(d) + \varepsilon_p} \Delta d \tag{10.75}$$

$$J_p(d) = qN_t \frac{\beta_n \beta_p (n(d)p(d) - n_0 p_0)}{\beta_n n(d) + \varepsilon_n + \beta_p p(d) + \varepsilon_p} \Delta d + qS_p(p(d) - p_0) \tag{10.76}$$

3. Infinite recombination of electrons

$$J_p(d) = qS_p(p(d) - p_0(d)) \tag{10.77}$$

$$n(d) = n_0 \tag{10.78}$$

In Figure 10.18 are shown the characteristics of the solar cell at open circuit in the case of generation across the whole active layer. When the diffusion length is higher than the device length ($L_n > d$), the concentration of electrons remains almost constant along the device, as already discussed in Section 10.5. In some cases shown in Figure 10.18, holes accumulate at the ESC. At open-circuit voltage, the bands are almost flat along the device and the carriers can flow to any direction. The holes (majority carriers in this semiconductor) are blocked at the ESC and, in addition, they cannot recombine due to the low recombination, thus, they accumulate at the ESC interface. This accumulation can be suppressed for different values of the parameters.

In Figure 10.19, a shorter penetration of the incident light occurs, which is reflected in the carrier distribution when the diffusion length is short, panel (b).

Figure 10.20 shows the characteristics of the p-type solar cell under different assumptions of light penetration and diffusion length for the parameters and results shown in Table 10.2. The diagrams obtained from the general simulation indicate the carrier distributions and the components of electron current, and they complement the previous calculation for the case of flat bands presented

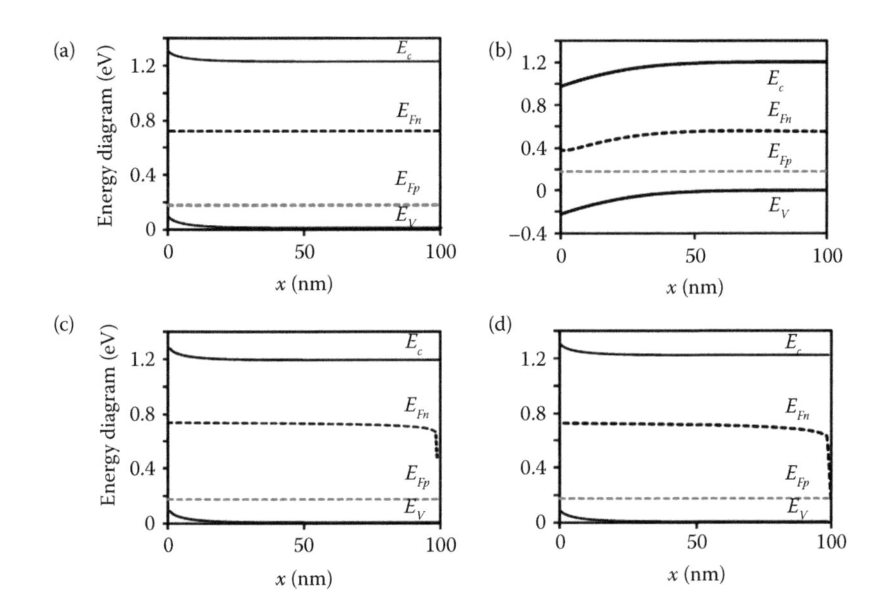

FIGURE 10.18 Drift–diffusion simulations of a p-type solar cell at open circuit voltage for different cases in which the boundary condition for electrons at the back ($x = d$) and the diffusion length, L_n, are modified. (a) HSC is a reflecting contact and the semiconductor has a long diffusion length; (b) HSC is a reflecting contact and the semiconductors have a short diffusion length. (c) HSC with SRH recombination and semiconductor with long diffusion length. (d) Ideal absorbing contact and semiconductor with long diffusion length. The ESC is always a reflecting contact. The physical parameters used in the simulation are $u_{n,p} = 1$ cm^2 V^{-1} s^{-1}, $\varepsilon_r = 10$, $N_A = 10^{17}$ cm^{-3}, $d = 100$ nm, $E_g = 1.2$ eV, N_C, $N_V = 10^{20}$ cm^{-3}, energy barrier at the hole selective contact for holes $\Phi_{B,a} = 0.2$ eV, energy barrier at the ESC for electrons $\Phi_{B,b} = 0.6$ eV, $\Phi_{ph} = 10^{16}$ cm^{-1} s^{-1}, and $\alpha d = 1$. The particular parameters are in (a, c, d): $\tau_n = (5d)^2/D_n = 3.5 \times 10^{-12}$ s and in (b) $\tau_n = (0.03d)^2/D_n = 10^{-7}$ s. For SRH: $\beta_n = \beta_p = 1$ cm^3 s^{-1}, $N_c = N_v = 10^{21}$ cm^{-3}, $\varepsilon_n = 10^{-6}$, and $\varepsilon_p = 10^{26}$. (Courtesy of Pilar Lopez-Varo.)

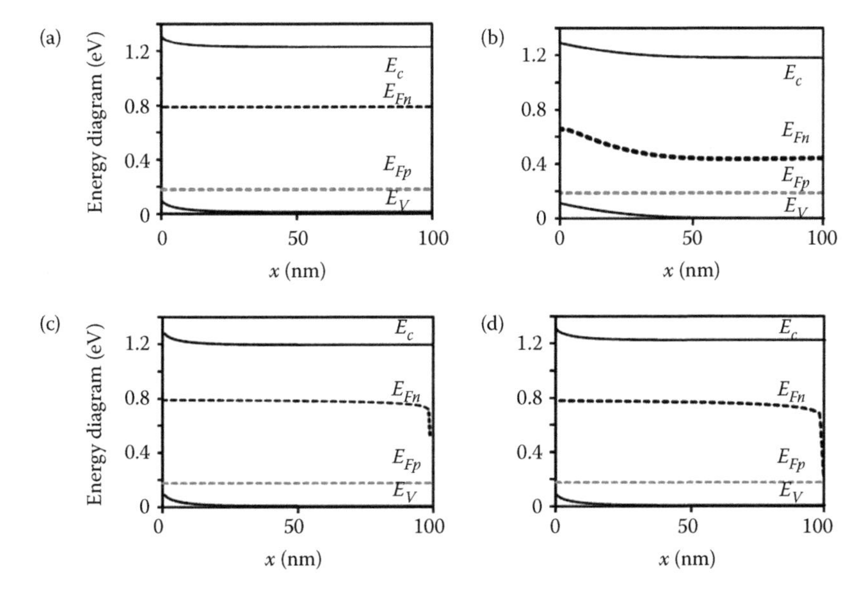

FIGURE 10.19 Drift–diffusion simulations for the same cases of Figure 10.18. The light penetration is the only parameter with a different value, $\alpha d = 10$. (Courtesy of Pilar Lopez-Varo.)

TABLE 10.2
Characteristics of Simulated Solar Cells of Figure 10.20

	(a)	(c)	(e)	(g)
L_n/d	0.03	0.03	5	5
αd	1	20	1	20
j_{ph} (mA cm^{-2})				
$q\Phi_{pho}(e^{-\alpha d} - 1)$ (mA cm^{-2})	10	16	10	16
j_{ph} (mA cm^{-2})	0.05	0.7	1	1.6
V_{oc} (V)	0.28	0.5	0.65	0.675
EQE_{PV}	0.03	0.4	0.66	1
IQE_{PV}	0.05	0.4	1	1

Parameters: $d = 100$ nm, $T = 300$ K, $\Phi_{ph} = 10^{16}$ cm^{-2} s^{-1}, $N_A = 10^{17}$ cm^{-3}, and $u_{n,p} = 1$ cm^2 V^{-1}s^{-1}.

in Figure 10.13. The internal current components in the second column indicate internal flues due to generation close to the left contact.

The effect of the surface states on solar cell characteristics is discussed with the example in Figure 10.12. The scheme and calculations represent an all-oxide solar cell based on Co_3O_4 absorber (Majhi et al., 2016), with TiO_2 ESC at the transparent side and Au layer as HSC. To improve the quality of the back contact, an additional MoO_3 layer is added as shown in Figure 10.12b, which reduces the density of surface states. The interface extraction rate of each carrier at the respective contact is modeled by a surface velocity constant S_n, S_p. On the right side, it is observed that surface traps reduce substantially the voltage and photocurrent for short thickness, while in the case of thicker absorber layer, the recombination at the back contact is less influential in the solar cell performance.

10.9 CLASSIFICATION OF SOLAR CELLS

Having analyzed in full detail the different steps that compose the photovoltaic operation, we now establish a comparison of the mechanisms that lead to charge collection (Kirchartz et al., 2015).

As mentioned earlier, charge extraction in a solar cell consists of a competition between charge transport and recombination. On their way to the contacts, electrons and holes encounter each other in the central layer, which can lead to recombination events that are quantified in a generic sense by a recombination lifetime τ. This lifetime may depend on impurities, dislocations and grain boundaries in inorganic solar cells, or on the local microstructure of the donor–acceptor network in organic solar cells. In polycrystalline solar cells such as CdTe or $CH_3NH_3PbI_3$ perovskite, one may take care and control the special properties of the grain boundaries, which may have electronic properties deleterious for the overall operation, as mentioned previously in Section PSC.9.3 and shown in Figure 10.21, where a large grain structure (left) forms conduction pathways that provoke a short circuit. The adequate recrystallization, annealing, and control of grain size procedures will provide a more compact film with optimized photovoltaic properties, as shown in the right column of Figure 10.21, and the passivation of grain boundaries reduces the deleterious recombination surface states, as indicated in Figure PSC.9.9 (Jaegermann et al., 2009). Controlling crystal growth and morphologies by solvent engineering has been a determinant step for the development of high-efficiency perovskite solar cells (Jeon et al., 2014; Ahn et al., 2015).

In Section PSC.4.5, we remarked on the construction of a built-in voltage in solar cell devices. In general, the selective contact to electrons must be ohmic and form a small injection barrier for electrons. Thus, materials with a low-work function are required. These will align well to the CB

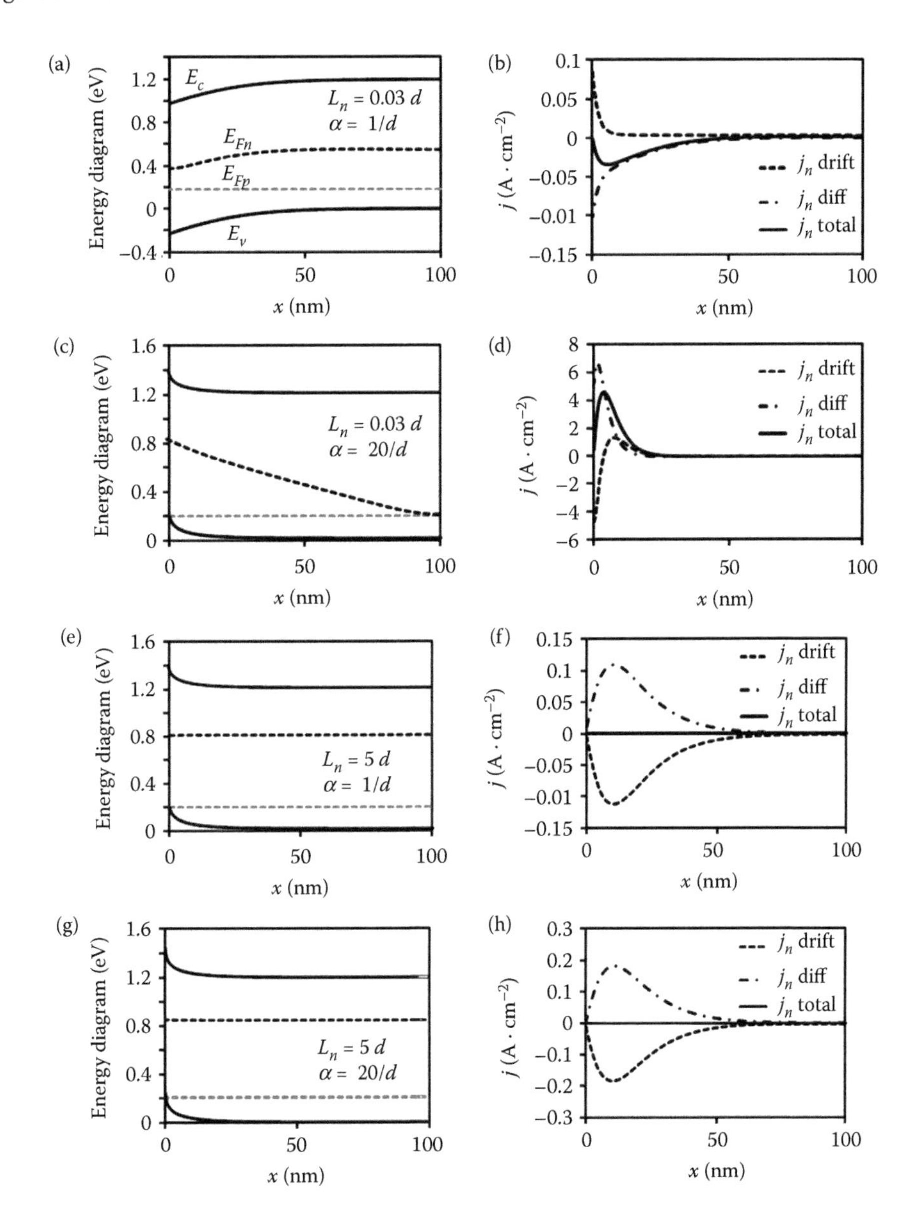

FIGURE 10.20 (a, c) Energy band diagrams including the quasi-Fermi levels in Schottky p-type solar cells at open circuit with different values of the absorption depth and the diffusion length. (b, d) Respective calculations of the drift, diffusion (diff), and total electron current densities. The simulation parameters are in Table 10.2. (Courtesy of Pilar Lopez-Varo.)

minimum energy level of the absorber layer. Oppositely, materials with a large work function may form an efficient hole extraction contact. In consequence, solar cell devices contain an intrinsic built-in potential between the outer contacts.

In some cases, the built-in field does not help much with charge extraction at short circuit. This is typically the case when the built-in field is confined to a dipole layer or to a narrow space-charge region, as in a DSC or crystalline Si (c-Si) solar cell, as discussed in Section PSC.9.3 and shown in Figures PSC.9.11 and PSC.9.12. Nevertheless, the built-in voltage V_{bi} ensures that the applied forward bias V (e.g., at the maximum power point of an illuminated solar cell) can drop somewhere. As long as $V_{bi} - V$ is still positive, no barriers for extraction will form. Thus, the built-in voltage

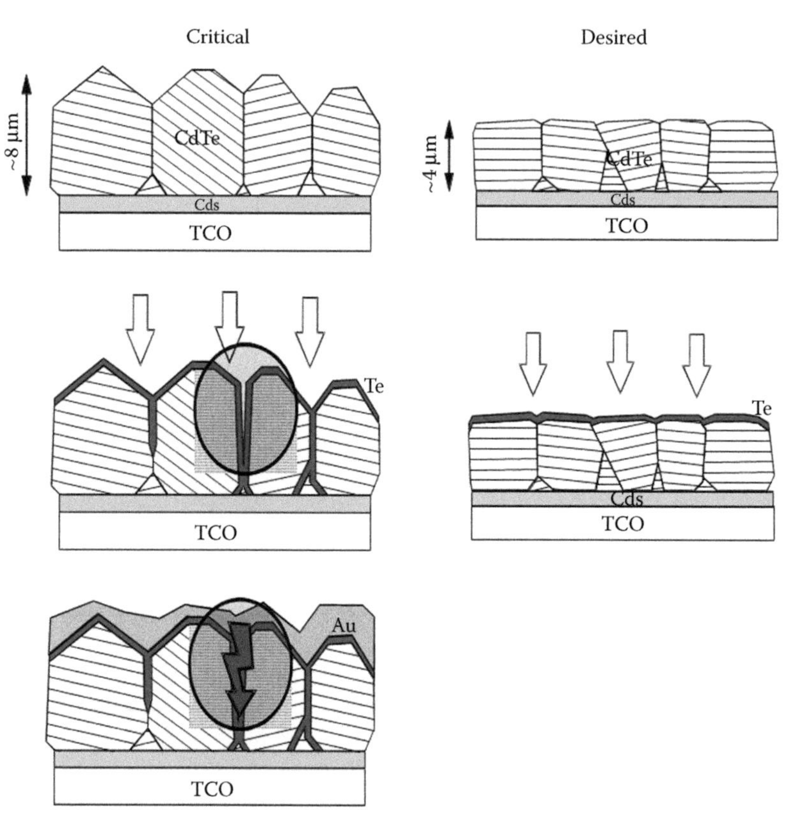

FIGURE 10.21 Schematic representation of recombination pathways producing low-shunt resistances along grain boundaries of three-dimensional, statistically oriented films (*left*), and the improved morphology of compact CdTe layers expected by better control of texture and nucleation (*right*). (Reproduced with permission from Jaegermann, W. et al. *Advanced Materials* 2009, 21, 4196–4206.)

serves an important role even if, like in c-Si solar cells, charge collection is nearly entirely driven by diffusion. Alternatively, the built-in field, obtained after initial device formation, may be extended over the absorber thickness and take a leading role in charge extraction. In fact, the device might be even fully depleted at short circuit in which case the built-in electric field extends over the whole absorber, as in amorphous silicon (a-Si) solar cells (Schiff, 2003), see the model in Figure PSC.4.11.

The ratio between the width w of the space-charge region and the absorber thickness d is therefore a useful criterion to classify solar cell types with regard to their way of separating charge carriers. Figure 10.22a compares important solar cell technologies in terms of their ratio w/d. Figure 10.22b shows the band diagram of a p-type semiconductor at short circuit in the dark. The doping is sufficiently high relative to the absorber thickness so that $w/d \ll 1$. c-Si as well as DSCs are typical examples of solar cells that have a tiny space-charge region relative to the total absorber width. Here, electron transport through the device will be mostly by diffusion as already explained above.

For typical inorganic thin film solar cells, the ratio w/d varies from around 1/10 in Cu(In,Ga)Se_2 (CIGS) to 1 in fully depleted devices like amorphous or microcrystalline Si (μc-Si). Figures 10.21c and d show the schematic band diagrams for devices with intermediate (c) and high (d) ratios of w/d. When going from small to large values of w/d, the way charge separation occurs in the solar cell changes. While devices with low w/d are controlled by diffusion, the larger the space-charge region becomes relative to the absorber thickness, the more drift will affect charge carrier collection. Alternate situations can exist, as in the case of organic bulk heterojunctions where both band bending in the absorber and contact dipole account for the total V_{bi} (Guerrero et al., 2012). Figure 10.23

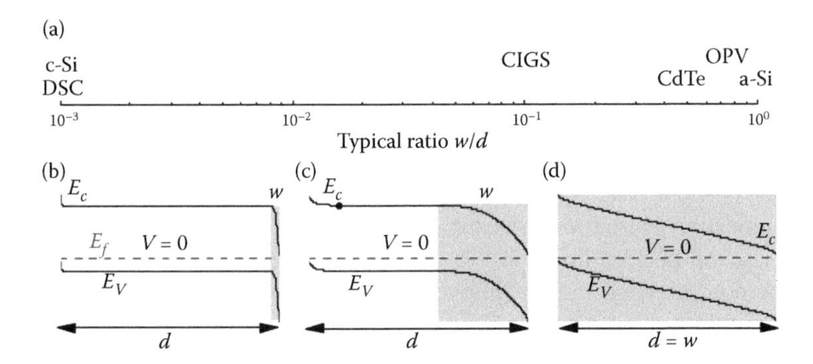

FIGURE 10.22 (a) The typical ratio of space-charge region width w and thickness d for different solar cell technologies. Band diagrams of solar cells with small w/d (b), intermediate w/d (c), and $w = d$ (d) at short circuit in the dark. The gray areas represent the main space-charge regions in the different diagrams at short circuit. (Reproduced with permission from Kirchartz, T. et al. *Physical Chemistry Chemical Physics* 2015, 17, 4007–4014.)

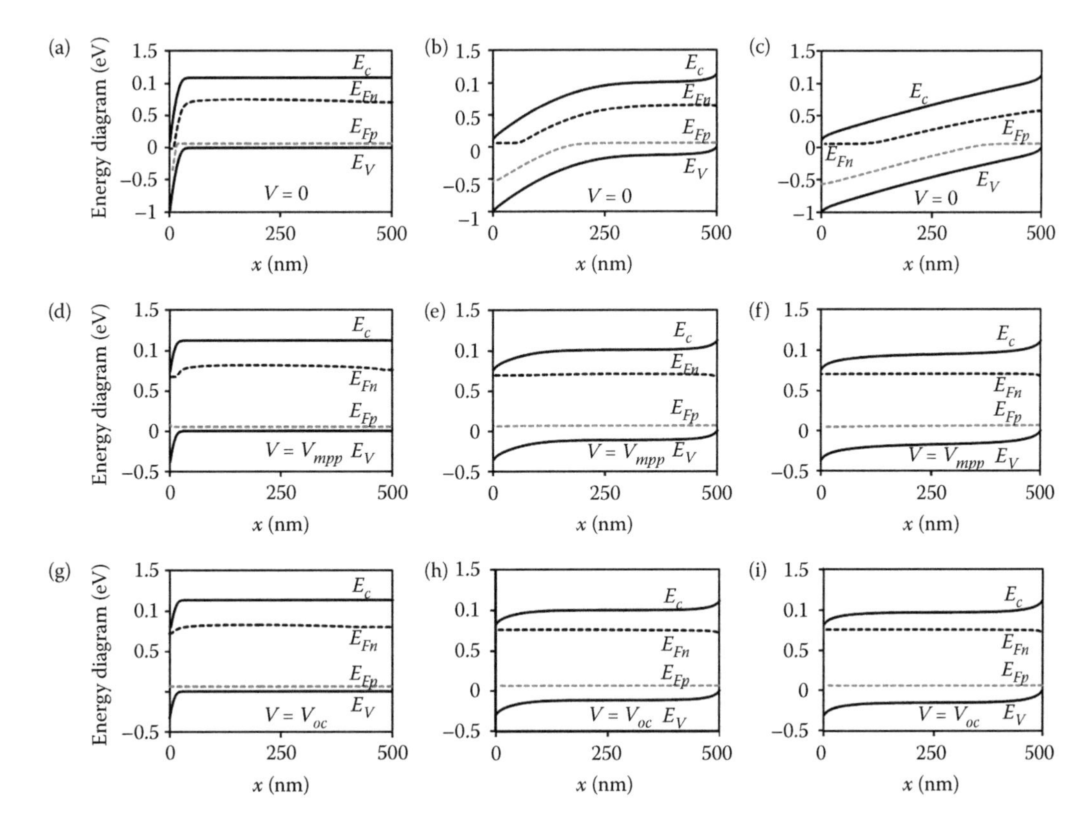

FIGURE 10.23 Solar cell simulations with different ratio between space charge region width w and device length d due to different doping ion acceptor concentrations. Band diagrams of the solar cells with small w/d (a), intermediate w/d (b), and $w = d$ (c) at short circuit, at maximum power point (d–f), and at open circuit (g–i). Parameters: $u_{n,p} = 0.1$ cm^2 V^{-1} s^{-1}, $\varepsilon_r = 10$, $d = 500$ nm, $E_g = 1.2$ eV, N_c, $N_v = 10^{19}$ cm^{-3}, barrier at hole selective contact for holes $\Phi_{B,a} = 0.05$ eV, barrier at ESC for electrons $\Phi_{B,b} = 0.05$ eV, $\alpha d = 1$, $\tau_n = (5d)^2/D_n$ and $\Phi_{ph} = 10^{16}$ cm^{-1} s^{-1}. The particular parameters are (a, d, g) $N_A = 10^{18}$ cm^{-3}, (b, e, h), $N_A = 10^{16}$ cm^{-3}, and (c, f, i) $N_A = 1 \times 10^{15}$ cm^{-3}. (Courtesy of Pilar Lopez-Varo.)

shows the distribution of the Fermi levels for the three cases of solar cells regarding the size of *w*/*d*, under illumination, from short- to open-circuit condition.

10.10 MEASURING AND REPORTING SOLAR CELL EFFICIENCIES

Investigating new types of solar cell materials requires adopting careful protocols of measurement to avoid communicating biased and unjustified claims, with devastating negative impact on the research field. The top performing cell of a given technology is routinely reported in efficiency tables and in the popular NREL chart (http://www.nrel.gov/ncpv). To become official, the device must have a minimum area of 1 cm^2 and the result must be certified by specialist labs. For solar cells still in their initial stages of development, the current published record may have been made with special materials that are not available to everyone, and with very unique skills of the staff or equipment in the lab, sometimes using sophisticated light management schemes. Although not widely recognized, some reported records have been achieved for very small area devices, which is not representative of practical solar cells.

Reporting the current density–voltage curve under standard conditions is obviously necessary when investigating new types of solar cells. Papers should also ideally report the EQE_{PV}, which provides an important test of calculation of the photocurrent as commented in Sections PSC.6.1 and 10.1. If the photocurrent is larger than the integral of EQE_{PV}, a major flaw has occurred. Figure PSC.8.2b shows the result of the integration that presents an excellent match with the photocurrent measured under 1 sun condition indicated in Figure PSC.8.2a. Note that such good match requires some precautions to avoid the difference between the solar cell conditions in dark and light (Jennings et al., 2010). Li et al. (2016, Figure PSC.8.8) remark that the measurements of EQE_{PV} were taken with chopped monochromatic light under a white light bias corresponding to 5% solar intensity.

Larger cells always lower the efficiency to a significant extent with respect to small ones. The best module efficiency is typically about 60% the best small cell efficiency (Nayak et al., 2011). Therefore, the active area of the device and the illumination procedure should be clearly stated. Standard procedures of characterization require avoiding the extra incident light from lateral reflection and also contributions from the material indirectly illuminated through light scattering. When reporting efficiencies, masks with (preferably) the same size as the active area of the device should be used. The calibration of the set-up with a solar cell that has similar spectral characteristics to the one measured is important to avoid mistakes due to the spectral features of a given light source (Doscher et al., 2016).

In this chapter, we have emphasized the power conversion efficiency under normal incident light, which is the standard for calibration. In real applications, there are other figures of merit such as the annual energy yield that depends on the light absorption at different incident angles and indirect light absorption across the year (Reale et al., 2014).

Perovskite solar cells and many other solution-processed photovoltaic devices, at their early stage of development, demonstrated an enormous variability of efficiency when manually prepared in the lab. The dispersion was huge until robust procedures and consistent protocols of materials and device making were established. It has happened often that a lab produces a large batch of samples and picks the best one for preparing a publication. Obviously, such a device may not be considered as representative of the state-of-art. The authors should always present histograms of the performance parameters, and average as well as best values. The stability of the reported devices under prolonged simulated sunlight should also be stated.

GENERAL REFERENCES

Simulation of organic and thin film solar cells: Fonash (1981), Schiff (2003), Kirchartz and Rau (2011), Kirchartz et al. (2012), Tress et al. (2012), and Tress (2014).

Measuring and reporting solar cells: Smestad et al. (2008), Snaith (2012), Zimmermann et al. (2014), and Christians et al. (2015).

Stabilized measurement of power conversion efficiency: Christians et al. (2015), Zimmermann et al. (2016), and Bardizza et al. (2017).

Measuring the diffusion length: Werner et al. (1993), Södergren et al. (1994), Brendel and Rau (1999), Halme et al. (2008), Jennings et al. (2010), and Pala et al. (2014).

Distribution of electrons in a DSC: Lobato and Peter (2006), Lobato et al. (2006), and Jennings and Peter (2007).

Diffusion–recombination model: Gärtner (1959), Reichman (1980), Brendel et al. (1995), and Bae et al. (2015).

Effect of built-in voltage: Rau et al. (2003), Turrión et al. (2003), and Kirchartz and Rau (2011).

Perovskite Solar Cells: Como et al. (2016) and Park et al. (2016).

REFERENCES

Ahn, N.; Son, D.-Y.; Jang, I.-H.; Kang, S. M.; Choi, M.; Park, N.-G. Highly reproducible perovskite solar cells with average efficiency of 18.3% and best efficiency of 19.7% fabricated via Lewis base adduct of lead(II) iodide. *Journal of the American Chemical Society* 2015, 137, 8696–8699.

Bae, D.; Pedersen, T.; Seger, B.; Malizia, M.; Kuznetsov, A.; Hansen, O.; Chorkendorff, I.; Vesborg, P. C. K. Back-illuminated Si photocathode: A combined experimental and theoretical study for photocatalytic hydrogen evolution. *Energy & Environmental Science* 2015, 8, 650–660.

Bardizza, G.; Pavanello, D.; Galleano, R.; Sample, T.; Müllejans, H. Calibration procedure for Solar Cells exhibiting slow response and application to a dye-sensitized photovoltaic device. *Solar Energy Materials and Solar Cells* 2017, 160, 418–424.

Barea, E. M.; Gonzalez-Pedro, V.; Ripolles-Sanchis, T.; Wu, H.-P.; Li, L.-L.; Yeh, C.-Y.; Diau, E. W.-G.; Bisquert, J. Porphyrin dyes with high injection and low recombination for highly efficient mesoscopic dye-sensitized solar cells. *The Journal of Physical Chemistry C* 2011, 115, 10898–10902.

Beaumont, J. H.; Jacobs, P. W. M. Polarization in potassium chloride crystals. *Journal of Physical Chemistry of Solids* 1967, 28, 657.

Bisquert, J.; Garcia-Belmonte, G.; Munar, A.; Sessolo, M.; Soriano, A.; Bolink, H. J. Band unpinning and photovoltaic model for P3HT:PCBM organic bulk heterojunctions under illumination. *Chemical Physics Letters* 2008, 465, 57–62.

Boix, P. P.; Ajuria, J.; Etxebarria, I.; Pacios, R.; Garcia-Belmonte, G.; Bisquert, J. Role of ZnO electron-selective layers in regular and inverted bulk heterojunction solar cells. *Journal of Physical Chemistry Letters* 2011, 2, 407–411.

Brendel, R.; Hirsch, M.; Stemmer, M.; Rau, U.; Werner, J. H. Internal quantum efficiency of thin epitaxial silicon solar cells. *Applied Physics Letters* 1995, 66, 1261–1263.

Brendel, R.; Rau, U. Effective diffusion lengths for minority carriers in solar cells as determined from internal quantum efficiency analysis. *Journal of Applied Physics* 1999, 85, 3634–3637.

Cendula, P.; Tilley, S. D.; Gimenez, S.; Bisquert, J.; Schmid, M.; Grätzel, M.; Schumacher, J. O. Calculation of the energy band diagram of a photoelectrochemical water splitting cell. *The Journal of Physical Chemistry C* 2014, 118, 29599–29607.

Christians, J. A.; Manser, J. S.; Kamat, P. V. Best practices in perovskite solar cell efficiency measurements. Avoiding the error of making bad cells look good. *The Journal of Physical Chemistry Letters* 2015, 6, 852–857.

Como, E. D.; Angelis, F. D.; Snaith, H.; Walker, A. (Eds.). *Unconventional Thin Film Photovoltaics*; Royal Society of Chemistry: London, 2016.

Donolato, C. A reciprocity theorem for charge collection. *Applied Physics Letters* 1985, 46, 270–272.

Donolato, C. An alternative proof of the generalized reciprocity theorem for charge collection. *Journal of Applied Physics* 1989, 66, 4524–4525.

Doscher, H.; Young, J. L.; Geisz, J. F.; Turner, J. A.; Deutsch, T. G. Solar-to-hydrogen efficiency: Shining light on photoelectrochemical device performance. *Energy & Environmental Science* 2016, 9, 74–80.

Dotan, H.; Sivula, K.; Gratzel, M.; Rothschild, A.; Warren, S. C. Probing the photoelectrochemical properties of hematite (a-Fe_2O_3) electrodes using hydrogen peroxide as a hole scavenger. *Energy & Environmental Science* 2011, 4, 958–964.

Fonash, S. J. *Solar Cell Device Physics*; Academic Press: New York, 1981.

Gärtner, W. Depletion-layer photoeffects in semiconductors. *Physical Review* 1959, 116, 84–87.

Gimenez, S.; Bisquert, J. *Photoelectrochemical Solar Fuel Production. From Basic Principles to Advanced Devices*; Springer: Switzerland, 2016.

Green, M. A. Do built-in fields improve solar cell performance? *Progress in Photovoltaics: Research and Applications* 2009, 17, 57–66.

Green, M. A.; Emery, K.; Hishikawa, Y.; Warta, W.; Dunlop, E. D. Solar cell efficiency tables (version 39). *Progress in Photovoltaics: Research and Applications* 2012, 20, 12–20.

Guerrero, A.; Marchesi, L. F.; Boix, P. P.; Ruiz-Raga, S.; Ripolles-Sanchis, T.; Garcia-Belmonte, G.; Bisquert, J. How the charge-neutrality level of interface states controls energy level alignment in cathode contacts of organic bulk-heterojunction solar cells. *ACS Nano* 2012, 6, 3453–3460.

Halme, J.; Boschloo, G.; Hagfeldt, A.; Lund, P. Spectral characteristics of light harvesting, electron injection, and steady-state charge collection in pressed TiO_2 dye solar cells. *The Journal of Physical Chemistry C* 2008, 112, 5623–5637.

Jaegermann, W.; Klein, A.; Mayer, T. Interface engineering of inorganic thin-film solar cells—Materials-science challenges for advanced physical concepts. *Advanced Materials* 2009, 21, 4196–4206.

Jennings, J. R.; Li, F.; Wang, Q. Reliable determination of electron diffusion length and charge separation efficiency in dye-sensitized solar cells. *The Journal of Physical Chemistry C* 2010, 114, 14665–14674.

Jennings, J. R.; Peter, L. M. A reappraisal of the electron diffusion length in solid-state dye-sensitized solar cells. *The Journal of Physical Chemistry C* 2007, 111, 16100–16104.

Jeon, N. J.; Noh, J. H.; Kim, Y. C.; Yang, W. S.; Ryu, S.; Seok, S. I. Solvent engineering for high-performance inorganic-organic hybrid perovskite solar cells. *Nature Materials* 2014, 13, 897–903.

Kirchartz, T.; Agostinelli, T.; Campoy-Quiles, M.; Gong, W.; Nelson, J. Understanding the thickness-dependent performance of organic bulk heterojunction solar cells: The influence of mobility, lifetime and space charge. *Journal of Physical Chemistry Letters* 2012, 3, 3470–3475.

Kirchartz, T.; Bisquert, J.; Mora-Sero, I.; Garcia-Belmonte, G. Classification of solar cells according to mechanisms of charge separation and charge collection. *Physical Chemistry Chemical Physics* 2015, 17, 4007–4014.

Kirchartz, T.; Rau, U. In *Advanced Characterization Techniques for Thin Film Solar Cells*; Abou-Ras, D., Kirchartz, T., Rau, U. (Eds.); Wiley: Berlin, 2011; p. 14.

Klahr, B.; Gimenez, S.; Fabregat-Santiago, F.; Hamann, T.; Bisquert, J. Water oxidation at hematite photoelectrodes: The role of surface states. *Journal of the American Chemical Society* 2012, 134, 4294–4302.

Klahr, B. M.; Hamann, T. W. Current and voltage limiting processes in thin film hematite electrodes. *The Journal of Physical Chemistry C* 2011, 115, 8393–8399.

Koster, L. J. A.; Smits, E. C. P.; Mihailetchi, V. D.; Blom, P. W. M. Device model for the operation of polymer/fullerene bulk heterojunction solar cells. *Physical Review B* 2005, 72, 085205.

Li, L.-L.; Diau, E. W.-G. Porphyrin-sensitized solar cells. *Chemical Society Reviews* 2012, 42, 291–304.

Li, X.; Bi, D.; Yi, C.; Décoppet, J.-D.; Luo, J.; Zakeeruddin, S. M.; Hagfeldt, A.; Grätzel, M. A vacuum flash-assisted solution process for high-efficiency large-area perovskite solar cells. *Science* 2016, 10.1126/science.aaf8060.

Lobato, K.; Peter, L. M. Direct measurement of the temperature coefficient of the electron quasi-fermi level in dye-sensitized nanocrystalline solar cells using a titanium sensor electrode. *The Journal of Physical Chemistry B* 2006, 110, 21920–21923.

Lobato, K.; Peter, L. M.; Wurfel, U. Direct measurement of the internal electron quasi-fermi level in dye sensitized solar cells using a titanium secondary electrode. *The Journal of Physical Chemistry B* 2006, 110, 16201–16204.

Majhi, K.; Bertoluzzi, L.; Rietwyk, K. J.; Ginsburg, A.; Keller, D. A.; Lopez-Varo, P.; Anderson, A. Y.; Bisquert, J.; Zaban, A. Combinatorial investigation and modelling of MoO_3 hole-selective contact in $TiO_2|Co_3O_4|MoO_3$ all-oxide solar cells. *Advanced Materials Interfaces* 2016, 3, 10.1002/admi.201500405.

Maldonado, S.; Knapp, D.; Lewis, N. S. Near-ideal photodiodes from sintered gold nanoparticle films on methyl-terminated Si(111) surfaces. *Journal of the American Chemical Society* 2008, 130, 3300–3301.

Nayak, P. K.; Bisquert, J.; Cahen, D. Assessing possibilities and limits for solar cells. *Advanced Materials* 2011, 23, 2870–2876.

Pala, R. A.; Leenheer, A. J.; Lichterman, M.; Atwater, H. A.; Lewis, N. S. Measurement of minority-carrier diffusion lengths using wedge-shaped semiconductor photoelectrodes. *Energy & Environmental Science* 2014, 7, 3424–3430.

Park, N.-G.; Grätzel, M.; Miyasaka, T. (Eds.). *Organic-Inorganic Halide Perovskite Photovoltaics: From Fundamentals to Device Architectures*; Springer: Switzerland, 2016.

Rau, W.; Kron, G.; Werner, J. H. Reply to comments on "Electronic transport in dye-sensitized nanoporous TiO_2 solar cells—Comparison of electrolyte and solid-state devices." *The Journal of Physical Chemistry B* 2003, 107, 13547.

Reale, A.; Cinà, L.; Malatesta, A.; De Marco, R.; Brown, T. M.; Di Carlo, A. Estimation of energy production of dye-sensitized solar cell modules for building-integrated photovoltaic applications. *Energy Technology* 2014, 2, 531–541.

Reichman, J. The current-voltage characteristics of semiconductor-electrolyte junction photovoltaic cells. *Applied Physics Letters* 1980, 36, 574–577.

Ritenour, A. J.; Cramer, R. C.; Levinrad, S.; Boettcher, S. W. Efficient n-GaAs photoelectrodes grown by close-spaced vapor transport from a solid source. *ACS Applied Materials & Interfaces* 2012, 4, 69–73.

Schiff, E. A. Low-mobility solar cells: A device physics primer with applications to amorphous silicon. *Solar Energy Materials and Solar Cells* 2003, 78, 567–595.

Smestad, G. P.; Krebs, F. C.; Lampert, C. M.; Granqvist, C.-G.; Chopra, K. L.; Mathew, X.; Takakura, H. Editorial: Reporting solar cell efficiencies in solar energy materials and solar cells. *Solar Energy Materials and Solar Cells* 2008, 92, 371–373.

Snaith, H. J. The perils of solar cell efficiency measurements. *Nature Photonics* 2012, 6, 337–340.

Södergren, S.; Hagfeldt, A.; Olsson, J.; Lindquist, S. E. Theoretical models for the action spectrum and the current-voltage characteristics of microporous semiconductor films in photoelectrochemical cells. *The Journal of Physical Chemistry* 1994, 98, 5552–5556.

Tilley, S. D.; Cornuz, M.; Sivula, K.; Grätzel, M. Light-induced water splitting with hematite: Improved nanostructure and iridium oxide catalysis. *Angewandte Chemie International Edition* 2010, 49, 6405–6408.

Tress, W. *Organic Solar Cells. Theory, Experiment, and Device Simulation*; Springer: Switzerland, 2014.

Tress, W.; Leo, K.; Riede, M. Influence of hole-transport layers and donor materials on open-circuit voltage and shape of I–V curves of organic solar Cells. *Advanced Functional Materials* 2011, 21, 2140–2149.

Tress, W.; Leo, K.; Riede, M. Optimum mobility, contact properties, and open-circuit voltage of organic solar cells: A drift-diffusion simulation study. *Physical Review B* 2012, 85, 155201.

Turrión, M.; Bisquert, J.; Salvador, P. Flatband potential of $F:SnO_2$ in a TiO_2 dye-sensitized solar cell: An interference reflection study. *The Journal of Physical Chemistry B* 2003, 107, 9397–9403.

Werner, J. H.; Kolodinski, S.; Rau, U.; Arch, J. K.; Bauser, E. Silicon solar cell of 16.8 μm thickness and 14.7% efficiency. *Applied Physics Letters* 1993, 62, 2998–3000.

Zimmermann, E.; Ehrenreich, P.; Pfadler, T.; Dorman, J. A.; Weickert, J.; Schmidt-Mende, L. Erroneous efficiency reports harm organic solar cell research. *Nature Photonics* 2014, 8, 669–672.

Zimmermann, E.; Wong, K. K.; Müller, M.; Hu, H.; Ehrenreich, P.; Kohlstädt, M.; Würfel, U. et al. Characterization of perovskite solar cells: Toward a reliable measurement protocol. *APL Mater* 2016, 4, 091901.

11 Solar Energy Conversion Concepts

The main feature that hinders high efficiency of energy collection in solar cells is the poor match of the optical absorption of semiconductors to the solar spectrum. These effects cause large energy losses by thermalization of carriers and subbandgap transparency. Here we describe different schemes to overcome the limitations of solar energy harvesting. We first discuss the conversion of the spectral characteristics by use of luminescent layers and molecules that perform different processes of thermalization, splitting, and fusion of photons to significantly change the wavelength of the emission. These processes result in the spectral conversion methods of down-shifting using fluorescent collectors (FCs), and upconversion of long wavelength photons via excited state absorption, energy transfer, or triplet–triplet annihilation. We discuss the tandem solar cell that combines different semiconductor layers of complementary bandgaps to absorb different portions of the solar spectrum for more efficient photon utilization. Finally, we review the main methods of conversion of sunlight to chemical fuel by application of phototelectrochemical cells and photovoltaic cells combined with catalyst layers.

11.1 CONVERSION OF PHOTON FREQUENCIES FOR SOLAR ENERGY HARVESTING

From the studies of the physical limitations to photovoltaic energy conversion presented in Chapters PSC.6 through PSC.8, it emerges that one central condition for efficient solar energy harvesting is the spectral match of the semiconductor optical properties with the solar radiation. As the latter is distributed over a very wide range of energies, any single semiconductor device is doomed to substantial losses, either by reduction of the photon energy or by the failure to collect the photons in the transparent range of the semiconductor. In order to improve the harvesting of the energy of the solar photons, we can either transform the incoming spectrum to adapt it to the semiconductor features, as discussed in the following paragraphs, or we can combine several semiconductors toward a better spectral match. The second strategy consists of the tandem solar cell that will be discussed in Section 11.2.

Solar photon conversion schemes consist of luminescent layers that produce a spectral change in a relevant part of the incoming spectrum. Three different luminescent processes can be used to increase the efficiency of single bandgap solar cells: *Down-shifting*, *quantum-cutting*, and *upconversion*. The conversion processes are indicated in Figure 11.1. The potential gain of these photon conversion methods is summarized in Figure 11.2, which shows the part of the solar spectrum absorbed by a typical silicon-based solar cell and the spectral regions that can be utilized through quantum-cutting and upconversion processes (Huang et al., 2013).

Down-shifting is a single photon conversion process, which consists of the transformation of a high-energy photon to a low-energy photon. Downconversion is usually investigated in the application of luminescent solar concentrators (LSC), also called fluorescent collectors (FCs). The aim of the collector is to capture light over a large area, which is then delivered to a smaller area solar cell reducing the cost of the overall energy conversion system. FCs are also useful to complement solar cells that show poor spectral response to short wavelength electromagnetic radiation.

The FC consists of a transparent plastic doped with luminescent species, normally either organic dye molecules or inorganic quantum dots (Erickson et al., 2014). The incident sunlight is absorbed

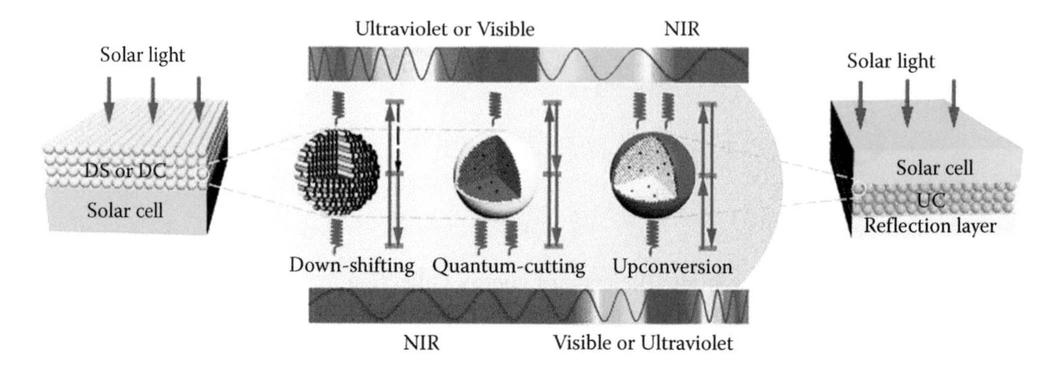

FIGURE 11.1 Spectral conversion design for PV applications involving down-shifting, quantum-cutting, and upconversion luminescent materials. In a typical down-shifting process, upon excitation with a high-energy photon, nonradiative relaxation takes place followed by radiative relaxation, thereby resulting in the emission of a lower-energy photon. In contrast, two-step radiative relaxation occurs in the quantum-cutting process upon excitation with a high-energy photon, leading to the emission of two (or more) lower-energy photons. The upconversion process can convert two (or more) incident low-energy photons into a single higher-energy photon. Note that the down-shifting and quantum-cutting materials are generally placed on the front surface of a monofacial solar cell, allowing the downconverted photons to be absorbed by the solar cell. The upconversion material is typically placed in between a bifacial solar cell and a light-reflection layer to harvest the sub-bandgap spectrum of sunlight. (Reproduced with permission from Huang, X. et al. *Chemical Society Reviews* 2013, 42, 173–201.)

by the dyes that perform isotropical reemission with high-quantum efficiency. By total internal reflection, a large part of the light is transmitted to the side of the plastic where it is collected by a solar cell, as shown in Figure 11.3. The emission is tuned to be just above the bandgap of the solar cell, as indicated in Figure 11.4, to ensure near-unity conversion efficiency of the radiation that is trapped and wave-guided to the edges of the slab. An optimum choice of the Stokes shift of the dye is an important factor for solar energy harvesting. If the Stokes shift is small, then a large part of the emitted light is reabsorbed producing photon transport losses. If, on the other hand, the Stokes

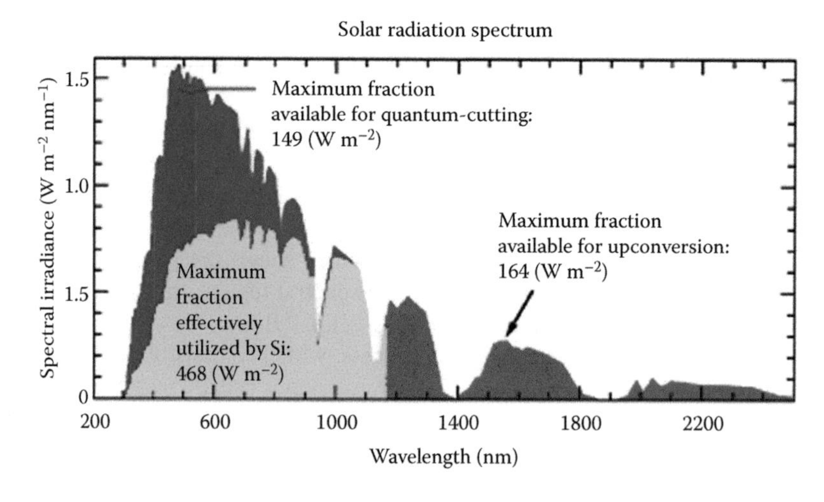

FIGURE 11.2 *AM1.5G* spectrum showing the fraction (highlighted in *green*) absorbed by a typical silicon-based PV cell and the spectral regions that can be utilized through quantum-cutting and upconversion processes (highlighted in *purple* and *red*, respectively). (Reproduced with permission from Huang, X. et al. *Chemical Society Reviews* 2013, 42, 173–201.)

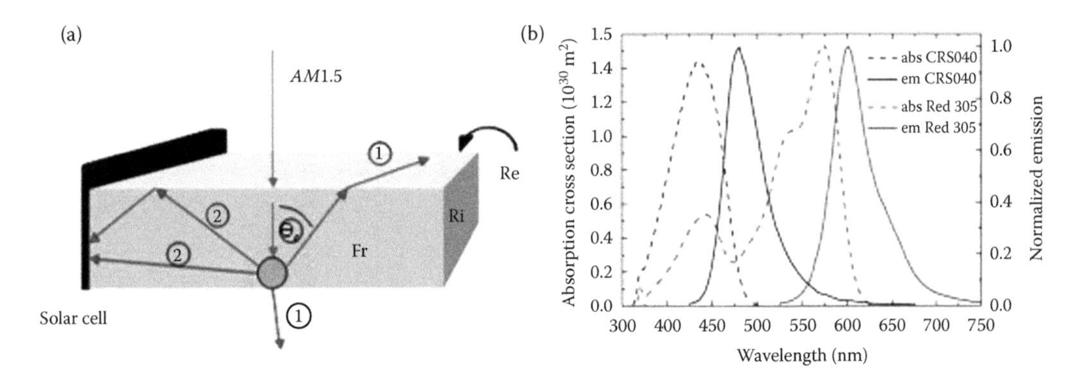

FIGURE 11.3 (a) Schematic three-dimensional view of a luminescent concentrator. *AM*1.5 light is incident on the top. The light is absorbed by a luminescent particle, and its luminescence is randomly emitted. Part of the emission falls within the escape cone and is lost from the luminescent concentrator at the surfaces (1). The other part (2) is guided to the solar cell by total internal reflection. (b) Absorption cross section and normalized emission spectra of two fluorescent dyes, illustrating Stokes' shift. (Reproduced with permission from van Sark, W. G. J. H. M. *Renewable Energy* 2013, 49, 207–210.)

shift is large, then a substantial part of the photons in the spectral range between the absorption and fluorescence maxima are not collected.

The FC provides a practical example of radiation at nonzero chemical potential that we discussed in Section PSC.6.4. The excitation by the surrounding radiation promotes the dye to the excited state and produces a separation of the Fermi levels in the dye molecules, which has the value μ_{dye} when the radiation and the molecules come to equilibrium by absorption and reemission. The spectral flux emitted by the FC achieves the form of Equation PSC.6.32, corresponding to a flux of photons at chemical potential $\mu_{flux} = \mu_{dye} > 0$.

Figure 11.5 shows the scheme of an optimized FC used to funnel the collected light to a solar cell, which covers the surface of the FC. The window open to light capture poses a problem in that light generated in the escape cone will be lost, decreasing the efficiency of effective light harvesting for the production of electricity in the solar cell. This is prevented with the use of a photonic crystal that suppresses reemission, as shown in Figure 11.5. Note the ideal characteristics of absorption and emission used in this model, as in Figure PSC.6.6. A real example of radiation produced in the FC is shown in Figure 11.6. The dye does not have a sharp absorption edge as in the model of Figure 11.5, but nonetheless at energies larger than the emission peak, Equation PSC.6.35 is well satisfied. The radiation in Figure 11.6 is characterized by a chemical potential $\mu_{flux} \approx 1.7$ eV > 0.

Quantum cutting is a form of downconversion in which one incident high-energy photon is split into two or more low-energy photons with a quantum yield larger than 100%.

Long-wavelength photons of the solar spectrum are not suitable for direct photoelectric conversion since they are transmitted through usual semiconductors, as shown in Figure 11.2. Upconversion involves the fusion of photons to significantly change the longer wavelength photons of the sunlight spectrum toward shorter wavelengths. In the scheme of photon upconversion or anti-Stokes shift, two photons of low energy are combined to produce one photon of higher energy. Upconversion is a considerable challenge as it requires the simultaneous or sequential absorption of two or more photons with lower energy than that of the emitted photon.

There are two main methods for the combination of two excitations. Excited state absorption (ESA) consists of sequential absorption of two photons in a single ion using an intermediate level and then causing the promotion of the excited state to a higher level, see Figure 11.7c. In energy transfer upconversion (ETU), two neighboring ions are excited by pump photons. Then, one ion passes the excitation by energy transfer to the neighbor, which becomes promoted to a higher-excited

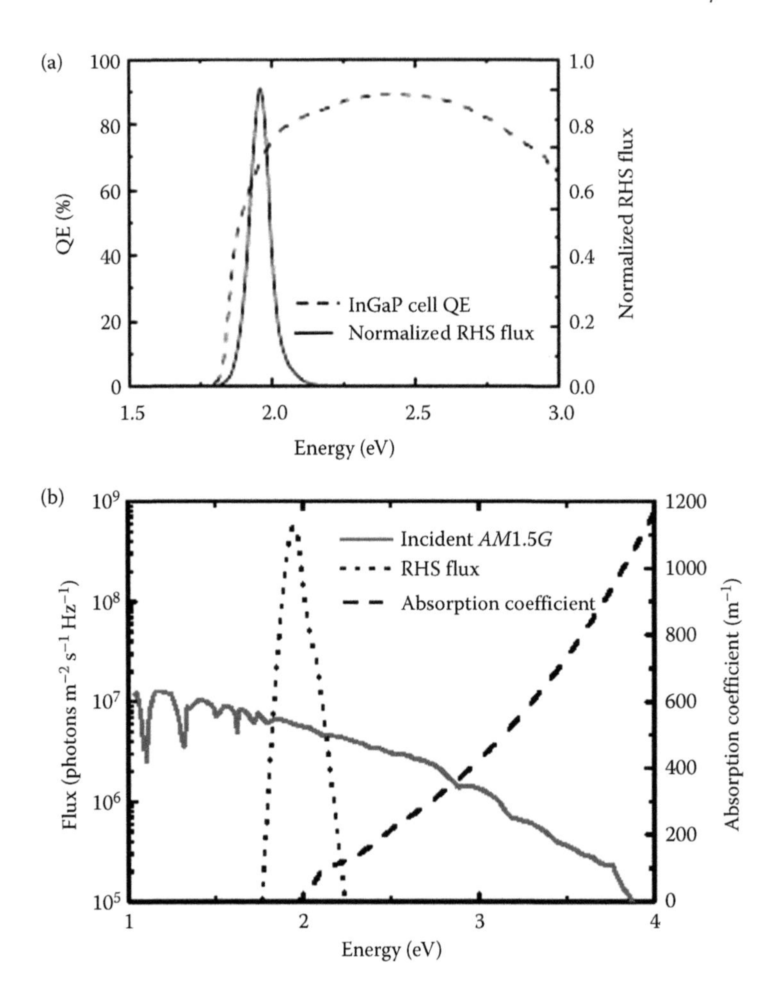

FIGURE 11.4 (a) Quantum efficiency (spectral response) of a GaInP cell used in modeling the idealized LSC together with the modeled luminescence escaping the right-hand surface (RHS) of the LSC that would be coupled into the cell. (b) Absorption of the LSC material used in the calculations for the idealized system together with the flux incident on the top surface and the predicted concentrated average luminescent flux escaping the right-hand surface of the idealized LSC. (Reproduced with permission from Van Sark, W. G. et al. *Optics Express* 2008, 16, 21773–21792.)

state (Figure 11.7d). These upconversion processes are possible with lanthanide ions that were mentioned in Section PSC.2.2 as phosphor materials. These ions possess metastable intermediate levels that are able to store the first excitation until the arrival of a second excitation, which provides sufficient energy for fluorescence at shorter wavelength. Actual energy transfer schemes for the couple Er^{3+} – Yb^{3+} are shown in Figure 11.8. Here Er^{3+} emits in the green and red after upconversion (de Wild et al., 2011).

Another approach to upconversion of low-energy photons is based on triplet–triplet annihilation (TTA). In this method, the triplet state of the sensitizer is excited by singlet excitation, Figure 11.9, or metal-to-ligand charge transfer (MLCT) excitation (Figure 11.10). The excitation and emission wavelengths of TTA upconversion can be selected by independent choice of the triplet sensitizer and triplet acceptor. Heavy metal containing porphyrins and phthalocyanines as well as metallated sensitizers are the compounds most studied as sensitizers since the presence of the π-conjugated aromatic rings place their absorption and emission maxima in the red and NIR (near-infrared)

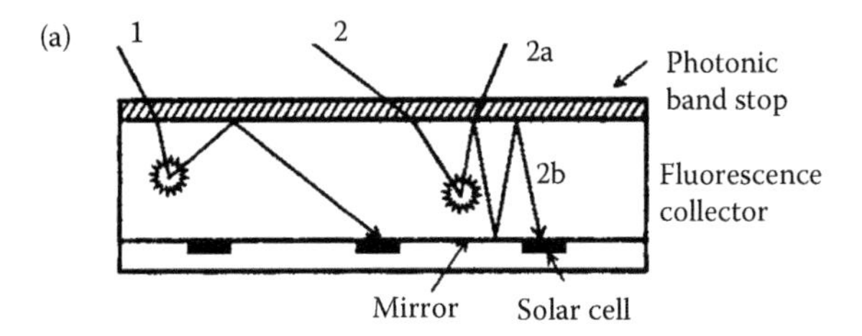

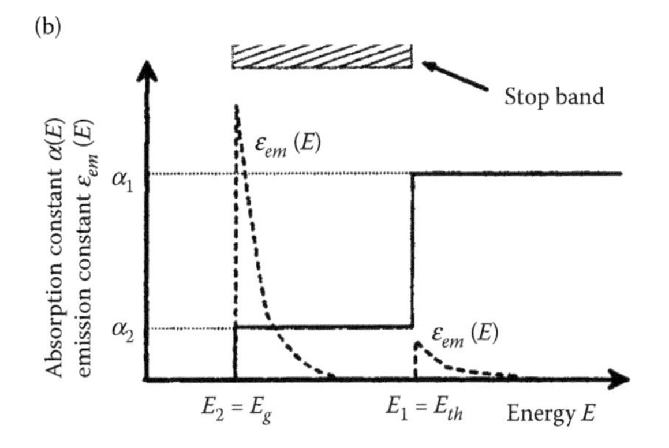

FIGURE 11.5 (a) Schematic drawing of an FC with solar cells at its bottom. The fluorescent layer absorbs incoming photons (1) and emits them at lower photon energy. Without photonic band stop some of these photons leave the collector (2a). The photonic band reflects these photons back into the collector (2b). (b) Spectral dependence of the absorption and emission coefficients, α and e, of the ideal fluorescent dye with $\alpha = \alpha_2$ for photon energies, E, in the range $E_2 \leq E \leq E_1$ and $\alpha = \alpha_1$ for $E_1 \leq E$. The photonic band has an omnidirectional reflectance $R = 1$ for $E_2 \leq E \leq E_1$. (Reproduced with permission from Rau, U.; Einsele, F.; Glaeser, G. C. *Applied Physics Letters* 2005, 87, 171101.)

region of the spectrum (Singh-Rachford and Castellano, 2010). The first process in the chain after light absorption is the intersystem crossing (ISC), which produces the metastable triplet state in the sensitizer. The excitation is passed to the annihilator by triplet–triplet energy transfer. The efficient ISC within the sensitizer molecules ensures a large population of the sensitizer triplet level after single photon absorption. On the other hand, the depopulation of the excited emitter triplet states via phosphorescence is impeded by the very weak ISC of the emitter molecules. The created triplet population of the emitter is, therefore, preserved for the process of TTA. Then, two triplets in two sensitizers recombine to form one singlet, which emits by fluorescence emission that is blueshifted to shorter wavelengths with respect to the incoming light.

11.2 TANDEM SOLAR CELLS

The tandem solar cell is another solution to the problem of optimal match to the spectral characteristics of sunlight. It consists of a group of different semiconductor absorbers with complementary spectral characteristics that are suitably connected electrically to combine their energy production rates. A device that uses a single absorber is termed a single-junction solar cell, and then one can use two junctions, three junctions, and so on.

The simplest possibility of a tandem device, the two-junction device, is composed of a wide bandgap material as top absorber that will produce a high voltage, followed by a smaller bandgap bottom absorber material that utilizes the photons that pass through the top absorber. In the monolithically integrated or two-terminal (2T) device shown in Figure 11.11, two solar cells are connected in series so that the current must be the same and the voltages of the individual cells are

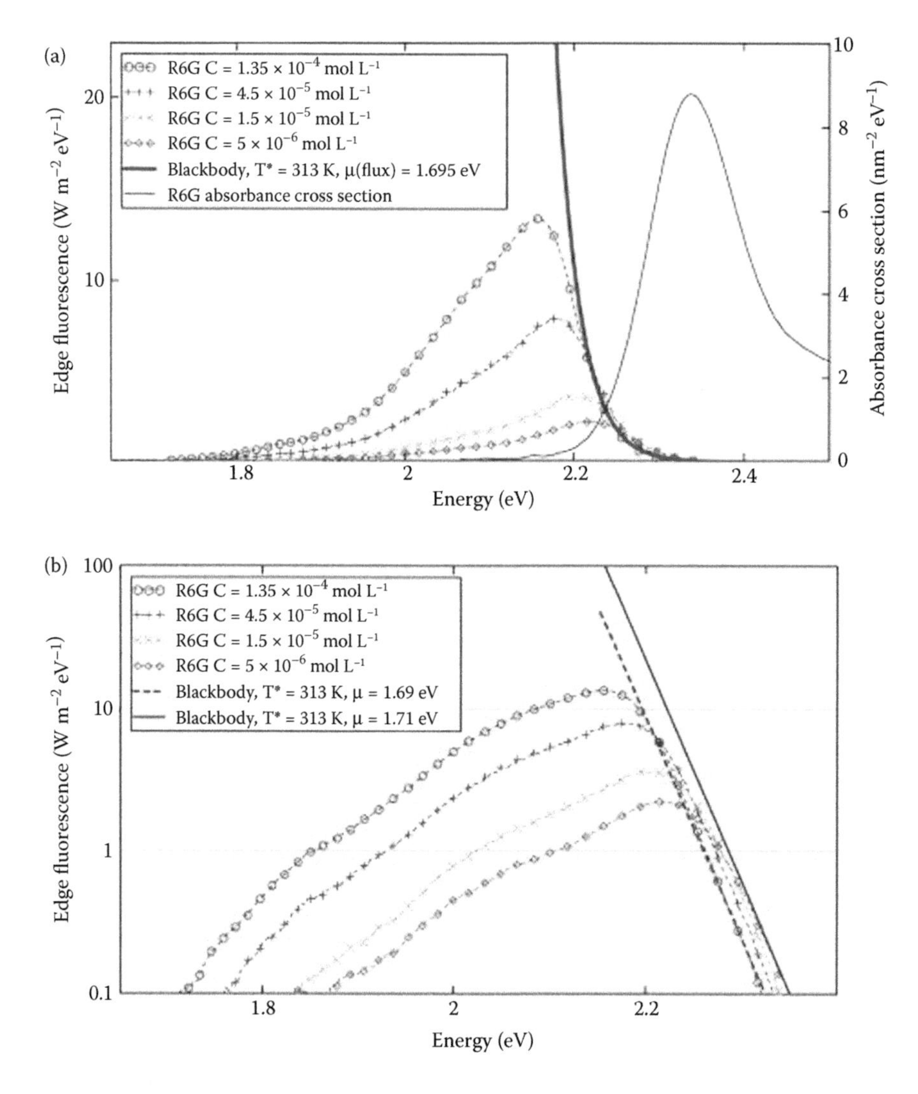

FIGURE 11.6 (a) Edge fluorescence spectra of Rhodamine 6G (R6G) compared with a blackbody function at $T = 313$ K and $\mu_{flux} = 1.695$ eV. (b) Log plot of (a), showing the fit lines to Equation PSC.6.35. The numbers in the legend give the dye concentration in the solvent. (From Meyer, T. J. J.; Markvart, T. *Journal of Applied Physics* 2009, 105, 063110.)

added. In the 2T device, it is required to build a stack of two, very high-efficiency solar cells including an intermediate, highly transparent recombination contact that aligns the Fermi levels. The technological aspects of the fabrication of these cells are rather demanding (Bush, 2017). Another type of tandem solar cell is a mechanically integrated device also called a four-terminal (4T) device, in which individual cells are operated while electrically separated.

By combining absorbers of approximately 1 and 1.6 eV, an optimal PCE of 44%–45% can be obtained under SQ maximal energy conversion conditions, as shown in Figure 11.12 (Henry, 1980). Figure 11.13 shows the use of organic polymers of complementary absorption ranges. When using inorganic semiconductors, the short wavelength absorption region of the bottom cell is obscured by the wide bandgap top cell, as in a perovskite-silicon tandem solar cell shown in Figure 11.14. The advantage of converting the photons to a larger voltage at the front cell must compensate the darkening of the bottom cell, as observed in the efficiency losses in Figure 11.11. The optical management

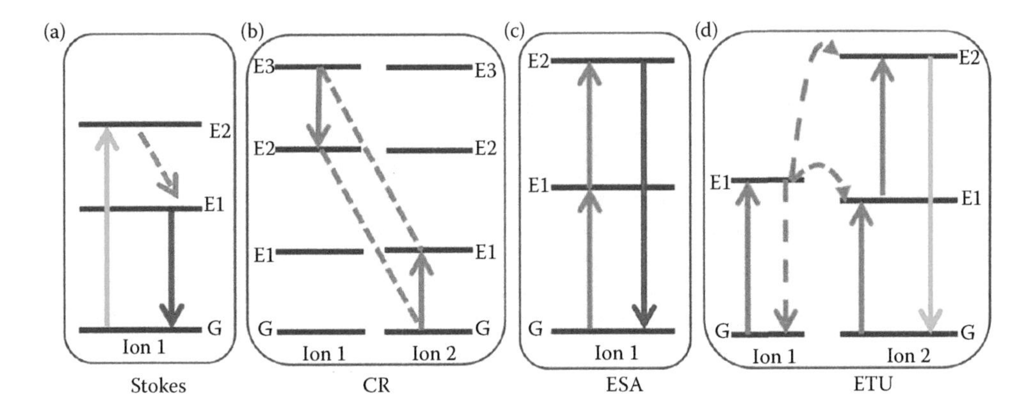

FIGURE 11.7 Mechanisms of (a) downconversion, (b) cross relaxation, (c) excited state absorption, and (d) energy transfer upconversion. (Reproduced with permission from Chen, G. et al. *Accounts of Chemical Research* 2012, 46, 1474–1486.)

of the incoming photons is crucial for an effective operation of the device, as indicated by the use of an antireflective coating in Figure 11.14. The features of a four-junction record device achieving 44% PCE under concentrated sunlight are shown in Figure 11.15.

11.3 SOLAR FUEL GENERATION

The generation of fuels with semiconductor materials offers a versatile strategy to efficiently capture and store the solar energy incident on the Earth's crust. One of the most interesting approaches involves the utilization of solar photons to realize the reduction of water to H_2 or CO_2 to carbon-based molecules. In order to efficiently carry out these processes, an energy-conversion device based on visible light-absorbing semiconductors that will capture solar photons and use them to

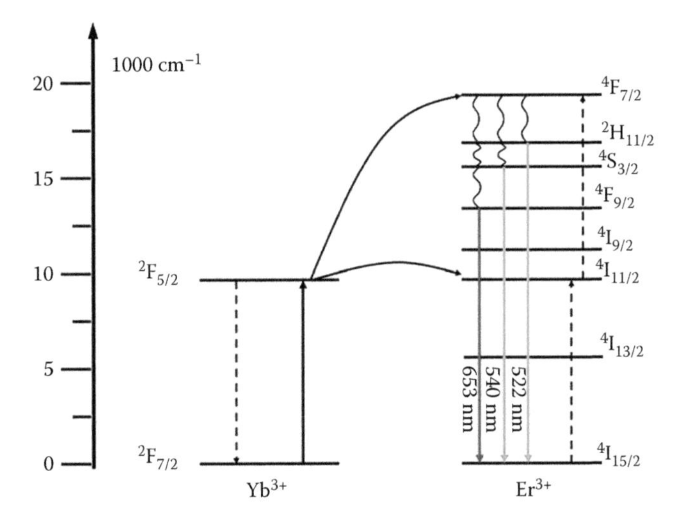

FIGURE 11.8 Schematic energy level scheme for the Yb/Er couple. The Yb_{2+} ion absorbs around 980 nm and transfers the energy from the $^2F_{5/2}$ level to the $^4I_{11/2}$ level of Er^{3+}. Subsequent energy transfer from a second excited Yb^{3+} ion to Er^{3+} ($^4I_{11/2}$) excites Er^{3+} ion to the $^4F_{7/2}$ excited state. After multiphonon relaxation to the lower lying $^4S_{3/2}$ and $^2F_{9/2}$ states, green and red emissions are observed, as indicated. (Reproduced with permission from de Wild, J. et al. *Energy & Environmental Science* 2011, 4, 4835–4848.)

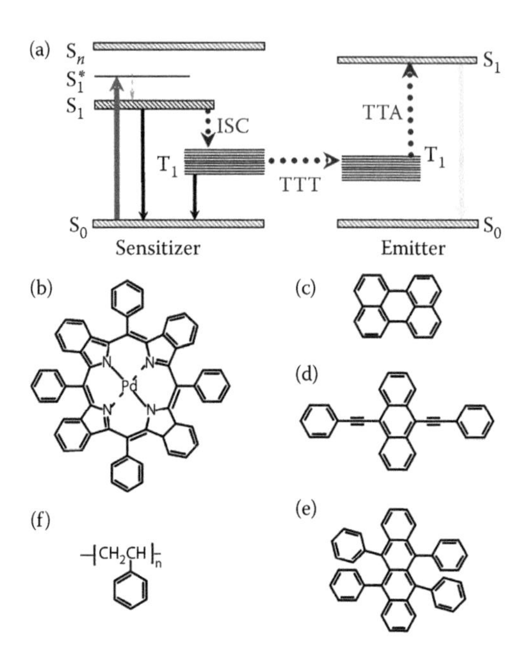

FIGURE 11.9 (a) Energetic scheme of the TTA-supported upconversion process. The structures of sensitizer PdPh4TBP (b), and the emitters: perylene (c), BPEA (d), rubrene (e), and the matrix-styrene oligomers (f). (Reprinted with permission from Miteva, T. et al. *New Journal of Physics* 2008, 10, 103002.)

carry out the required electrochemical reactions must be developed, for example, the oxygen evolution reaction (OER) for water oxidation and hydrogen evolution reaction (HER) for H^+ reduction to gas molecules. This topic is treated in the monography (Gimenez and Bisquert, 2016).

There are aspects of contrast in the application of semiconductor light absorbers for either photovoltaic or solar fuel conversion devices. For a photovoltaic material, a larger bandgap increases the open-circuit voltage, but decreases the absorption range and hence the photocurrent, as discussed in Section PSC.6.1. The benchmark method to calculate the optimal semiconductor property for

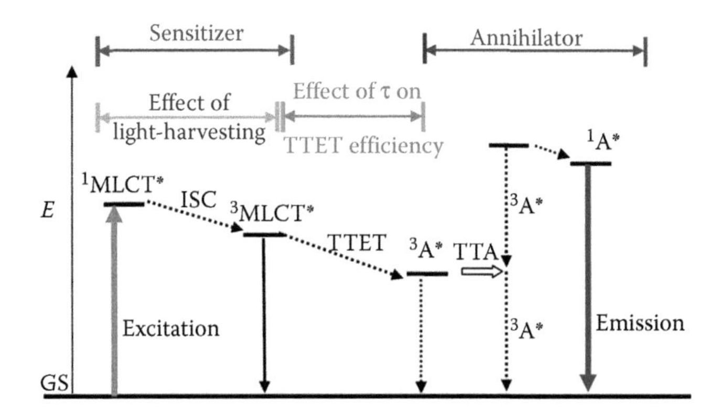

FIGURE 11.10 Qualitative diagram illustrating the sensitized TTA upconversion process between triplet sensitizer and acceptor (annihilator/emitter). The effect of the light-harvesting ability and the excited state lifetime of the sensitizer on the efficiency of the TTA upconversion is also shown. *E* is energy. GS is ground state (S_0). ^{3}MLCT* is the metal-to-ligand-charge-transfer triplet-excited state. TTET is triplet–triplet energy transfer. ^{3}A* is the triplet-excited state of annihilator. ^{1}A* is the singlet-excited state of annihilator. (Reproduced with permission from Zhao, J. et al. *RSC Advances* 2011, 1, 937–950.)

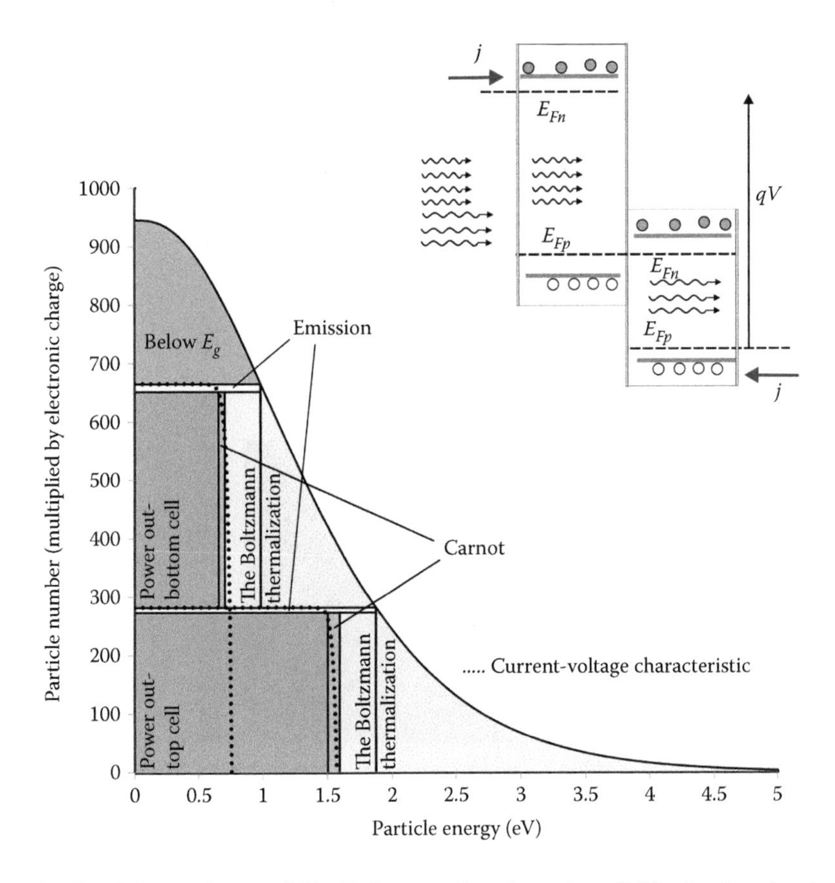

FIGURE 11.11 *Top*: Scheme of a monolithically integrated tandem solar cell. The first large bandgap material absorbs short-wavelength photons in the solar spectrum, while the long-wavelength photons are not absorbed. These are collected in a second absorber layer consisting of a low bandgap material. *Bottom*: Intrinsic losses occurring in an unconstrained double junction device with optimal bandgaps (0.98 and 1.87 eV) under 1 sun illumination. The Carnot, Boltzmann, and thermalization losses reduce the optimal-operating voltage in each absorber. (Reproduced with permission from Hirst, L. C. et al. *Progress in Photovoltaics: Research and Applications* 2011, 19, 286–293.)

maximal electrical power output is the SQ approach that provides a value of about 1.1 eV, as shown in Section PSC.8.4. However, a solar fuel generator must deliver a voltage to exceed the difference of free energies of the proposed reaction scheme, including the overvoltage required by the surface catalysts for the specific redox reactions, as shown in Figures ECK.6.29 and 11.16. For water splitting resulting in hydrogen production, a potential of 1.23 V is needed thermodynamically, Figure ECK.3.11, plus 0.2 V for each overvoltage, which adds up to about 1.8 V depending on the catalyst quality. This voltage must occur at the operating point of the device so that, even with an excellent fill factor, the photovoltage should be larger.

The second essential requirement that marks a contrast with electricity-producing photovoltaics is the necessary contact of the system with the active electrolyte in which effective catalysts must carry out the desired reactions. This feature poses the need for stability at the reaction site. This is a particularly critical aspect that determines the chosen approach among a range of alternatives that we discuss in the following (Guerrero and Bisquert, 2017).

The photoelectrochemical cell (PEC), already discussed in Section ECK.6.8 and shown in Figure 11.16a, is among the simplest arrangements to produce the water-splitting reaction and generation of hydrogen as fuel. Here, a semiconductor electrode realizes the functions of light absorption, charge separation in the space charge region, and catalytic OER. The complementary

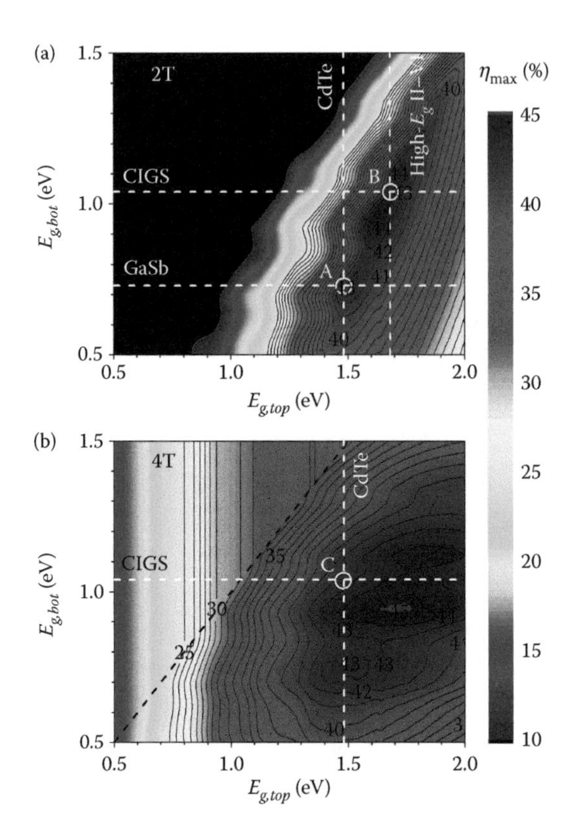

FIGURE 11.12 Two-junction radiatively limited efficiency for (a) series-connected monolithic 2T tandem and (b) mechanically stacked 4T tandem solar cells. In the 4T architecture, for $E_{g,top} < E_{g,bot}$, there is no contribution from the bottom cell. The bandgap of CdTe, GaSb, CIGS, and high-E_g II–VI are marked with white dashed lines while different architectures A, B, and C are marked using the white circle marks. (Reproduced with permission from Mailoa, J. P. et al. *Energy & Environmental Science* 2016, 9, 2644–2653.)

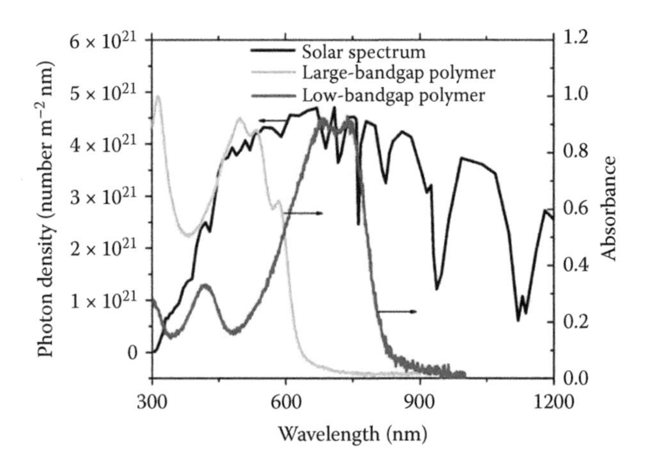

FIGURE 11.13 Complementary absorbing organic materials: absorbance of P3HT, a large bandgap polymer (*green*) and PSBTBT (Poly[(4,4′-bis(2-ethylhexyl)dithieno[3,2-b:2′,3′-d]silole)-2,6-diyl-alt-(2,1,3-benzothiadiazole)-4,7-diyl]), a low bandgap polymer (*red*), compared to the solar spectrum. (Reproduced with permission from Sista, S. et al. *Energy & Environmental Science* 2011, 4, 1606–1620.)

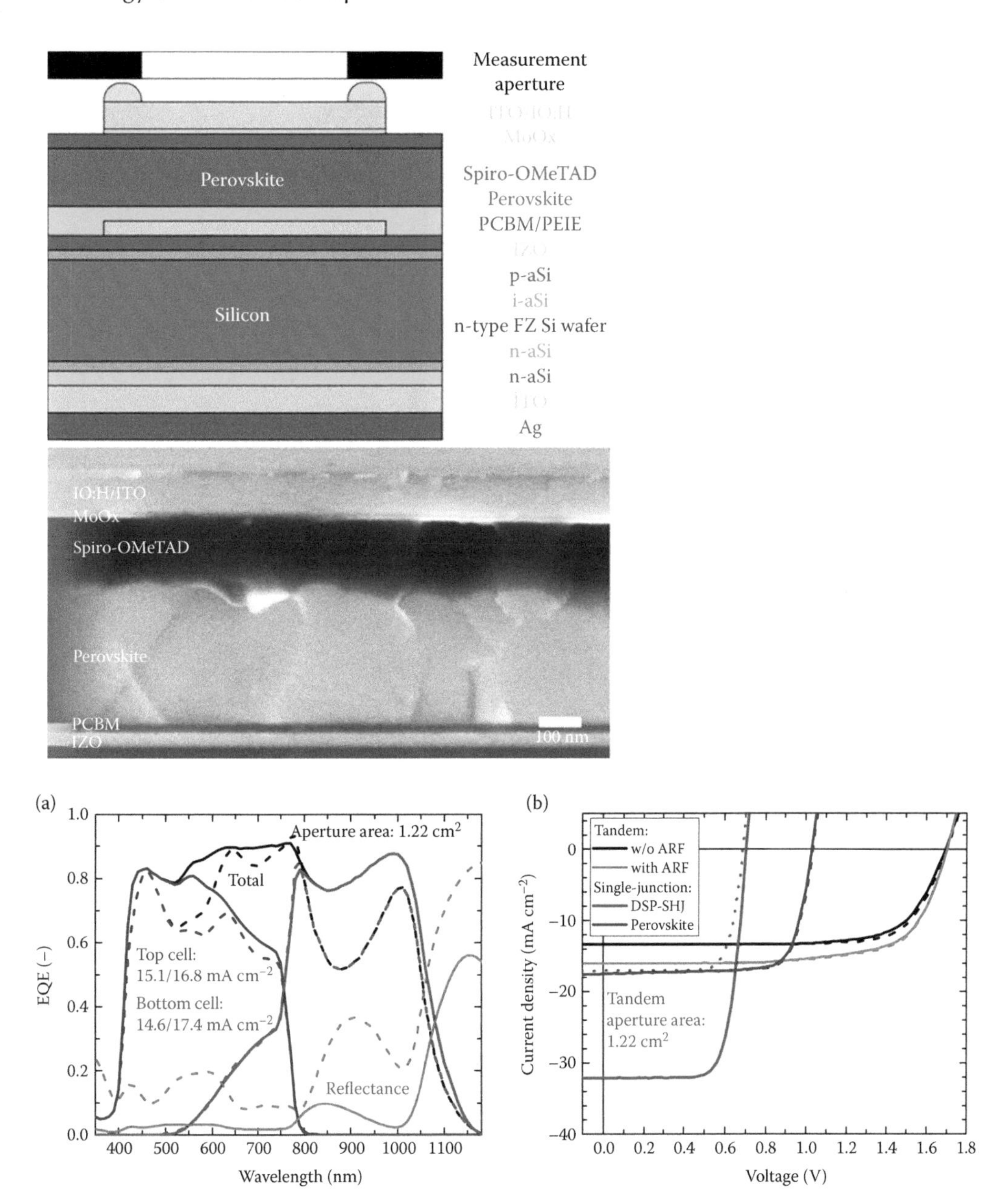

FIGURE 11.14 *Top*: Schematic drawing of a planar monolithic perovskite/(amorphous/crystalline silicon heterojunction) tandem cell layer stack and a SEM cross-sectional view of the perovskite top cell. *Bottom*: (a) EQE spectra of a perovskite/Si monolithic tandem with (*solid lines*) and without (*dashed lines*) anti-reflective foil (ARF) as well as the corresponding reflectance (*green curves*). The integrated j_{sc} for both top and bottom cells is given in the legend (without ARF/with ARF). (b) j–V measurements of the best perovskite/Si monolithic tandem with 1.22 cm^2 aperture area and of the single junction perovskite and Si cells. Reverse (*solid lines*) and forward (*dashed lines*) scans are shown for perovskite single-junction and tandem cells. The *dotted red curve* shows the j–V curve of the Si cell when illuminated at an intensity of 0.53 suns. (Reproduced with permission from Werner, J. et al. *The Journal of Physical Chemistry Letters* 2016, 7, 161–166.)

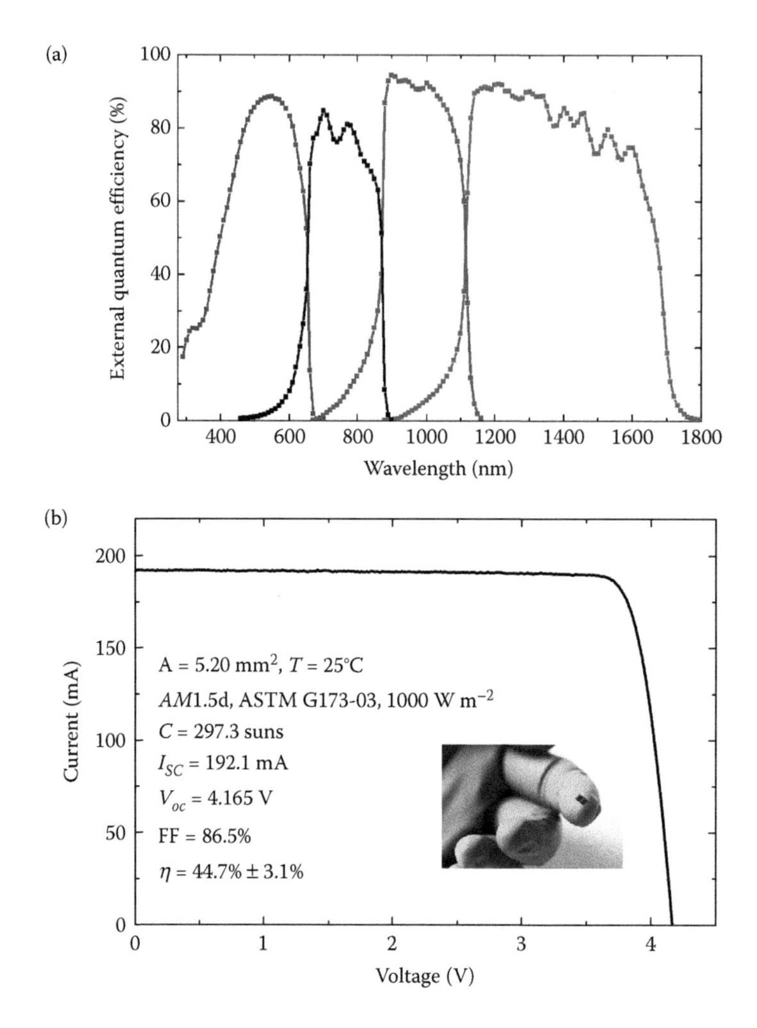

FIGURE 11.15 (a) EQE of the four-junction solar cell prepared jointly by Fraunhofer Institute for Solar Energy Systems ISE, Soitec, CEA-Leti and the Helmholtz Center Berlin. (b) Current density–voltage characteristic under *AM*1.5d ASTM *G*173–03 spectrum at a concentration of 297 suns. The inset shows an image of the cell.

reaction occurs at a counterelectrode. Suitable semiconductor materials must satisfy stringent conditions in terms of light absorption in the visible range, adequate alignment of band edges with the relevant redox potentials, overall efficiency, cost, and stability under operating conditions. The main visible-light-driven junction water-splitting photo(electro)catalysts reported include $BiVO_4$, Fe_2O_3, Cu_2O, and C_3N_4, but they show important shortcomings, which result in relatively low efficiencies. For example, oxide materials such as WO_3, $BiVO_4$, or Fe_2O_3 are suitable for OER generation but cannot produce hydrogen on the surface even after absorbing visible light because their conduction band edges are more positive (in the electrochemical scale, see Figure ECK.2.7) than the hydrogen evolution potential. Furthermore, traditional oxide materials for water splitting based on a surface SB show very poor fill factor characteristics due to low mobilities and large recombination close to flatband conditions, see Figure PSC.10.16. Alternatively, chalcogenides containing CuI ions, such as $CuGaSe_2$, Cu_2ZnSnS_4, and Cu(Ga,In)$(S,Se)_2$, are suitable for HER but not for water oxidation due to photocorrosion and an insufficient bandgap. For this reason, in order to accomplish water splitting using these oxides or chalcogenides, a combination of semiconducting materials and an electric power supply are needed,

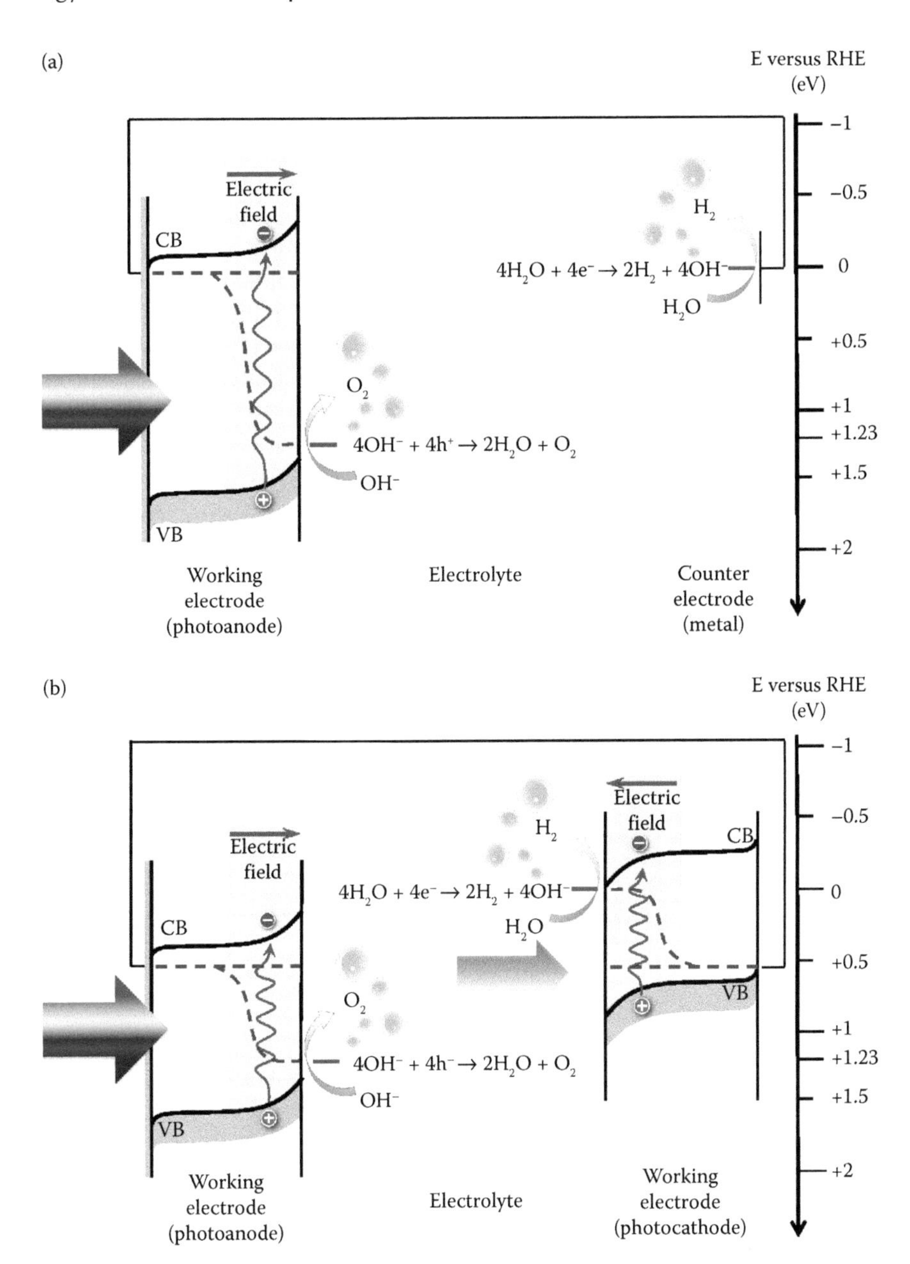

FIGURE 11.16 Schematic illustration of (a) a single absorber PEC with a photoanode and a metal cathode, and (b) a dual absorber system consisting of a photoanode and a photocathode in a tandem configuration. *Blue dashed lines* represent Fermi levels under illumination. CB and VB indicate the conduction and valence bands, respectively. (Reproduced with permission from Smith, W. A. et al. *Energy & Environmental Science* 2015, 8, 2851–2862.)

similar to the tandem solar cell, in PEC water-splitting arrangement, shown in Figure 11.16b. A tandem configuration of semiconductor–electrolyte junctions using the earth's abundant elements that realize the complementary redox reactions is a suitable option for the competitive conversion of water to hydrogen (competing with US\$2–3 kg^{-1} for the steam reforming of natural gas) (Prévot and Sivula, 2013). Optimized nanostructured catalysts are often deposited in order to promote the desired reactions at the surface.

In order to relax some of these highly demanding material specifications, other configurations of solar fuel converters have been developed. The photovoltaic part can be placed outside the electrolytic solution while the catalytic fragment will be in direct contact with water, both portions being connected by a wire. By using this approach, the photovoltaic part does not need to be water resistant and the catalytic fragment of the system does not need to offer good optical properties, that is, fulfill all the bandgap requirements. In Figure 11.17, we show the use of a tandem solar cell configuration achieving stable photocurrents of $\approx$10 mA cm^{-2} for the production of H_2/O_2 (Luo et al., 2014) with solar-to-hydrogen (STH) efficiency of 12.3%. In this setup, two hybrid perovskite ($CH_3NH_3PbI_3$) solar cells are connected in series in a tandem configuration. An electrolyzer is connected by an external wire to an electrode containing earth-abundant catalysts such as nickel–iron-layered double hydroxide NiFe-LDH, which carries out the desired electrochemical reaction. Similarly, monolithic tandem solar cells based on lead halide perovskites and $BiVO_4$ have also been described with photocurrents of $\approx$2 mA cm^{-2} for the production of H_2/O_2 (Chen et al., 2015). The main disadvantage of this approach is the high complexity of the devices, which obviously will pose a negative impact by increasing the final cost of the device.

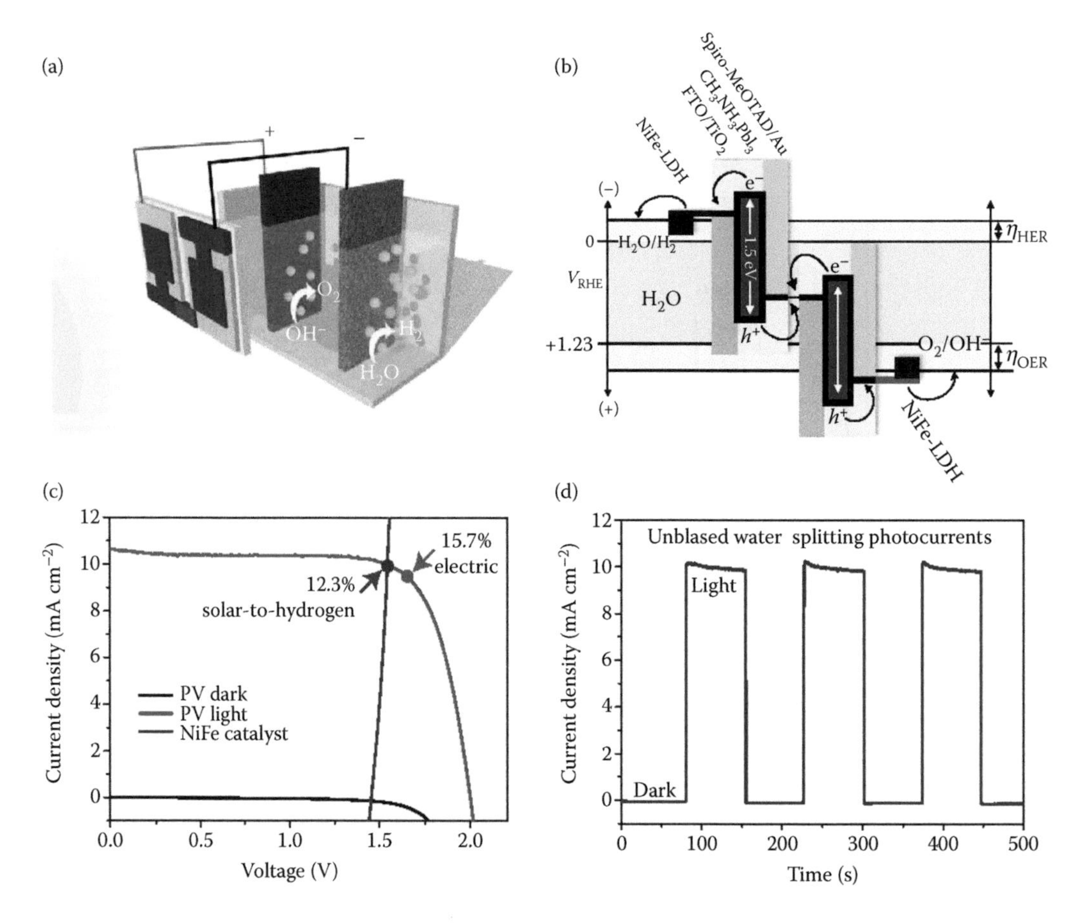

FIGURE 11.17 (a) Schematic diagram of the water-splitting device. (b) A generalized energy schematic of the perovskite tandem cell for water splitting. (c) j–V curves of the perovskite tandem cell under dark and simulated $AM1.5G$ 100 mW cm^{-2} illumination, and the NiFe/Ni foam electrodes in a two-electrode configuration. The illuminated surface area of the perovskite cell was 0.318 cm^2 and the catalyst electrode areas (geometric) were ~5 cm^2 each. (d) Current density–time curve of the integrated water-splitting device without external bias under chopped simulated $AM1.5G$ 100 mW cm^{-2} illumination. (Reproduced with permission from Luo, J. et al. *Science* 2014, 345, 1593–1596.)

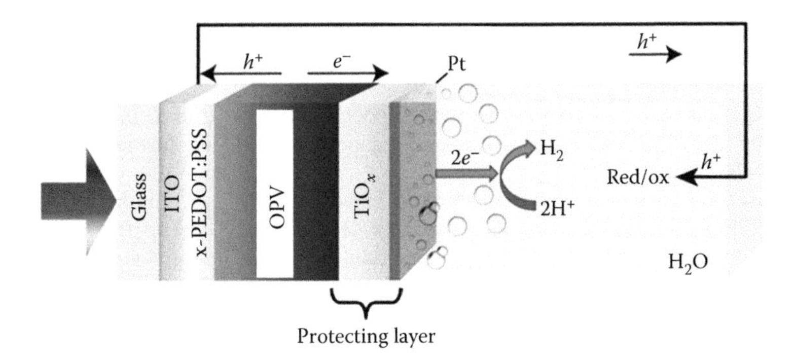

FIGURE 11.18 Organic PEC. (Reproduced with permission from Haro, M. et al. *The Journal of Physical Chemistry C* 2015, 119, 6488–6494.)

A third approach aims at converting a photovoltaic device into a PEC by using the protecting layers to avoid the chemical reaction of the liquid solution with the materials used in the photovoltaic device. Khaselev and Turner (1998) converted GaAs/GaInAs tandem photovoltaic devices into PECs for the production O_2/H_2 in the absence of protecting layers. Although highly efficient, this type of configuration lacks long-term stability and the photocurrent decreased by 15% over the initial 20 h. Similarly, photovoltaic devices based on CIGS have been turned into PEC and stability in the liquid solution has been gained by the use of a Ti protective layer (Kumagai et al., 2015). Indeed, it was observed that a configuration such as CIGS/Ti/Mo/Pt could improve the stability over a period of 10 days with gradual decrease in efficiency. With current low-voltage photovoltaic technologies such as silicon, a multiple tandem is required to attain the required voltage (Nocera, 2012; Cox et al., 2014). STH efficiency of 10%–14% has been demonstrated with extremely expensive III–V PV materials (May et al., 2015; Verlage et al., 2015). Similarly, the conversion of organic photovoltaic devices into PECs has been conferred by the use of protective TiO_2 layers that enhance protection against the aqueous solution and electrically communicates the organic layer with a Pt catalyst, see Figure 11.18 (Haro et al., 2015).

GENERAL REFERENCES

Luminescent solar concentrators: Yablonovitch (1980), Rau et al. (2005), Markvart et al. (2012), van Sark (2013), and Klimov et al. (2016).

Upconversion: Trupke et al. (2002), Auzel (2003), and Huang et al. (2013).

Triplet–triplet annihilation: Miteva et al. (2008), Singh-Rachford and Castellano (2010), Ji et al. (2011), and Zhao et al. (2011).

Solar fuel production and photoelectrochemistry: Morrison (1980), Nozik and Memming (1996), Memming (2001), van de Krol et al. (2008), Walter et al. (2010), Prévot and Sivula (2013), Hisatomi et al. (2014), Nielander et al. (2015), and Gimenez and Bisquert (2016).

Solar fuel conversion efficiency: Weber and Dignam (1984), Chen (2010), Ager et al. (2015), Coridan et al. (2015), McCrory et al. (2015), Smith et al. (2015), Doscher et al. (2016), and Guerrero and Bisquert (2017).

REFERENCES

Ager, J. W.; Shaner, M. R.; Walczak, K. A.; Sharp, I. D.; Ardo, S. Experimental demonstrations of spontaneous, solar-driven photoelectrochemical water splitting. *Energy & Environmental Science* 2015, 8, 2811–2824.

Auzel, F. Upconversion and anti-Stokes processes with f and d ions in solids. *Chemical Reviews* 2003, 104, 139–174.

Bush, K. A.; Palmstrom, A. F.; Yu, Z. J.; Boccard, M.; Cheacharoen, R.; Mailoa, J. P.; McMeekin, D. P. et al. 23.6%-efficient monolithic perovskite/silicon tandem solar cells with improved stability. *Nature Energy* 2017, 2, 17009–17014.

Chen, G.; Yang, C.; Prasad, P. N. Nanophotonics and nanochemistry: Controlling the excitation dynamics for frequency up- and down-conversion in lanthanide-doped nanoparticles. *Accounts of Chemical Research* 2012, 46, 1474–1486.

Chen, Y.-S.; Manser, J. S.; Kamat, P. V. All solution-processed lead halide perovskite-$BiVO_4$ tandem assembly for photolytic solar fuels production. *Journal of the American Chemical Society* 2015, 137, 974–981.

Chen, Z.; Jaramillo, T.; Deutsch, T.; Kleiman-Shwarsctein, A.; Forman, A.; Gaillard, N., ... Dinh, H. Accelerating materials development for photoelectrochemical hydrogen production: Standards for methods, definitions, and reporting protocol. *Journal of Materials Research* 2010, 25, 3–16.

Coridan, R. H.; Nielander, A. C.; Francis, S. A.; McDowell, M. T.; Dix, V.; Chatman, S. M.; Lewis, N. S. Methods for comparing the performance of energy-conversion systems for use in solar fuels and solar electricity generation. *Energy & Environmental Science* 2015, 8, 2886–2901.

Cox, C. R.; Lee, J. Z.; Nocera, D. G.; Buonassisi, T. Ten-percent solar-to-fuel conversion with nonprecious materials. *Proceedings of the National Academy of Sciences* 2014, 111, 14057–14061.

de Wild, J.; Meijerink, A.; Rath, J. K.; van Sark, W. G. J. H. M.; Schropp, R. E. I. Upconverter solar cells: Materials and applications. *Energy & Environmental Science* 2011, 4, 4835–4848.

Doscher, H.; Young, J. L.; Geisz, J. F.; Turner, J. A.; Deutsch, T. G. Solar-to-hydrogen efficiency: Shining light on photoelectrochemical device performance. *Energy & Environmental Science* 2016, 9, 74–80.

Erickson, C. S.; Bradshaw, L. R.; McDowall, S.; Gilbertson, J. D.; Gamelin, D. R.; Patrick, D. L. Zero-reabsorption doped-nanocrystal luminescent solar concentrators. *ACS Nano* 2014, 8, 3461–3467.

Gimenez, S.; Bisquert, J. *Photoelectrochemical Solar Fuel Production. From Basic Principles to Advanced Devices*; Springer: Switzerland, 2016.

Guerrero, A.; Bisquert, J. Perovskite semiconductors for photoelectrochemical water splitting applications. *Current Opinion in Electrochemistry*, 2017, doi.org/10.1016/j.coelec.2017.04.003.

Haro, M.; Solis, C.; Molina, G.; Otero, L.; Bisquert, J.; Gimenez, S.; Guerrero, A. Toward stable solar hydrogen generation using organic photoelectrochemical cells. *The Journal of Physical Chemistry C* 2015, 119, 6488–6494.

Henry, C. H. Limiting efficiencies of ideal single and multiple energy gap terrestrial solar cells. *Journal of Applied Physics* 1980, 51, 4494–4500.

Hirst, L. C.; Ekins-Daukes, N. J. Fundamental losses in solar cells. *Progress in Photovoltaics: Research and Applications* 2011, 19, 286–293.

Hisatomi, T.; Kubota, J.; Domen, K. Recent advances in semiconductors for photocatalytic and photoelectrochemical water splitting. *Chemical Society Reviews* 2014, 43, 7520–7535.

Huang, X.; Han, S.; Huang, W.; Liu, X. Enhancing solar cell efficiency: The search for luminescent materials as spectral converters. *Chemical Society Reviews* 2013, 42, 173–201.

Ji, S.; Guo, H.; Wu, W.; Wu, W.; Zhao, J. Ruthenium(II) polyimine–coumarin dyad with non-emissive 3IL excited state as sensitizer for triplet–triplet annihilation based upconversion. *Angewandte Chemie International Edition* 2011, 50, 8283–8286.

Khaselev, O.; Turner, J. A. A monolithic photovoltaic-photoelectrochemical device for hydrogen production via water splitting. *Science* 1998, 280, 425–427.

Klimov, V. I.; Baker, T. A.; Lim, J.; Velizhanin, K. A.; McDaniel, H. Quality factor of luminescent solar concentrators and practical concentration limits attainable with semiconductor quantum dots. *ACS Photonics* 2016, 3, 1138–1148.

Kumagai, H.; Minegishi, T.; Sato, N.; Yamada, T.; Kubota, J.; Domen, K. Efficient solar hydrogen production from neutral electrolytes using surface-modified $Cu(In,Ga)Se_2$ photocathodes. *Journal of Materials Chemistry A* 2015, 3, 8300–8307.

Luo, J.; Im, J.-H.; Mayer, M. T.; Schreier, M.; Nazeeruddin, M. K.; Park, N.-G.; Tilley, S. D.; Fan, H. J.; Grätzel, M. Water photolysis at 12.3% efficiency via perovskite photovoltaics and Earth-abundant catalysts. *Science* 2014, 345, 1593–1596.

Mailoa, J. P.; Lee, M.; Peters, I. M.; Buonassisi, T.; Panchula, A.; Weiss, D. N. Energy-yield prediction for II-VI-based thin-film tandem solar cells. *Energy & Environmental Science* 2016, 9, 2644–2653.

Markvart, T.; Danos, L.; Fang, L.; Parel, T.; Soleimani, N. Photon frequency management for trapping & concentration of sunlight. *RSC Advances* 2012, 2, 3173–3179.

May, M. M.; Lewerenz, H.-J.; Lackner, D.; Dimroth, F.; Hannappel, T. Efficient direct solar-to-hydrogen conversion by *in situ* interface transformation of a tandem structure. *Nature Communications* 2015, 6, 8286.

McCrory, C. C. L.; Jung, S.; Ferrer, I. M.; Chatman, S. M.; Peters, J. C.; Jaramillo, T. F. Benchmarking hydrogen evolving reaction and oxygen evolving reaction electrocatalysts for solar water splitting devices. *Journal of the American Chemical Society* 2015, 137, 4347–4357.

Memming, R. *Semiconductor Electrochemistry*; Wiley-VCH: Weinheim, 2001.

Meyer, T. J. J.; Markvart, T. The chemical potential of light in fluorescent solar collectors. *Journal of Applied Physics* 2009, 105, 063110.

Miteva, T.; Yakutkin, V.; Nelles, G.; Baluschev, S. Annihilation assisted upconversion: All-organic, flexible and transparent multicolour display. *New Journal of Physics* 2008, 10, 103002.

Morrison, S. R.: *Electrochemistry at Semiconductor and Oxidized Metal Electrodes*; Plenum Press: New York, 1980.

Nielander, A. C.; Shaner, M. R.; Papadantonakis, K. M.; Francis, S. A.; Lewis, N. S. A taxonomy for solar fuels generators. *Energy & Environmental Science* 2015, 8, 16–25.

Nocera, D. G. The artificial leaf. *Accounts of Chemical Research* 2012, 45, 767–776.

Nozik, A. J.; Memming, R. Physical chemistry of semiconductor-liquid interfaces. *The Journal of Physical Chemistry* 1996, 100, 13061–13078.

Prévot, M. S.; Sivula, K. Photoelectrochemical tandem cells for solar water splitting. *The Journal of Physical Chemistry C* 2013, 117, 17879–17893.

Rau, U.; Einsele, F.; Glaeser, G. C. Efficiency limits of photovoltaic fluorescent collectors. *Applied Physics Letters* 2005, 87, 171101.

Singh-Rachford, T. N.; Castellano, F. N. Photon upconversion based on sensitized triplet-triplet annihilation. *Coordination Chemistry Reviews* 2010, 254, 2560–2573.

Sista, S.; Hong, Z.; Chen, L.-M.; Yang, Y. Tandem polymer photovoltaic cells-current status, challenges and future outlook. *Energy & Environmental Science* 2011, 4, 1606–1620.

Smith, W. A.; Sharp, I. D.; Strandwitz, N. C.; Bisquert, J. Interfacial band-edge energetics for solar fuels production. *Energy & Environmental Science* 2015, 8, 2851–2862.

Trupke, T.; Green, M. A.; Wurfel, P. Improving solar cell efficiencies by up-conversion of sub-band-gap light. *Journal of Applied Physics* 2002, 92, 4117–4122.

van de Krol, R.; Liang, Y. Q.; Schoonman, J. Solar hydrogen production with nanostructured metal oxides. *Journal of Materials Chemistry* 2008, 18, 2311–2320.

van Sark, W. G.; Barnham, K. W.; Slooff, L. H.; Chatten, A. J.; Büchtemann, A.; Meyer, A.; Mc.Cormack, S. J.; Koole, R.; Farrell, D. J.; Bose, R. et al. Luminescent solar concentrators. A review of recent results. *Optics Express* 2008, 16, 21773–21792.

van Sark, W. G. J. H. M. Luminescent solar concentrators. A low cost photovoltaics alternative. *Renewable Energy* 2013, 49, 207–210.

Verlage, E.; Hu, S.; Liu, R.; Jones, R. J. R.; Sun, K.; Xiang, C.; Lewis, N. S.; Atwater, H. A. A monolithically integrated, intrinsically safe, 10% efficient, solar-driven water-splitting system based on active, stable earth-abundant electrocatalysts in conjunction with tandem III-V light absorbers protected by amorphous TiO_2 films. *Energy & Environmental Science* 2015, 8, 3166–3172.

Walter, M. G.; Warren, E. L.; McKone, J. R.; Boettcher, S. W.; Mi, Q.; Santori, E. A.; Lewis, N. S. Solar water splitting cells. *Chemical Reviews* 2010, 110, 6446–6473.

Weber, M. F.; Dignam, M. J. Efficiency of splitting water with semiconducting electrodes. *Journal of the Electrochemical Society* 1984, 131, 1258–1265.

Werner, J.; Weng, C.-H.; Walter, A.; Fesquet, L.; Seif, J. P.; De Wolf, S.; Niesen, B.; Ballif, C. Efficient monolithic perovskite/silicon tandem solar cell with cell area >1 cm^2. *The Journal of Physical Chemistry Letters* 2016, 7, 161–166.

Yablonovitch, E. Thermodynamics of the fluorescent planar concentrator. *Journal of the Optical Society of America* 1980, 70, 1362–1363.

Zhao, J.; Ji, S.; Guo, H. Triplet-triplet annihilation based upconversion: From triplet sensitizers and triplet acceptors to upconversion quantum yields. *RSC Advances* 2011, 1, 937–950.

Appendix

Physical Constants and Conversion Factors

Quantity	Symbol, Equation	Value[a]
Speed of light in vacuum	c	2.998×10^{8} m s^{-1}
Electron charge magnitude	q	1.602×10^{-19} C
Planck's constant	h	6.626×10^{-34} J s = 4.136×10^{-15} eV s
Planck's constant, reduced	$\hbar = h/2\pi$	1.0546×10^{-34} J s
Permittivity of free space	ε_0	8.85×10^{-12} F m^{-1} = 8.85×10^{-14} F cm^{-1}
Permeability of free space	μ_0	$4\pi \times 10^{-7}$ N A^{-2}
Electron mass	m_0	9.11×10^{-31} kg
Proton mass	m_p	1.67×10^{-27} kg
Unified atomic mass unit (u)	(mass ^{12}C atom)/12 = (1 g)/(N_A mol)	1.66×10^{-27} kg
(e^- Compton wavelength)/2π	$\lambda\!\!\!^{-}_e = \hbar/m_e c$	3.86×10^{-13} m
Bohr's radius ($m_{nucleus} = \infty$)	$a_\infty = 4\pi\varepsilon_0\hbar^2/m_e e^2$	0.529×10^{-10} m
Rydberg's energy	$hcR_\infty = m_e e^4/2(4\pi\varepsilon_0)^2\hbar^2$	13.6 eV
Bohr's magneton	$\mu_B = e\hbar/2m_e$	9.274×10^{-24} J T^{-1}
Nuclear magneton	$\mu_N = e\hbar/2m_p$	5.051×10^{-27} J T^{-1}
Avogadro's constant	N_A	6.02×10^{23} mol^{-1}
Boltzmann's constant	k_B	1.3806×10^{-23} J K^{-1} = 8.617×10^{-5} eV K^{-1}
Gas constant	$R = N_A k$	8.314 J K^{-1} mol^{-1}
Thermal energy at 300 K	$k_B T$	0.0259 eV
Wien displacement law constant	$b = \lambda_{max}{}^{T}$	2.87×10^{-3} m K
Faraday's constant	$F = N_A q$	96.490 C mol^{-1}
Étendue of solar disc	ε_S	6.8×10^{-5}
Constant Blackbody flux for radiation to the hemisphere	$b_\pi = \dfrac{2\pi}{h^3c^2}$	9.883×10^{26} eV^{-3} m^{-2} s^{-1}
Stefan–Boltzmann's constant	$\sigma = \pi^2 k^4/60\hbar^3 c^2$	5.67×10^{-8} W m^{-2} K^{-4}

[a] Complete values with uncertainties at www.physics.nist.gov/constants

0°C = 273.15 K	1 dina = 10^{-5} N	1 eV = 1.602×10^{-19} J
1 Å = 10^{-10} m	1 erg = 10^{-7} J	1 Wh = 3600 J

1 eV corresponds to 96.48 kJ mol^{-1} = 23.05 kcal mol^{-1} or to the energy of a quantum of wavelength 1240 nm.

$$\hbar\omega\lambda = hc_{vac} = 1240 \text{ eV nm}$$

$$1 \text{ debye} = 3.33564 \times 10^{-30} \text{ cm}$$

General List of Acronyms and Abbreviations

4T	Four terminal device
APCE	Absorbed photon-to-collected-electron efficiency
ARF	Antireflective foil
CB	Conduction band
CCT	Correlated color temperature
CRI	Color rendering index
CT	Charge transfer
CTC	Charge transfer complex
DFT	Density functional theory
DOS	Density of states
DSC	Dye-sensitized solar cells
ECK	Equilibrium concepts and kinetics (accompanying volume of this book)
EL	Electroluminescence
EQE	External quantum efficiency
ESA	Excited state absorption
ESC	Electron selecting contact
ETU	Energy transfer upconversion
FC	Fluorescent collector
FET	Field-effect transistor
FRET	Förster resonant energy transfer
FTPS	Fourier transform photocurrent spectroscopy
HER	Hydrogen evolution reaction
HOMO	Highest occupied molecular orbital
HSC	Hole selecting contact
IPCE	Incident photon-to-current collected electron efficiency
IPES	Inverse photoemission spectroscopy
IQE	Internal quantum efficiency
IR	Infrared
IS	Impedance spectroscopy
ISC	Intersystem crossing
KMC	Kinetic Monte Carlo
LED	Light-emitting diode
LHE	Light-harvesting efficiency
LSC	Luminescent solar concentrators
LUMO	Lowest unoccupied molecular orbital
MIM	Metal–insulator–metal
MLCT	Metal-to-ligand charge transfer
mpp	Maximum power point
NIR	Near infrared
OER	Oxygen evolution reaction
OET	Optoelectronics and transport (this volume)
OFET	Organic field-effect transistor
OLED	Organic light-emitting diode
PA	Pair approximation
PCE	Power conversion efficiency
PEC	Photoelectrochemical cell
PL	Photoluminescence
QD	Quantum dot
QE	Quantum efficiency
QY	Quantum yield

(*Continued*)

General List of Acronyms and Abbreviations (*Continued*)

SCLC	Space-charge limited current
SCR	Space-charge region
STC	Standard test conditions
STH	Solar to hydrogen
TAS	Transient absorption spectroscopy
TOF	Time-of-flight
TRMC	Time-resolved microwave conductivity
TRTS	Time-resolved Terahertz spectroscopy
TTA	Triplet–triplet annihilation
UPS	Ultraviolet Photoelectron Spectroscopy
UV	Ultraviolet
VB	Valence band
VL	Vacuum level
VLA	Vacuum level alignment

Nomenclature

FTO	Fluor-doped tin oxide
Alq3	Tris(8-hydroxyquinolino)aluminum
α-NPD	*N,N′*-diphenyl-*N,N′*-bis(1-naphthyl)-1,1′-biphenyl-4,4′-diamine
a-Si	Amorphous silicon
BPEA	9,10-Bis(phenylethynyl)anthracene
CIGS	$Cu(In,Ga)Se_2$
c-Si	Crystalline silicon
CZTSe	$Cu_2ZnSnSe_4$
DPM6	4,4′-dihexyloxydiphenylmethano[60]fullerene
DSB	Distyrylbenzene
F8T2	poly(9,9-dioctylfluorene-alt-bithiophene)
FA	Formamidinium
ITO	Indium-doped tin oxide
MA	Methylammonium
μc-Si	Microcrystalline silicon
MEH-PPV	Poly[2-methoxy-5-(2-ethylhexyloxy)-1,4-phenylenevinylene]
OF	Oligofluorenes
OMETAD	[2,2′,7,7′-tetrakis(*N,N*-di-*p*-methoxyphenyl-amine)9,9′-spirobifluorene]
PBDTTT	Poly[4,8-bis(5-(2-ethylhexyl)thiophen-2-yl)benzo[1,2-b;4,5-b′]dithiophene-2,6-diyl-alt-(4-(2-ethylhexyl)-3-fluorothieno[3,4-b]thiophene-)-2-carboxylate-2–6-diyl]
P3HT	Poly(3-hexylthiophene)
$PC_{60}BM$	[6,6]-phenyl-C60-butyric acidmethyl ester
$PdPh_4TBP$	Meso-tetraphenyl-tetrabenzoporphyrin palladium
PEDOT	Poly(3,4-ethylenedioxythiophene)
PSBTBT	Poly[(4,4′-bis(2-ethylhexyl)dithieno[3,2-b:2′,3′-d]silole)-2,6-diyl-alt-(2,1,3-benzothiadiazole)-4,7-diyl]
PTB7	Poly[[4,8-bis[(2-ethylhexyl)oxy]benzo[1,2-b:4,5-b′]dithiophene-2,6-diyl][3-fluoro-2-[(2-ethylhexyl)carbonyl]thieno[3,4-b]thiophenediyl]]
TCO	Transparent conducting oxide

Index

T - #0280 - 160425 - C238 - 254/178/11 - PB - 9781138099968 - Gloss Lamination